D0712643

FORGING AMERICA

FORGING

IRONWORKERS, ADVENTURERS, AND THE INDUSTRIOUS REVOLUTION

AMERICA

JOHN BEZÍS-SELFA

CORNELL UNIVERSITY PRESS

Ithaca and London

First published 2004 by Cornell University Press

Printed in the United States of America

Library of Congress Cataloging-in-Publication Data

Bezís-Selfa, John, 1966–
Forging America : ironworkers, adventurers, and the industrious revolution / John Bezís-Selfa.
p. cm.
Includes bibliographical references and index.
ISBN 0-8014-3993-0 (cloth)
1. Iron industry and trade—United States—History—18th century. 2. United States—History—Revolution, 1775–1783. 3. Work ethic—United States—History. I. Title.

HD9515.B478 2003
338.4'76691'09730903—dc21

2003055168

Cornell University Press strives to use environmentally responsible suppliers and materials to the fullest extent possible in the publishing of its books. Such materials include vegetable-based, low-VOC inks and acid-free papers that are recycled, totally chlorine-free, or partly composed of nonwood fibers. For further information, visit our website at www.cornellpress.cornell.edu.

Cloth printing 10 9 8 7 6 5 4 3 2 1

Paola,

antes de você chegar,

era tudo saudade.

CONTENTS

ACKNOWLEDGMENTS

Like many adventurers and even more ironworkers, I have accumulated more debts than I can ever repay. Research for this project was supported by grants from the Scholars-in-Residence Program of the Pennsylvania Historical and Museum Commission, the National Endowment for the Humanities, the Virginia Historical Society, and the Hagley Museum and Library. A year as a fellow at the Charles Warren Center for Studies in American History at Harvard University gave me time to reflect and write, as did a semester of leave from Wheaton College. Material in chapter 4 previously appeared in print in *Pennsylvania History: A Journal of Mid-Atlantic Studies,* as did material in chapters 5 and 6 in *The William and Mary Quarterly*. I thank both journals for allowing me to draw upon those articles.

Everywhere I went archivists graciously accommodated me. I especially want to thank the Pennsylvania State Archives, the Hagley Museum and Library, the Virginia Historical Society, the Maryland Historical Society, the Maryland State Archives, and the Historical Society of Pennsylvania. Marcia Grimes and Martha Mitchell at Wheaton's Wallace Library pursued my obscure requests through Inter-Library Loan.

I cannot imagine a better place to attend graduate school than Penn, especially with what has become the McNeill Center for Early American Studies. Many friends helped to make my time there so enjoyable, particularly Rose Beiler, Harold Van Lonkhuyzen, Alison Games, Marion Nelson Winship, Celia Cussen, Todd Barnett, Liam Riordan, John Majewski, Simon Newman, and Camilla Townsend. Ann Little and Chris Moore took me in when I visited Baltimore. Marion and Liam both read portions of this book and Liam graciously shared some of his forthcoming work with me.

I also had an excellent dissertation committee: Walter Licht, Drew Gilpin Faust, Michael Zuckerman, and Richard S. Dunn. Mike pushed me to think harder and to remember that the ghosts trapped within documents were once human. I owe the most to Richard. He was and is the ideal mentor. He read and improved all of the book manuscript that I got to him. As I have said before, if I should ever become half the teacher that Richard has been for me, I would be very proud.

I benefited from the counsel of many other scholars. Rachel Klein, Charles Dew, Douglas Monroy, Cathy Matson, Peter Onuf, and Joe William Trotter, Jr., all commented on presentations of parts of this project. Nina Dayton, Phil Morgan, Laurel Thatcher Ulrich, Hal Hansen, Cathy Matson, Jonathan Prude, and David Waldstreicher each read one or more chapters. I am especially grateful to David for sharing his unpublished work on Benjamin Franklin with me and for much else. Anne Kelly Knowles read the entire manuscript. So did Joyce Chaplin, who read the "complete" draft, patiently awaited a conclusion, and lucidly advised me on how to improve both. Bruce Acker provided expert copyediting and Karen Laun patiently and skillfully shepherded this manuscript into print. I also owe many thanks to my Cornell editor, Sheri Englund, who adroitly switched between providing a sympathetic ear and keeping an often wayward author on task and sometimes on message.

Wheaton College has been a wonderful place for me. I received invaluable advice from Susanne Woods, Bill Goldbloom-Bloch, Eric Denton, Leo Cabranes-Grant, Anni Baker, Candace Quinn, Paul Helmreich, Travis Crosby, Vipan Chandra, Alex Bloom, Terri McCandies, Eduardo Fichera, Michelle Harris, Hyun Kim, Brenda Wyss, Deyonne Bryant, Matthew Allen, Ann Sears, Javier Treviño, Frank Guridy, and, above all, from Dolita Cathcart and Kathryn Tomasek.

The Wheaton students whom I have had the privilege to teach have probably shaped this book the most. Two of the best—Sean Britt and Chris Benedetto—helped with research. I have aired most of the book's major themes in classes. My senior seminars of 2000 and 2001 read portions of the manuscript and told me candidly how to improve them. I hope that they, and the others who have endured my courses, will read this book and recognize what they have given to it and to me. I wrote it, as best I could, for them.

My parents John and Joanne, my sister Jennifer, and my brother Justin contributed mostly by asking, "Are you *still* working on that book?" So did my Argentine family, especially Carlos Zucchi, *padre e hijo.* My brother Jason sheltered me on research trips and never got glazed eyes when I talked about what I was up to.

Neither did Paola, most of the time. Her industry and her love made my passage to and through graduate school possible. She happily accompanied me as we drove backroads looking at old furnace stacks. She has heard more of adventurers and ironworkers, and of my obsession about committing their world to paper, than anyone should. That makes her at least as happy to see me finish this book as I am. With help from Caetano Veloso's composition "Sou Você," I dedicate this story to you, *coisa linda.*

FORGING AMERICA

INTRODUCTION

Visitors to the Historical Society of Berks County in Reading, Pennsylvania, will encounter Daniel Udree's image gazing down on the main meeting hall. Udree holds an ore-bearing rock in his right hand. An Oley Furnace account book, either a daybook or a journal, rests on a stand behind him to his left, quill and inkwell ready to record transactions. The book lies open to a blank page designated for business on the day of July 4, 1799. Udree's portrait bears witness to three intertwined revolutions: (1) the early stage of what most people call the industrial revolution; (2) an armed struggle in which the United States of America won its political independence from the British Empire; and (3) an *industrious* revolution—a gradual transformation in how peoples of the early modern North Atlantic world organized, conducted, and valued work—on which both the industrial revolution and the American Revolution rested.[1]

This book tells a story about the three revolutions that unfolded in eastern British North America between the early seventeenth and the early nineteenth centuries. It claims that close consideration of how and why Americans made iron allows us to understand better the unique and complex relationship between work, economic development, politics, and identity that emerged in early Anglo America. I argue that slavery, as much as gendered and religiously inspired "work ethics," enabled the United States to become the first of the early modern colonies to industrialize and join the developed world.

Figure 1. Daniel Udree, c. 1800. Courtesy of the Historical Society of Berks County, Reading, Pennsylvania.

Daniel Udree's life testified to early Anglo America's revolutions. He presented part of that life in his portrait, through which he told viewers about himself, his enterprise, and how both helped to forge America. By cradling the ore-bearing stone and displaying an Oley Furnace account book, Udree had himself depicted as an ironmaster—someone who owned and oversaw one of the largest integrated manufacturing operations of his time. It made hundreds of tons of iron each year and might employ over a hundred men, some around the clock for months, to do it. When he entered the iron business in the 1760s, Udree joined the biggest industry in British North America, one that was about to become the third largest iron producer in the world. Udree was a product of the iron industry. He began his career as a clerk. The account book symbolized how he rose in the world and how far he had risen. Someone kept accounts for Udree long before he commissioned his portrait. He continued to rise after it was painted. Udree served as a major general in the War of 1812 and was elected to Congress three times. He died Berks County's wealthiest taxpayer in 1828.[2]

Daniel Udree's ascent and that of the United States, his portrait argues, went hand-in-hand. He reminded viewers that he and his enterprise served the nation. Oley Furnace produced some of the iron that had enabled the United States to revolt against the British Empire and to preserve its political independence thereafter. Udree took a personal stake in that revolution; he was a militia colonel at the Battle of Brandywine. Udree went farther—to identify himself and his enterprise with an emerging American nationalism. Making iron was patriotic; the American iron industry provided the foundation upon which the United States could develop a vibrant economy. It needed and deserved government's protection, ironmasters like Udree had already begun to assert. It was probably no coincidence that the account book page in his portrait is dated July 4, 1799. For Udree, at least as he chose to represent himself in paint, the fashioning of iron, self, and nation were nearly inseparable.[3]

An industrious revolution made possible all that Udree's portrait celebrates. It rested on the ability of household heads and entrepreneurs to organize their own labor and that of others to produce more goods and provide more services. Udree, his painted image tells us, held the knowledge and the power to extract iron from otherwise worthless stone. In the background lay a key instrument of the industrious revolution, the account book. In it Udree tracked what he earned and what he spent by making iron. With it he displayed his business acumen and moral fitness to fellow entrepreneurs. In account books Udree (or, more accurately, his clerk) inscribed his relationship to workers and his assessment of them.

He exercised authority over and assigned value to the men (and through them over women and children in their households) whose energy and knowledge converted rock and wood into kettles, stove plates, and pig iron. Oley accounts also recorded what drew many workers to Udree's employ—the chance to earn income to buy what they could not make or obtain through barter. Consumption drove the industrious revolution forward in early Anglo America as in early modern Europe.

Most significant to the making of early Anglo America's industrious revolution was men's desire to achieve economic independence. An independent man was his own boss. A truly independent man secured a comfortable life for himself and his dependents, and he ensured that his sons would also enjoy independence. The promise of independence lured most men who emigrated from Britain and German-speaking Europe to colonial Anglo America. There, independence, or a "competency," as some called it, had became inseparable from personal liberty, political identity, and manhood—a man was neither truly free nor fully a man if he spent his life working for someone else. Employment, unless it was temporary, rendered him dependent and potentially servile, analogous to children, servants, slaves, and most women. Many men who had business with Oley Furnace pursued a competency or preserved that of their fathers by hiring themselves out to Daniel Udree. Others remained ironworkers their entire lives, which forced them to reconcile the mainstream ideal of independence with their own dependence on adventurers.[4]

They chose to serve the iron industry; Tim and thousands of others never had that option. Tim was around forty years old when Udree registered him as his slave in 1780. He reminds us that it took slaves and slavery to make early Anglo America's three revolutions. Enslaved men like Tim reminded free men of the value of independence and of why ownership of one's own work mattered so much in their world.[5]

Work Matters

In this book I try to explain what work meant to its subjects, in part so that we may understand better what work and working people mean to Americans today. Work structures people's lives and it shapes how they view themselves and others, whether as individuals or as groups. Who does which jobs, on what terms, for what purpose, and for whom reflect and influence economic imperatives, social hierarchies, cultural values, and political priorities. Divisions of labor serve to demarcate who is who; they indicate the relative value that those who constitute a society, especially those who are most powerful, attach to different tasks and to the people who carry them out. In other words, how societies organize, conduct, and reward work mirrors and largely determines who "counts" and who does not, who "belongs" and who does not, who rules and who follows. "Work" matters because it produces goods and services. It matters even more because it shapes what people envision as possible and what they consider impermissible, especially for those who must work for others.[6]

What work is and what it means is never fixed. Work is situationally and culturally defined. My "hobby" may be your "chore." The same task takes on different meanings depending on who does it and on the context in which it is done. Consider a comparison which one could make from this book. Two forgemen, one an African American slave and the other a free white man, have mastered the same skill and do the same job. But that job means something different to each of them, largely because the society in which they live regards them differently. Both may earn money for iron that they make, but only the white man gets something approaching full value for his time, energy, and expertise. Each may pass his trade on to a son, though the white forgeman sees a future for his son that lies beyond the dreams of the slave. The slave forgeman may garner more respect from other slaves than the white forgeman does from his peers and neighbors, both for reasons involving legacies of African beliefs about metal and metalworkers and because he has risen as high as a slave can. He remains an outsider, someone who may or may not be socially dead but who is civically dead, principally because he is a commodity.[7]

In early Anglo America, work—how it was organized, performed, understood, and valued—mattered because people spent most of their lives working.[8] What gave work particular meaning for early Anglo Americans was that control of others paved the way to economic independence, social advancement, and political power, all almost exclusively for white men. To become and to remain independent, a man needed dependents—women and children in nearly all cases and bound laborers (indentured servants or slaves) where his faith, finances, and local law permitted.[9] Most immigrants to colonial British North America came as servants or slaves. While at work they learned and they influenced the cultures of their masters, their neighbors, and their fellow laborers. Race, ethnicity, or legal status did not necessarily determine who performed which jobs. The scarcity of skilled hands for hire encouraged masters to train servants and slaves or to purchase those who already practiced a trade. Whatever the division of labor within a household, farm, shop, plantation, or enterprise, its head needed to keep dependents as dutiful and diligent as possible. His reputation and status rested on their industry.[10]

The American Revolution politicized work and gradually collapsed formal divisions of labor into "slave" and "free." Revolutionary rhetoric afforded some white women a heightened profile as guardians and teachers of republican virtue but confined them to the status of wife and mother, and it hardly freed them from the unrecognized and unpaid drudgery of maintaining households.[11] The revolutionary era saw thousands of African Americans struggle to obtain their freedom, use it to establish autonomous institutions, and become marginalized, excluded from nearly all skilled or industrial jobs and gradually denied the right to vote in most of the North. Their fate was tied to the demise of white indentured servitude and the reinforcement of slavery in the South. Indentured servitude, an institution through which many journeyed from slavery to freedom, slowly withered in the face of legal challenges and the growing availability of workers for hire. Southern masters enshrined their right to enslave others in the Constitution and presided over slavery's expansion, which sundered thousands of families as it accorded more slaves a wider range of job opportunities that were increasingly denied free northern blacks.[12] Contests over what the American Revolution meant clarified for whites the distinction between independence and dependence, between freedom and slavery. Some male artisans and female millhands modified and deployed such ideas to challenge their employers; some politicians mustered them to create the Republican Party, the only successful third party in U.S. history, which provoked secession and led to the war that restored the Union and abolished slavery. For African Americans, freedom's coming sharpened the color line, what W.E.B. DuBois identified as the main challenge that would confront the nation in the twentieth century.[13]

Making Iron as an American "Crucible"

I have chosen the iron industry as a window through which to view and understand the industrious and industrial revolutions as they unfolded in early Anglo America. I have done so partly because iron manufacturing created a microcosm of early Anglo American society. It brought together tens of thousands of workers—African and African American slaves, British and Irish convicts, indentured servants from the British Isles and central Europe, and free white immigrant and American-born tradesmen and laborers. Ironworks, like crucibles, functioned as melting pots of a sort in which capital, people, and ideas from Europe, Africa, and North America met, collided, and melded to form something new and uniquely American. Work helped to acculturate ironworkers by introducing them to each other and to many of the dominant values of the society in which they lived, much as factories and other workplaces have for immigrants since. In this sense early ironworks foreshadowed the industrial capitalism that would employ millions.[14]

What makes iron manufacturing so useful for understanding what work meant to early Anglo Americans is that it evoked tensions between their industrious revolution and economic development. Ironworks, though tiny by our standards, dwarfed other manufacturing enterprises of their time and place. They were rural factories which demanded thousands of pounds to finance them and dozens to hundreds of workers to operate them in a place where capital and labor were often scarce.[15] Entrepreneurs who owned ironworks really were adventurers—the word that English on both sides of the Atlantic often used to describe them during the colonial era. Many dreamed grandly and some failed spectacularly. Circumstances unique to early Anglo America—men's desire for independence and especially their ability to attain it—rendered adventurers uncomfortably dependent on ironworkers, the very people they thought should be dependent on them. As a result, adventurers made the United States the only developed nation in the world to industrialize on the backs of black slaves—an exceptional case indeed.[16]

The direct roles that slavery and race played in the industrious and industrial revolutions seldom figure in accounts of economic development in the United States. Most studies of industrialization have focused on textile mills in New England and the Delaware Valley or on trades in northeastern cities, sectors in which relatively few blacks worked (thanks largely to discrimination) in areas where relatively few African Americans lived.[17] The residual bias in favor of free labor that most economists and historians share has led most to assume that, although slavery as a labor system may have fostered economic growth, it inherently inhibited economic development, technological innovation, and industrialization where it existed. In this scenario, plantation slaves may have experienced their own industrious revolution under the direction of profit-minded planters, but the creation of "factories in the fields" powered development elsewhere in the North Atlantic world by creating commodities that millions consumed, by providing stimulants and cheap calories that enabled factory hands to work longer hours, or by generating capital that merchants invested in mills and factories.[18]

The colonial iron industry's expansion owed directly to unfree labor, principally slavery. Despite abundant natural resources, iron production proceeded fitfully during the seventeenth century and grew rapidly after 1715. By 1775, Britain's North American colonies turned out one-seventh of the world's iron. This surpassed England and Wales and made the colonies the globe's third largest producer of raw iron.[19] The most rapid growth occurred within a broad arc which swept south from northern New Jersey through Pennsylvania to central Virginia. Most of those furnaces and forges depended principally or heavily on slave labor. Iron production in the South remained largely or mostly in slaves' hands into the Civil War. By 1830, slavery had nearly become a memory in the North, the industry's center. But its legacies still influenced adventurers' relations to employees, and the lives, aspirations, and attitudes of ironworkers white and black.

Reckoning Accounts

Historians construct history from the stories our subjects leave behind. Both adventurers and ironworkers shaped the industrious revolution, yet adventurers largely get to tell their story of it. Ironworkers have left us few accounts of themselves by their own hands. Most of their handiwork is now dust, soot, and rust. Trees, highways, or suburban subdivisions cover most of their landscapes. Iron objects preserved and displayed reveal how they were made but not who made them. Museums and collectors treasure the works of individual silversmiths and furniture makers. Moulders and forgemen remain anonymous—tradesmen denied authorship of their work largely because of its utility. Most ironworkers who were literate wrote sparingly—a signature on a contract or in a receipt book more often than not. Occasionally ironworkers put their names on petitions. Some—mostly tradesmen—composed letters, usually to seek employment. I use such sources when I can, though the ledger will never balance.

This is especially true when one considers most of the tales that adventurers left behind. They are accountants' views of life—hundreds of daybooks, journals, and ledgers, generally neatly written and painstakingly compiled (which makes them pleasant to work with) but myopic in so many ways. Double-entry accounts froze the transactions that swirled around them, classified them as credits or debits, and categorized people mostly as creditors or debtors. It is difficult to avoid seeing the world through their creators' eyes and in effect becoming their captive. For them, women and their work were often invisible or subsumed under men of their households. Slaves seldom appeared by name unless they completed additional work, had something distributed to them, resisted their bondage, or had a calamity befall them. In every case the value of what they did or who they were was rendered in pounds, shillings, and pence or in dollars and cents. In 1791, "1 Superannuated Neg[r]o named Frank" died at Northampton Furnace's mine. The furnace's clerk dutifully recorded a debit of six pence to "Profit and Loss." Accounts skew our view of the people whose transactions they distill and record. Imagine how someone might reconstruct our lives and our values from business we did at one store, where we received most of our earnings in credit and where we had to purchase most of what we could not make or grow ourselves.[20]

Yet despite their limitations, account books tell compelling stories. They testify clearly to the values and perspectives of the people who created them. Account books are instruments of power through which their compilers quantified work, assigned value to it, and structured workers' activities. When Robert Pinion broke into the New Haven ironworks' counting house in 1666 and tore out the ledger pages that summarized his accounts, he acknowledged and struck at the power that bookkeeping accorded his employer. Daybooks, journals, and ledgers reveal how adventurers and ironworkers related to one another, the circumstances which shaped ironworkers' choices, and how ironworkers chose to enjoy the fruits of their labor. Account books are cause and product of the changing world of work in early Anglo America—evidence of and contributors to the industrious revolution. When studied in conjunction with other documents and official public records, they permit us to trace the changing meanings of work for the entrepreneurs who kept them and for the workers whose activities they measured.[21]

I should say a few words about the story that I tell in the following pages. First, I have tried to write it as a story—or perhaps more accurately, as a series of linked stories—to reach the widest possible audience. This means, for example, that I offer in the text few details about quantified data or about the methods by which I compiled them. Readers who wish to know more may consult my endnotes. Similarly, explicit discussion of historiography generally can be found in the endnotes. I have, however, tried to let adventurers and ironworkers speak in their own words whenever I could. I have mostly left their emphasis, their spellings, their syntax, and their punctuation as I found them, except when doing so would, in my judgment, prove too confusing.

I should also note that this is not the story that I thought I would tell. I began this project certain that I would go to archives and tap a stream of tales about ironworkers who resisted adventurers and industrial capitalism at every turn—the pioneer moulders and forgemen of an American industrial working class, if you will. To be sure, I encountered such stories. But I found more of ironworkers who mostly played by adventurers' rules, partly because they shaped those rules just enough to follow them willingly most of the time and partly because circumstances, be they years invested in mastering a trade, necessity, or enslavement's stark realities, left them no better alternative.

My story about making iron during the industrious revolution is told in two parts with six chapters. Chapter 1 presents an overview of the iron industry from 1620 to 1830 and of the processes by which people made iron. Ironworks were expensive and risky ventures which required the marshalling of capital and labor on a scale and scope seen nowhere else in early Anglo America save on sugar plantations. Adventurers sometimes reaped considerable profits, but the iron business was never easy. Iron production exploded, particularly in the Middle and Chesapeake Colonies after 1715, to meet surging British and local demand.[22] The Revolution created a truly "American" industry, especially when adventurers began to demand high tariffs to protect a growing but unstable domestic market. Ironmasters needed reliable workers; how well these workers carried out their duties largely determined how much iron they made and how well it sold. Making iron was dirty, backbreaking, and dangerous work, though for tradesmen, especially those who handled and shaped hot metal, it promised a comfortable and respectable living. The high stakes involved in the iron business and the influence that ironworkers exercised over what they made and how they made it compelled adventurers to do all they could to create a disciplined work force. That was exceedingly difficult in a place where labor was often scarce and where men expected work to make them and keep them free, so adventurers invented an American version of industrial bondage.

Part 1, "Iron and Empire: The Colonial Era," begins with Chapter 2, which examines the creation of an iron industry in Virginia and in New England between 1619 and 1685. Adventurers aspired to serve God by using their enterprises to lift up people and societies. Anglo Virginians sought to fund the conversion of Natives into anglicized Christians by making iron; their venture perished at the hands of those they claimed to save. In New England, adventurers and authorities dreamed that ironworks would save their godly societies. They tried to realize that dream by recasting unruly and blasphemous ironworkers into sober and dutiful men through a "culture of discipline" founded in Puritan values and centered on a legal system that punished "ungodly" behavior. Puritans largely succeeded at acculturating the ironworkers whom they invited among them, mostly because workers made use of the institutions that policed them for their own purposes. New England's ironworks soon failed, though not before a different vision of industrial work emerged, one which never sought to redeem those who were charged to do it. It emphasized coercion of unfree outsiders as vital to industrial enterprise in a land where free ironworkers seemed too expensive, too ambitious, and too powerful.

The subsequent two chapters examine the Chesapeake and the Middle Colonies between 1715 and the American Revolution. Chapter 3 explores how and why Chesapeake ironmasters turned to a system dominated by unfree labor in which they readily resorted to the whip to control convict servants and to acculturate African slaves and their children to colonial society and to the needs of iron production. Whips alone could not make good ironworkers of slaves, so adventurers had to modify bondage—especially for tradesmen—by emphasizing incentives and inviting all slave ironworkers to participate in the industrious revolution. Enough slaves did so to improve their lives and thus helped to forge a stronger, more flexible form of slavery. Chapter 4 discusses iron production and the industrious revolution in New Jersey and Pennsylvania. There, adventurers became some of the region's largest slaveowners. They selectively deployed slave labor as one of many techniques designed to discipline free white ironworkers and make them work harder and better. Adventurers largely succeeded in defining the contours of free labor and of the industrious revolution as their region became the colonial iron industry's center.

Part 2, "Iron and Nation: The Early Republic," examines the iron industry from the American Revolution to 1830. The Revolution and the new nation that it created recast both the relationship between adventurers and ironworkers and the industry's relationship to the industrious revolution. North and South headed down similar paths with different destinations. Chapter 5 charts industrial slavery's transformation in the Chesapeake. The death of white indentured servitude, the expansion of slave hiring, and more slave ironworkers with families encouraged southern adventurers to adopt and even celebrate a form of industrial paternalism akin to that which planters had begun to practice. Slave ironworkers manipulated that paternalism to help them establish more limits on their bondage. But they remained slaves, which ultimately left them nearly powerless to protect what their industry and their love had built, and which also complicated how their masters related to northern adventurers and participated in national politics. Chapter 6 follows the transformation of work and life within the iron industry of the mid-Atlantic states. Gradual emancipation slowly ended slavery and confined most black ironworkers first to servitude and then to low-paying unskilled jobs. The region's adventurers, eager to mold free labor in a new nation and a new business climate, began to view white ironworkers as family and as potential political allies. They tried to reform employees by fining them and by promoting evangelical Christianity, education, and temperance, because they hoped for more industrious hands and because they believed that what they were doing was right. Adventurers also sought to mobilize their employees to join their battle against imported iron and against opponents of high tariffs. Ironmasters mostly enjoyed victorious political campaigns and peaceful labor relations, partly because they accommodated enough ironworkers' aspirations enough of the time.

This book examines how and why early Anglo Americans made iron between 1620 and 1830. It presents the iron industry as an American "crucible" which combined many of the aspirations, achievements, contradictions, and cruelties of the industrious and industrial revolutions. It tells how the United States became the first early modern European colony to join the developed world—a story which includes the familiar tale that so many Americans like to tell themselves, each other, and the world. Despite great hardships, entrepreneurs and employees motivated by profit, self-improvement, and an internalized work ethic together harnessed abundant natural resources to make a rich, modern democratic republic in which men, or at least most white men, were roughly equal. But that is not the whole story. It also took slaves and whips to forge early Anglo America's three revolutions, a legacy most Americans have never learned, or have chosen to forget.

1

MASTERED BY THE FURNACE

In 1749, Joshua Hempstead stopped at Principio Iron Works while travelling through Maryland. It impressed him. "There is," he remarked,

> 30 piggs now cast that ly hot in the sand as they Run out of a hole in the bottom or Lower End of the furnace. . . . The furnace . . . is fed with oar and coal &c at the Top as if it were the Top of a Chimney. . . . There they bring Horse Carts [with] the oar the Coal & oyster Shels and there Stayd two men Day and night. The top of ye furnace is about breast high from the floor where they Stand to Tend it & ye flame Jets out Continually, is extinguished by the oar Coal & Shels as they feed it. Each Couple Tend 24 hours in which time they Run or Cast twice. They have Small Baskets that hold about a peck & a half & they put in a Cart [a] number of Baskets full of oar & a Certain Number of Baskets of Coal and a Certain Number of Baskets of oyster Shels, all in exact Proportion, and as the materials Consume below in the furnace they filled up at the Top and out at the Bottom. Besides the Iron that is drawn off near a day there is a vast Quantity of Glass that Runs out Every now & then. . . . [1]

The "piggs," oblong bars of cast iron formed by trenches in the furnace floor, headed to a refinery forge. There forgemen melted and hammered them into bar iron, out of which smiths beat tools, horseshoes, and other goods.

A furnace in blast was an impressive sight. It lit the night sky and was visible several miles away. Indeed, the Principio Iron Works was a landmark when Hempstead visited; Benjamin Franklin included it in the first edition of *Poor Richard's Almanack* in 1733. Workers ministered to the furnace's needs around the clock for months. Scores of others, who Hempstead may have seen but never mentioned, labored year-round to supply it with ore and charcoal until the stack gave way or winter struck. The workers mastered by Principio served what was then the largest firm in the colonies.[2]

Making iron challenged everyone involved in it. Furnaces and forges required plenty of money and constant oversight. Principio's owners had risked thousands of pounds in buildings, equipment, land, and bound labor to produce a metal vital to daily life, commerce, and politics throughout the Atlantic world. They stood to lose their reputations as well. "I expect it will be nothing new to hear that we Iron Masters are in general a sett of Hungry, needy Beings," Henry Drinker began a 1790 request for payment, "frequently bare of Money & straining our credit." Nature, technology, and turbulent markets often upset adventurers' best-laid plans and sometimes badly humiliated them. So might ironworkers, whose skill, muscle, and will adventurers had to harness if their enterprises were to succeed.[3]

Ironworkers felt the demands of making iron even more keenly. Ironmasters ventured their money and their honor; ironworkers ventured their lives and their bodies. The iron industry promised artisans, especially those who knew how to mold or forge hot metal, a lucrative career and the honor of claiming mastery of a difficult trade. It offered other workers a living or a chance to secure or pursue economic independence. Such benefits often carried a high price. The work was arduous, dirty, and dangerous. Ironmasters recorded and quantified production costs in ledgers. Ironworkers could tally the costs in sore backs, blinded eyes, and burned skin.

Adventurers and ironworkers were the first to confront changes in work that few others experienced until the nineteenth century. They were engaged in the largest industrial enterprises in the colonies. Both reckoned with production routines that required coordination of discrete tasks, many of them time-dependent. For adventurers, this necessitated close attention to work, which some ironworkers experienced as more vigilant supervision and more accountability to machines and clocks. Conditions unique to early Anglo America intensified such imperatives for adventurers. A successful ironworks needed capital, labor, and skill—all were often hard to come by. For workers, this potentially cut both ways. They faced greater scrutiny from ironmasters, who worried about risk and sought to control production as best they could. Scarcity of labor and skill empowered workers, who already wielded significant influence over work. Adventurers' desire to lower costs, organize work as they wished, and resolve difficulties in finding and keeping reliable hands led them to invent an Anglo American version of industrial bondage.

Methods for Making Iron

Early North American ironmasters knew two ways to make iron: the direct and indirect methods. The former, known as the "bloomery" or "Catalan" method, was the older of the two and the most widely used until the early modern era. Bloomery forges converted ore directly into wrought iron by melting it in a hearth with charcoal and repeatedly hammering and reheating the pasty mass to expel as much slag as possible.[4] The indirect method split production into two stages. First, a blast furnace stoked by a water-powered draft smelted ore with charcoal and a fluxing agent (oyster shells at Principio, limestone most everywhere else) to make cast iron—either pig iron or moldings such as pots, kettles, or stove plates. Because pig iron was too brittle for most applications, it required processing at a refinery forge to toughen it.

Bloomeries produced most iron made in North America before 1720. They remained common in remote parts of New Jersey decades after the Revolution, largely because they ran inexpensively and served local markets well.[5] Bloomery forges dominated the iron industry during the seventeenth century and persisted because they suited the needs of proprietors and consumers. A bloomery demanded far less capital to build and to operate than did a furnace or a refinery forge. Processing of metal took place under one roof with relatively simple equipment. Colonial ironmasters sometimes built bloomeries to test the quality of ore deposits before they would commit to building a furnace. Bloomeries also offered adventurers another advantage; they required relatively little labor. It took fewer than ten workers to run one.[6]

Bloomeries gave way to furnaces and refinery forges—the indirect process—after 1720. The indirect process developed gradually out of bloomery techniques in medieval western Europe. During the fourteenth century, a rudimentary blast furnace emerged in the Low Countries, and from there the technology spread to northern France. French artisans introduced furnaces and refinery forges to England during the late fifteenth century. By 1600, blast furnaces were common throughout southern England and Wales.[7]

Bloomery iron was often superior to refined iron for many purposes. But the indirect process made more iron and it required less labor and probably less fuel for each ton of iron produced. Fuel efficiency mattered little to most colonial ironmasters, who considered wood a cheap and nearly inexhaustible resource. Labor, however, was neither cheap nor abundant to them. Blast furnaces mostly replaced bloomeries in northwestern Europe by the sixteenth century thanks to war—armies needed cast iron weapons and munitions. Refinery forges arose largely to process what military use did not claim. Growing consumer demand for hollowware and wrought iron, a product of the industrious revolution, sustained furnaces and forges during peacetime.[8]

Adventurers and the Explosion of the Colonial Iron Industry

As markets for cast and wrought iron goods grew in Britain's North American colonies, ironmaking ventures there looked more alluring to entrepreneurs on both sides of the Atlantic. Hope of earning steady profits drew them to the iron business. In 1764, Charles Carroll touted his son's marriage prospects by informing him that the one-fifth share of the Baltimore Iron Works that he would inherit earned £400 sterling annually. After visiting the Batsto Iron Works in 1799, Julian Niemcewicz concluded that, despite expenses for labor, livestock, and raw materials, "the net profit must be large." Exactly what the "net profit" was of Batsto or of any other ironworks is difficult to determine, but some ironmasters clearly did well. Philadelphia merchant Joseph M. Paul earned several thousand dollars between 1801 and 1807 from his one-sixth stake in a furnace and an ironworks in New Jersey.[9]

The cost of entering the iron business and staying in it was steep. Few colonists could afford either before 1730. In 1786, Henry Drinker warned a prospective ironmaster that he should expect to spend £6,000 to £10,000 to buy a suitable site; build a furnace, forge, storage facilities, and employees' dwellings; and stockpile enough raw materials and provisions for blast. Jeffrey Zabler has calculated that the average value of the land, buildings, and moveable property of a Pennsylvania furnace which operated between 1800 and 1830 was nearly $33,000. The tugs on adventurers' purse strings grew more insistent once ironmaking began. Waterwheels, mill races, furnace stacks, and forge helves and hammers demanded constant repair; a breakdown in any might halt production. To safeguard against drought, ironmasters frequently had dams built to store water and regulate their power supply.[10]

It also took managerial talent and technical expertise to build and run an ironworks. Until well into the eighteenth century, few colonists commanded either. English entrepreneurs and agents largely financed and oversaw the first ironworks in Massachusetts and Virginia in the seventeenth century. British capital and supervisors relaunched the Chesapeake's iron industry in the 1710s and played a key role in it until the Revolution.[11]

The indirect process and the colonial iron industry took off together after 1715, thanks partly to the crowning of King George I. The United Kingdom went to war against Sweden after George I learned that King Charles XII had plotted to support a rebellion against him. He persuaded Parliament to outlaw trade with Sweden in 1717. British iron manufacturers, dependent on Swedish bar iron, protested that the embargo would drive them out of business. Some petitioned the Crown to offer incentives to anyone seeking to start ironworks in North America. Trade and the full flow of Swedish iron into Britain resumed the following year. But reliance on an enemy for such a strategic commodity gave ammunition to men on both sides of the Atlantic who wanted imperial authorities to promote a colonial iron industry.[12]

Appointed and elected officials of the Chesapeake colonies and Pennsylvania tried to encourage iron production. Maryland enacted three laws between 1719 and 1750 which granted land to adventurers and exempted most, if not all, ironworkers from levies and from requirements to provide labor on roads and highways. In 1727 and 1732, Virginia passed similar legislation. Pennsylvania's government financed some ironworks through its General Loan Office, set up in 1723. Seventeen ironmasters borrowed to finance construction of new ironworks or improve existing ones. Adventurers William Branson and John Taylor were the only borrowers for whom the Loan Office waived its credit limit. Thomas Penn, one of the colony's proprietors, sought to help adventurers lower fuel costs. In 1764, he dispatched English colliers to Pennsylvania to teach "those employed at the Iron Works to make the coal with a much less quantity of wood than they had hitherto consumed."[13]

Penn had reason to be concerned; proprietary efforts to promote an iron industry had led to violence in New Jersey. Rioters attacked employees of the Union Iron Works in 1749 and 1754 to strike at its owners William Allen and Joseph Turner; both were closely connected to the colony's Board of Proprietors. Allen, Turner, and their associates, the rioters believed, had endangered their industrious revolution—fuelling the ironworks deprived them of trees which they logged and sold to purchase economic independence. They had a point—the Union Iron Works devastated the woods that surrounded it. In 1783, Johann David Schoepf reported that it had "exhausted a forest of nearly 20,000 acres in about twelve to fifteen years, and the works had to be abandoned for lack of wood." New Jersey's leaders partially agreed with the rioters; the iron industry's appetite for wood created problems. In 1752, the Board of Proprietors of eastern New Jersey noted that "a great number of iron works have been lately erected amongst the mountains" of Morris County "and that the owners of these works or persons employed under them are making great havock and destruction of the Proprietor's timber, to supply the said works with coal &c., and that the lands there, after the timber is cut off, will be of little or no value." Nine years later, they admitted that the problem had worsened. Ironworks' role in accelerating eastern North America's deforestation soon had enormous consequences for adventurers and their enterprises.[14]

Provincial leaders stood with adventurers to lobby for imperial bounties to encourage iron production and against British ironmasters' campaigns to constrain the colonial industry's growth and development. On the first score they failed; on the second they enjoyed mixed success. Adventurers and their supporters sank proposals to outlaw or restrict construction and operation of North American forges. By 1757, buyers in Britain paid no duties on colonial pig or bar iron. The Iron Act of 1750, however, decreed that all colonial pig and bar iron shipped to Britain had to bear a stamp indicating where it was made or be taxed as foreign iron. More important, the Iron Act outlawed new mills to manufacture finished iron or steel goods—each violation brought a fine of £200 and demolition of the works. Resentment of the Iron Act prompted many colonial adventurers to support the Revolution and likely spurred Benjamin Franklin to articulate a new vision of colonists' place within the British Empire.[15]

Colonial Anglo America never commanded more than a 10 percent share of Britain's imported iron market, far less than Sweden or Russia. But efforts to supply Britain with iron shaped the industry profoundly, particularly in Maryland and Virginia. Exports from the Chesapeake to Britain dwarfed those of the rest of the colonies.[16] Accounts of Chesapeake ironworks document that some annually shipped hundreds of tons of iron to British ports. Maryland and Virginia contained the largest concentration of British-owned firms, the first and biggest of which was the Principio Company. The majority of the region's ironworks were within ten miles of Chesapeake Bay or one of its navigable tributaries. The transatlantic iron trade followed the plantation economy's rhythms. Most Chesapeake adventurers who resided in the area were wealthy planter-merchants who exported tobacco; they saw ironworks as another way to diversify their estates. John Tayloe, for example, sometimes shipped tobacco and iron together, with pig iron serving as ballast. Dr. Charles Carroll explained to a London merchant in 1750 that "the scarcety of Tobacco has hitherto prevent'd" a ship "from Sailing hence with my Pigg Iron wch I hope shee will soon."[17]

Pennsylvanians began to invest heavily in ironworks in the 1720s. Within a generation, native-born ironmasters dominated the region's industry. Many were merchants, mostly headquartered in Philadelphia, who financed ironworks in southern New Jersey and southeastern Pennsylvania and either hired agents or formed partnerships with men with ironmaking experience. Many saw iron ventures as an opportunity to leave behind most of the pressures and uncertainties of transatlantic commerce. Making iron broadened their portfolios and gave them a commodity that they could exchange for imports. Older and wealthier merchants often considered ironworks relatively safe investments over which they could delegate most managerial responsibilities.[18]

Growing demand for iron propelled the rise of British North America's iron industry, especially in the Middle and Chesapeake Colonies. There were virtually no furnaces or refinery forges in the region in 1715. By 1750, adventurers had built at least twenty-eight furnaces and forty-six forges. Pennsylvania and New Jersey adventurers, who sold most of their product locally, opened most of the forges. Chesapeake ironmasters channeled most of their resources into furnaces focused on making pig iron. Ironworks construction accelerated between 1750 and 1776: forty-seven furnaces and fifty-nine forges started up in that period. In every colony except Delaware, at least as many and usually more furnaces and forges were built in the generation before the Revolution than before 1750.[19] Ten furnaces first went into blast in New Jersey between 1766 and 1776, thanks mostly to the American Iron Company and to Charles Read.[20] Forge construction expanded most sharply in New Jersey and grew considerably elsewhere.

Adventurers' new emphasis on making bar iron partly flowed from the confluence of imperial regulation and technological change in the British iron industry. The Iron Act of 1750 officially sanctioned colonial forges; removal of all duties on colonial bar iron also may have encouraged investment in them. By the 1750s, British pig iron made with coke was cheaper than charcoal pig iron. Once its quality was comparable, the Principio Company informed its agent, it would "Urge us upon other methods of Converting more Pigs into Barrs."[21]

The boom in forge construction owed principally to domestic markets. Pennsylvania and New Jersey adventurers had long sold mostly to local con-

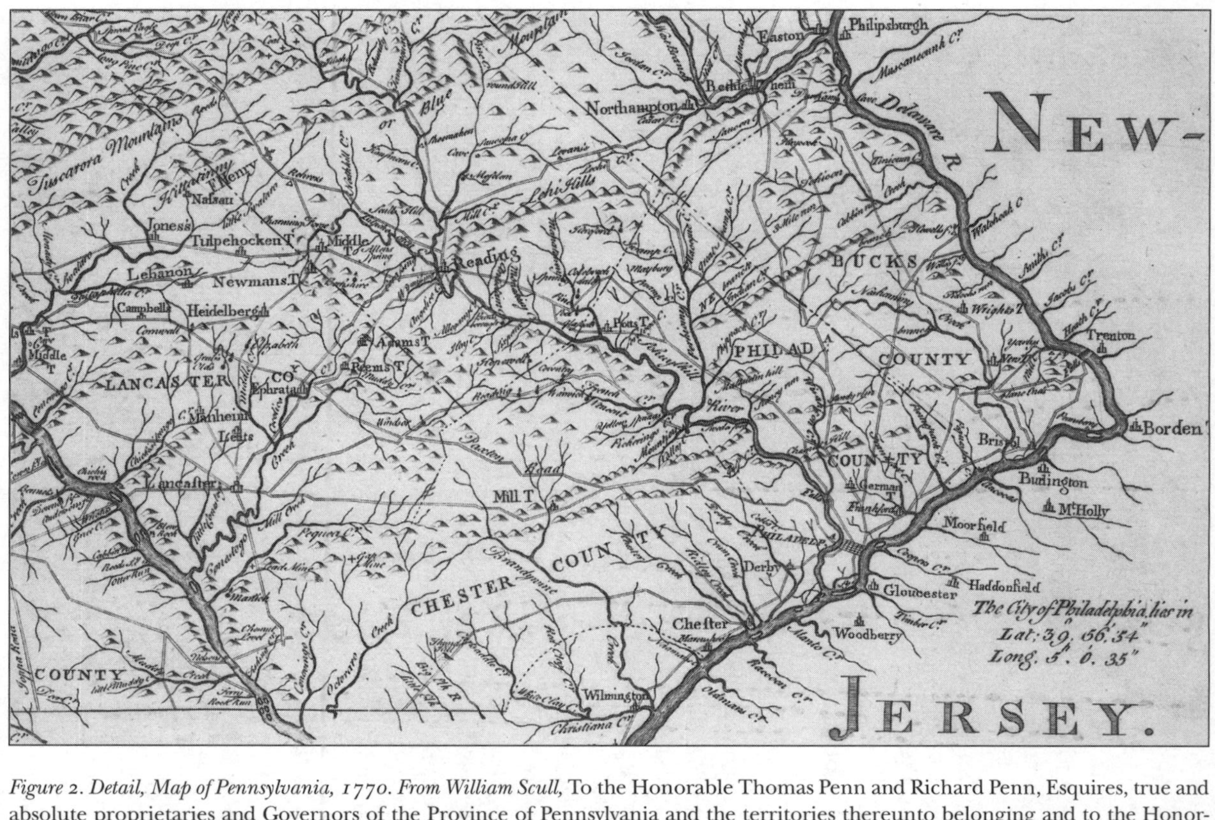

Figure 2. Detail, Map of Pennsylvania, 1770. From William Scull, To the Honorable Thomas Penn and Richard Penn, Esquires, true and absolute proprietaries and Governors of the Province of Pennsylvania and the territories thereunto belonging and to the Honorable John Penn, Esquire, Lieutenant-Governor of the same, this map of the Province of Pennsylvania *(Philadelphia, 1770). By permission of the Houghton Library, Harvard University. Scull's map locates and names furnaces and forges (such as Cornwall Furnace and Coventry Forge), an indication of their significance to Pennsylvania's economy and landscape.*

sumers. Population growth increased demand for hollowware such as pots, kettles, and skillets, as well as for bar iron that smiths reworked into finished products. Heavy German immigration created a large and expanding market for cast-iron stoves. By the Revolution, most colonial adventurers considered domestic buyers more important than British customers, and many, tired of dealing with faraway markets which often became glutted before their iron arrived and over which they had little influence, focused their efforts accordingly.[22] By 1774, ironmasters in Pennsylvania and New Jersey had even organized a cartel which tried to prop up domestic prices by requiring each furnace and forge to export a few tons to Britain.[23]

The boom in ironworks construction between 1763 and the Revolution and the local glut of iron that it created helped to ruin some adventurers. Peter Hasenclever, agent for the American Company, erected three furnaces and seven forges in New Jersey, staffed them with over 500 workers, and spent over £54,000 before his employers in London saw any return on their money. They fired him; Hasenclever wrote a tract to defend his conduct and salvage his reputation. John Semple, who arrived in the Chesapeake colonies as a tobacco buyer for a Glasgow firm, bought part of Virginia's Occoquan Iron Works in 1762 and overborrowed to restart production. In 1771, his creditors had him confined to the town of Dumfries until he died two years later. Henry William Stiegel, owner of Pennsylvania's Elizabeth Furnace and Charming Forge, bought more land than he needed or could afford. In 1774, he languished in debtor's prison until an act of clemency from the Assembly freed him on Christmas Eve. Stiegel hoped that the sale of munitions to Continental forces would save him. By 1778, he was penniless. Stiegel died a teacher and clerk at Charming Forge, the charge of his nephew George Ege.[24]

Stiegel was hardly alone in his great expectations of the Revolution. Most ironmasters thought that it promised a windfall. Continental forces needed cannon, shot, salt pans, pots, and skillets; furnaces geared up to supply them. Between July and December 1776, Northampton Furnace sold tens of thousands of pounds of shot to Maryland. "Had we ten Furnaces & Forges," Isaac Zane wrote from Virginia's Marlboro Furnace in 1777, "the demand is more than equal to what could be made."[25]

Zane and other adventurers had to temper their enthusiasm. Zane had complained in 1775 that debtors could not pay him because the "general convulsion of affairs have so stagnated commerce that cash & every standard seems gone from amongst us." Rampant inflation, especially toward the war's end, increased costs steeply and ate most, if not all, of the profits. After he celebrated wartime demand for iron, Zane noted that "the extravagance of Phila. has crept up here almost every thing is double price &

some things three times the usual prices." Worse, the iron industry's heartland became a battlefield as Continental and British forces contested northern New Jersey and southeastern Pennsylvania. The British Army destroyed a total of four ironmaking facilities in Pennsylvania, New Jersey, and Virginia, and it forced Continental soldiers to defend ironworks, as did George Washington's troops when they camped at Warwick Furnace and Reading Furnace after the Battle of Brandywine.[26] The British Army had another reason to target furnaces and forges. Five adventurers had signed the Declaration of Independence. At least twenty Pennsylvania ironmasters served as Continental or militia officers.[27]

Ecological damage also shuttered many ironworks during the revolutionary era. In 1783, Samuel Gustaf Hermelin identified several furnaces and forges which had recently closed because they had exhausted their timber reserves. Johann David Schoepf foresaw doom. "The business of the mines and foundries," he asserted,

> in New Jersey as well as throughout America, cannot be said to be on as firm a basis as in most parts of Europe, because nobody is concerned about forest preservation, and without an uninterrupted supply of fuel and timber many works must go to ruin, as indeed has already been the case here and there. Not the least economy is observed with regard to forests. The owners of furnaces and foundries possess for the most part great tracts of appurtenant woods, which are cut off, however, without any system or order.

"If it does not fortunately happen that rich coal mines are discovered," he predicted, "enabling such works to be carried on, as in England, with coal, it will go ill with many of them later on."

Schoepf was right. In 1809, Benjamin Henry Latrobe observed that "the disease of which all the small Iron works in Pennsylvania are dying, *is the scarcity of Charcoal.* In the Jersey this scarcity has also been fatal to some principal works." By 1850, many ironmasters practiced conservation more systematically, a legacy of which can be seen in the public forests or private reserves which surround furnace ruins in New Jersey, Pennsylvania, and Maryland.[28]

Adventurers and the Making of an "American" Industry

The Revolution marked a watershed for the iron industry. After national independence, it turned inward and more insular as adventurers marketed almost exclusively to domestic customers. By 1800, the United States had virtually stopped exporting bar or pig iron, forcing its ironmasters to compete with each other and with foreign producers for American buyers. To keep foreign competitors at bay, ironmasters lobbied state authorities and then the federal government for the highest tariffs they could get. In short, during the revolutionary and early national eras, an industry that had looked out to the Atlantic world became fiercely protectionist and self-consciously "American."[29]

The Americanization of the iron business owed partly to state actions taken during the Revolution. Pennsylvania and Maryland seized ironworks owned by Loyalists or by British firms. In 1779, the Commissioner of Forfeited Estates sold Loyalist Joseph Galloway's interest in Pennsylvania's Durham Iron Works to Richard Backhouse for nearly £13,000. In 1781, Maryland confiscated Daniel Dulany's share in the Baltimore Iron Works and nearly all the property of the Principio and Nottingham companies, which included hundred of slaves, and auctioned them off.[30]

Confiscation was part of wider changes in ownership within the industry during the revolutionary era, particularly in the Chesapeake. Most prominent planters abandoned ironmaking by 1790. Northern migrants, many of whom had personal or familial ties to ironmasters in Pennsylvania and New Jersey, largely replaced them. By contrast, most of the prominent adventurers in Pennsylvania and New Jersey had either entered the iron business just before the Revolution or descended from the principal ironmaking families of the colonial era. Some of their sons and grandsons played an influential role within the industry through the Civil War. Merchants, especially in Philadelphia, also continued to invest heavily in ironworks well into the nineteenth century.[31]

Iron production grew rapidly, particularly in Pennsylvania, after the Revolution. Adventurers from New Jersey south to Virginia built a total of 113 furnaces and forges from 1784 to 1800. In Pennsylvania, ironmasters opened more furnaces (twenty-seven) and more forges (forty-nine) between 1784 and 1800 than they did during the entire colonial era.[32] Growth continued after 1800, with Pennsylvania leading the way. By 1840, when federal officials assembled the first comprehensive and reliable census of manufacturing, there were more than eight hundred furnaces and nearly eight hundred forges, bloomeries, and rolling mills operating in the United States. More than one-quarter of the furnaces and more than one-fifth of the forges were in Pennsylvania. Iron production increased in every state in the mid-Atlantic and Chesapeake regions between the Revolution and 1840.[33]

The industry's center shifted north and west. Adventurers built most new furnaces and forges in areas that the Revolution and subsequent conquest of native peoples had recently opened to white settlement: upstate New York, central and western Pennsylvania, Ohio, western Virginia, Kentucky, and Tennessee. By 1760, production had already begun to shift south in New Jersey and west in the Chesapeake and Pennsylvania. Twenty-three ironworks opened between 1784 and 1827 just in New Jersey's Pine Barrens. Virginia had developed a substantial iron industry in the Shenandoah Valley by the Revolution; production spread beyond the Alleghenies in the late eighteenth century. Nearly half of the furnaces and forges built in Pennsylvania between 1783 and 1800 were west of the Susquehanna River, mostly in the Juniata and Ohio valleys.[34]

The early republic's iron industry grew thanks to four developments: explosive demand for iron, new products, better marketing techniques, and government protection. An expanding nation, growing cities, and new industries needed more iron and a wider variety of cast and finished iron goods. Adventurers could supply them more easily because they had better information. By the 1820s, specialized commission merchants handled most iron that entered Baltimore, New York, and Philadelphia. They extended ironmasters credit and told them which goods sold best.[35] Forges and furnaces learned to make new products, such as cast iron ploughshares, equipment for mills and factories, and cast iron pipe for urban water and gas works. For example, between 1822 and 1830, New Jersey furnaces owned by Samuel Richards sold pipe for utility systems in Philadelphia, Richmond, and New Orleans.[36]

To judge from adventurers' public statements, an activist federal government mattered most in determining the industry's fortunes. Most adventurers supported ratification of the Constitution because it empowered the federal government to provide a more stable financial system, create a truly national market by giving it the sole power to regulate interstate commerce, and most important, enact and collect duties on imports.[37] Ironmasters of the early republic focused their political energies on obtaining, preserving, or raising tariffs. They formed associations to collect data on production and on how many people they employed so that they could make a stronger case that their interests and the nation's were one. They also insisted, when asked, that unchecked imports would push them under. Louis McLane's investigation of manufacturing, released in 1833, reported that thirty-nine of fifty-three ironmasters claimed that reducing the tariff would cause them to leave the iron business or stop production. This was not just posturing. That same year, Edward B. Grubb told his brother that he feared sinking any more money into Mt. Hope Furnace because he foresaw lower tariffs, "which in all probability will shut us up or if we do go on it will probably be at a loss we then find ourselves possessed of a large and unproductive landed estate very much reduced in value and on which we shall have to pay high state taxes."[38]

The early republic's adventurers knew that the wild business cycles that they faced correlated closely with how much iron the United States imported—the less iron that entered the country, the better they did. Indeed, U.S. ironmasters fared best between 1808 and 1815, when an embargo and the War of 1812 shut out all foreign-made iron. Otherwise, uneven demand and foreign competition forced iron prices down or made them stagnate for most of the early nineteenth century. When imports resumed after the War of 1812, the iron industry slumped and adventurers swamped Congress with petitions for relief.[39]

The Panic of 1819 sliced demand for iron and restricted access to credit, which only pinched ironmasters harder. The price of bar iron fell more than 20 percent in Baltimore between 1820 and 1821, prompting Jacob M. Haldeman to stop production at Cumberland Forge until the market recovered. Other ironmasters bailed out. John Anderson, part-owner of Colerain Forge, asked Haldeman to recommend him for a managerial position in New England. He would accept a demotion because "a good salary nowadays is better than an ownership in Huntingdon County," as long as he could sell his interest "without sacrificing too much." Weeks later, Anderson dumped his share in Colerain at considerable "*sacrifice.*"[40] The next few years brought slight and uncertain recovery. Virginia forges had trouble selling bar iron in Richmond and their owners had trouble getting credit. John Donihoe informed Buffalo Forge owner William Weaver in 1830 that a Lynchburg bank's directors were reluctant to lend to ironmasters after they "discovered that the Iron *Business* is not *Profitable* & it is likely that some of the Ironmen have not payd up their paper." A year earlier, John W. Schoolfield had wryly summed up the financial predicament of adventurers: "I expect an Iron master out of debt would feel something like a fish out of water."[41]

Because they could control so little and stood to lose so much, adventurers tried to limit their risk. Direct control over supplies of raw materials such as timber and iron ore promised greater predictability. Many Maryland and Pennsylvania adventurers, particularly during the colonial era, sought to curb expenditures on provisions by growing what they could on what Arthur Cecil Bining called their "iron plantations." Rather than depend on someone else for pig iron, forge owners like William Weaver sometimes acquired furnaces. The relative isolation that the need for large wooded tracts imposed on ironworks often reinforced many adventurers' desire for self-sufficiency.[42]

Reluctance to assume risk also led early Anglo American ironmasters to embrace technological innovation cautiously. They introduced only two major changes between 1620 and 1830, one geared to increasing furnace output, the other to diversifying it. The first involved replacing bellows with blowing tubs, a development which began in the mid-eighteenth century. The other surfaced in Pennsylvania in the 1730s and spread rapidly as colonial ironmasters adopted new methods to cast iron. Otherwise, the basic techniques for making iron changed little for over two centuries.

Adventurers were hesitant to innovate because they saw little reason to do so. Unlike their British counterparts, many of whom suffered from timber shortages and who abandoned charcoal for coke by 1775, most Anglo American ironmasters saw nearly limitless forests. Moreover, most sold to buyers who preferred charcoal iron to that made with mineral coal. Even if they had wanted to convert to coal, ironmasters had no economical way to get it to their furnaces until the advent of canals and railroads. Deposits of bituminous coal lay near some ironworks, but its high sulfur content yielded pig iron that was brittle and difficult to forge. Ironmasters who wanted to use anthracite, the cleanest and most efficient coal, had to await techniques that would heat air before it entered the furnace—stacks that relied on ambient drafts could not sustain the temperatures that anthracite needed in order to burn. Only in the 1830s, when hot-blast technology, canals, and railroads arrived together to the Pennsylvania countryside, did many ironmasters begin to adopt mineral fuel.[43]

Besides changes in furnace draft and in molding technology, most American adventurers restricted themselves to minor improvements. Some devised ways to extend a furnace's blast. In the 1760s, Peter Hasenclever buried pipes that fed the American Company's waterwheel to "secure them from the frost in Winter-time." In 1794, John Potts expected that Virginia's Keeptryst Furnace would run well into winter because he had "built a House over the Wheel." Ironmasters also tried to power their ironworks more efficiently. By the early nineteenth century, most had replaced undershot wheels—which turn as water flows through and under them—with breast or overshot wheels onto which a mill race dumps water, enabling the wheel to tap the power of the stream's flow and gravity.[44]

Minimal innovation also enabled adventurers to avert risk; they lost little if a minor adjustment failed to yield good results. Why invest scarce capital in a new technology when the old one had functioned well for as long as anyone could remember? Why bother converting to mineral fuel when most forges preferred charcoal pig iron? The relative isolation of many ironworks could only have reinforced such sentiments, especially when they had a captive local market to which they sold most of what they produced. High tariffs and political and economic uncertainty also combined to discourage major innovation during the early nineteenth century. As a result, basic work routines within early Anglo America's iron industry, with one notable exception, changed little during its first two centuries.

Ironworkers Make Iron

For adventurers, the stakes were clear. They had gambled that they would profit handsomely by making and selling iron. To win, they needed the labor of dozens to hundreds of workers. Cutting wood, mining ore, and making charcoal commanded the attention of most. A furnace needed at most a dozen hands to operate, a forge considerably fewer. Workers, especially the tradesmen who shaped iron, supplied the muscle and the knowledge that transformed rock and wood into valuable metal. From their observations, expertise, and willingness to experiment came most improvements in technique. How well ironworkers performed their duties often meant the difference between success and failure.

The stakes were, if anything, higher for workers. For all who chose to do it, making iron offered a living and even a career. For many it provided a way to attain or to preserve their economic independence as landowners. For tradesmen it promised a relatively lucrative job in which they enjoyed considerable autonomy and security. All ironworkers took risks that most adventurers never faced. Their jobs left them covered in grime and sweat. Most chanced serious injury every time they reported to work.

Although the process of making iron presented workers with several challenges, they shared few of them. The specialized tasks and distinct stages of iron production divided them by occupation. Furnace and forge operations followed seasonal rhythms. Winter stopped metal production and provided time to make repairs and to lay in a supply of wood. Seasonal staffing changes, as well as the expanse and hilly terrain of many ironworks, meant that many ironworkers seldom saw one another. The coordination of discrete tasks, particularly at furnaces, rendered some workers subject to clock discipline and night toil that few outside the iron industry experienced, while others remained mostly free to set their own pace. Tradesmen, especially artisans who handled metal, enjoyed considerable autonomy; many other workers came under tighter scrutiny.

Of all the activities associated with making iron, felling trees and chopping wood for coaling demanded the most labor. Ironworks consumed so much charcoal that a plurality, and often a majority, of those employed at an ironworks spent time clearing woods. Except for clerks, woodcutters probably had the least dangerous job at an ironworks. Unless they were servants or slaves, they experienced the least supervision. Few ironmasters could oversee woodcutting, especially on huge estates. Besides, most woodchoppers knew what they were doing. Clearing forests was one of the most common tasks in early Anglo America; most farmers grudgingly had to do it to open new lands for cultivation.[45]

Cutting wood was tiresome but not complicated. After felling a tree, woodcutters removed small branches and chopped the wood into pieces four feet long. They then piled it to determine how many cords they had cut. It was at this point that ironmasters or their agents scrutinized their work. Hired woodcutters were paid by the cord; ironmasters wanted to verify that they got what they were paying for. They examined the stacks; choppers might try to deceive them by piling the wood over rocks or stumps, by placing it loosely, or by cutting the pieces too short. Some adventurers in colonial New Jersey and Pennsylvania passed the problem along to colliers by charging them for the wood that they coaled.[46]

Once the wood had dried, colliers began to convert it into charcoal. They burned part of the wood to generate enough heat to drive out most of the spirits, pitch, and moisture, leaving behind lumps of almost pure carbon. Colliers largely determined the quality and quantity of charcoal, and so influenced the productivity of a furnace or forge. Furnaces required strong and compact charcoal. Wood charred too hot or too quickly left only a few small and brittle coals that would probably crumble under the weight of a furnace charge. A slow, controlled burn would make plenty of light, sturdy charcoal, if colliers kept the pile tightly covered.

That was easier said than done. Coaling required skill, diligence, patience, and luck. Colliers had little control over several factors that would determine their success. The quality and quantity of charcoal depended largely on the age, type, and density of the wood from which it was made. Mature hardwoods, such as sugar maple, beech, and hickory, generally made the most and the best charcoal. Coaling also depended on the weather. Colliers usually constructed their first coaling pits in early spring and raked their last coals by late fall. Precipitation, humidity, or cold slowed burns and cut yields. Heavy rain or snow might flood charcoal pits and extinguish the burn, ruining the wood and wasting several days' labor. Wind was a collier's worst enemy; it could set his pile ablaze and reduce his work to ashes.[47]

Problems with handling and storing charcoal shaped how colliers worked. Ironworks could not keep charcoal outdoors; it deteriorates quickly when exposed to weather. It also readily absorbs water from the atmosphere. The more moisture that charcoal contained, the more of it a furnace had to burn. Charcoal is also fragile: long hauls or the stress of its own weight might reduce it to dust that could choke a blast. A greater worry was the threat of fire. Live coals could easily ignite and immediately engulf the wooden structures which ironworks used to store supplies and house equipment for casting and forging iron. Ironmasters liked to use charcoal quickly. Colliers seldom made it more than a few weeks before a furnace or forge needed it.

Small stockpiles left ironworks dependent on a steady and timely stream of charcoal. Insufficient reserves slowed iron production; a shortage could halt it. An orderly and sufficient flow of charcoal could not proceed without enough labor. In 1778, Baltimore Company manager Clement Brooke reported that Hockley Forge was "standing still for want of hands to cut wood & make coal; the other forge goes on but slowly for the same reason." The furnace, he added, had long stood "ready to go in blast, but for the want of a sufficient number of hands to coal I cannot put her to work."[48]

Adventurers needed colliers to execute their duties promptly and well. John Doyle, supervisor of Virginia's Bath Iron Works, considered supervision of the coaling grounds one of a manager's principal duties. Other ironmasters, particularly in eighteenth-century Pennsylvania, subjected colliers to considerably less oversight. They commonly negotiated agreements in which master colliers subcontracted work. This allowed a master collier to choose his helpers and to organize them as he saw fit. It also rendered him directly responsible and financially liable for their performance.[49]

The rapid turnover of charcoal potentially empowered colliers. It also sometimes forced them to work under constraints that most of their peers seldom faced. Generally the more time that colliers had, the better the charcoal they could produce. Time and sleep were luxuries that master colliers often did not enjoy. Each charcoal burn might demand their attention for at least seven consecutive days. When a furnace needed charcoal, ironmasters expected it. Colliers often paid when they could not provide it. In the 1760s, ironworks in Pennsylvania and New Jersey began to fine colliers regularly for failing to have enough charcoal ready when a teamster arrived for it. In 1810, New Jersey's Martha Furnace responded to a sporadic supply of charcoal by refusing to rehire colliers and by adjusting how their replacements were compensated.[50]

Colliers needed a dry, sandy spot, preferably nestled in the woods to protect against wind. Where a master collier constructed his pits mattered. In 1773, Patrick Campbell secured an "allowance" from Andover Furnace "on Acc^{t}. [of] the Badness of the Ground whereon he Coaled."

Few conditions impeded a burn more than damp earth. Delaware Furnace manager Derick Barnard complained that everything "has been done to a disadvantage in coaling ever since we went to the swamp. In fact it is a ruinous job at best," he continued, "except for a very short time, and even then ½ more expence than high ground coaling."[51]

After they had prepared the area, colliers piled the wood to form a coaling pit. Each pit contained twenty-two to fifty cords of wood and measured ten to fourteen feet high and thirty to forty feet in diameter. The workers stacked the wood carefully, tightened the pile with small limbs and wood chips, and topped it with about two inches of leaves and a thin coat of dust for extra protection against wind. They left one large hole at the pit's center, opened several small holes about a foot above the ground for ventilation, and then lit the stack. Weather permitting, colliers could leave the stack alone for twelve to eighteen hours before closing the top. They adjusted the side vents to draw the burn from the pile's middle to its sides. Until teamsters arrived to claim the charcoal, the pits required constant attention for up to two weeks. A crew of three to four men under the master collier's direction watched over several pits at once. The cover began to sink three to four days after lighting the pile. If the side vents failed to distribute heat evenly, the cover would settle haphazardly and create large openings through which air could enter to ignite the wood. The colliers tried to avoid such a disaster by walking atop the smoldering heap, raking dirt and leaves to seal holes, and jumping on the cover. They also carefully poked beneath the surface with a long iron rod to settle the coals.[52]

Coaling demanded sharp eyes, good balance, and sure feet. It also required high tolerance for dirt, smoke, and heat. Grime and soot coated colliers. A *Maryland Gazette* advertisement told readers that they might recognize three runaway servants because their clothes were "very black" after having recently worked with colliers. Methodist itinerant Benjamin Abbott recalled that after he had preached at an ironworks, "several of the collier's faces were all in streaks where the tears ran down their cheeks." The most dangerous part of a collier's job began once a pit started to sink. As he stood, walked, and jumped on the smoldering heap to settle the coals, he had to be vigilant and agile. Wind could instantly set a pile and anyone on it ablaze. A cover placed too tightly could trap gases which might explode and throw off anyone atop it. Colliers minded where they stepped; a thin layer of dirt and leaves separated them from the live coals below. Sometimes it failed to support their weight. In 1774, Charles Dowd "Burnt his foot" at Northampton Furnace's coaling grounds. Seven years earlier, a collier named Sam missed eleven days of work at Kingsbury Furnace for "being burnt."[53]

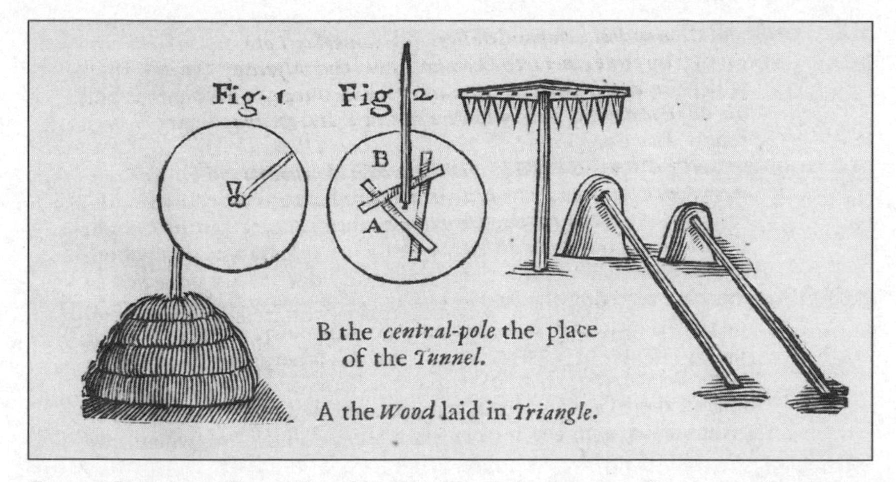

Figure 3. Preparing a Charcoal Pit, 1664. From John Evelyn, Sylva, or a Discourse of Forest-trees and the Propagation of Timber in His Majesties Dominions *(London, 1664), 103. By permission of the Houghton Library, Harvard University. Within an extended explanation of how to make charcoal to fuel an ironworks, Evelyn illustrated how to build a coaling pit chimney and how a pit should look once colliers had finished stacking the wood.*

When the pit stopped smoking, colliers covered it and let it cool a few days. They gathered the charcoal with wooden rakes, shovels, and baskets to limit breakage and lower the chance of causing sparks that could send their work up in flames. If the first coals that they drew seemed cold enough, colliers worked around the perimeter of the pit, remaining careful to expose no more than what would fill a wagon. Teamsters hauled the charcoal away and kept a close eye on their cargo. Their wagons often contained bottoms of loose boards which could be removed to dump the charcoal and spare the wagon from becoming kindling.[54]

Mining took less skill or care than making charcoal. Most miners dug ore from large pits with hand tools; Hermelin called most American iron mines "ore quarries." In 1748, Peter Kalm noted that in Pennsylvania "commonly the ore is here mined infinitely easier than our Swedish ore. For in many places with a pick ax, a crow-foot and a wooden club, it is obtained with the same ease with which a hole can be made in a hard soil." Often, he remarked, "the people know nothing of boring, blasting and firing." Johann David Schoepf agreed. "Any knowledge of mining is superfluous here," he recorded dismissively, "where there is neither shaft nor galley to be driven, all work being done at the surface or in great, wide trenches or pits." Furnaces that smelted harder ores used explosives to remove them.[55]

A few furnaces, concentrated in northern New Jersey and central Virginia, used shaft mines. Julian Niemcewicz observed four shafts at New Jersey's Mount Hope Furnace: "The deepest are 30 to 40 feet; two or three people work in each one." Subterranean mines required excavation of tunnels, caverns, and air shafts, all reinforced with timbered drifts, as well as a drainage system to prevent flooding. Their construction and maintenance demanded skill, labor, and money. Daniel Udree's Oley Furnace began to smelt ore from an underground mine just before the Revolution. The war made gunpowder scarce and expensive, and Udree could find few skilled miners. He decided to abandon the mine after it began to flood.[56]

Most furnaces dug ore from pits because it was the cheapest and most practical method. It required relatively few workers, simple hand tools, and little infrastructure. Henry Drinker calculated that three men could supply a furnace with ore. Schoepf reported that there were two miners at Hopewell Furnace and six at Oley Furnace. Their jobs required more strength and stamina than skill. Most used tools like those Accokeek Furnace issued a miner in 1751: "3 Mattocks steeled, 1 sledge, 1 shovel, 3 stone axes, 6 wedges, 6 points sharpt." Miners threw or shovelled ore into baskets or buckets, which were dumped into a wagon or, if the mine was deep and steep enough, hoisted to the surface with a rope or pulley. Few pits were deeper than fifty feet, but a sudden storm could stop work. In 1768, Kingsbury Furnace's ore bank flooded and "prevented the raising of 50 Tons ore this Last week."[57]

Most of Kingsbury's miners probably rejoiced. The flood gave them a break from hewing, breaking, and raising rock in a giant treeless pit. At least they avoided the dangers of shaft mines. Three slaves at the Oxford Iron Works died after a rock landed on them. Lower risks to life and limb hardly made pit mining more pleasant. Rain, seeping ground water, and poor drainage left stagnant pools, ideal breeding grounds for mosquitoes. Pit miners breathed dust constantly. Like colliers, their work stained them. When John Mathews fled the Principio Iron Works, *Pennsylvania Gazette* readers might have spotted him by "his dirty Clothes" which were "colour'd yellow by working amongst Iron Stone." His masters and his neighbors thought little of Mathews or his peers. English traveler William Eddis noted in 1770 that Chesapeake whites considered pit mining "the most laborious employment allotted to worthless servants." Mathews's bid to escape his job and his servitude suggests how much he hated both. For the slaves and especially for transported British felons who mined ore in the Chesapeake colonies, the iron business was a prison.[58]

When a furnace had assembled enough ore, charcoal, and limestone or oyster shells, the blast could begin. A charcoal furnace was a large stack of brick or stone up to forty feet high and twenty-four feet wide at its base.

Masons often built furnace stacks into the side of a hill to facilitate charging it. At the stack's center was a roundish core open at both ends, made of fire brick or other refractory material and protected by a lining of sand or broken stone about ten feet in diameter at the widest part called the bosh. The bosh and the inwalls that tapered down from it supported the weight of the charge as carbon monoxide and heat from the blast reduced the iron and separated it from gangue: the dirt, rock, and other unwanted materials in the ore. Molten iron dripped into the crucible, where it pooled. Atop the metal floated slag, which was mostly liquefied gangue. The tuyere, which supplied blasts of air to stoke the furnace, led directly into the crucible. At the stack's front were two taps, one for molten iron, the other for slag.

The only major change in furnace design resulted from efforts to make blasts more efficient by increasing and standardizing air flow into the stack. Bellows made of leather and wood stoked furnaces well into the eighteenth century, but they did not provide a steady supply of air, wore out quickly, and were difficult and expensive to replace. In the 1760s, furnaces in England began to use blowing tubs, made of two wooden cylinders which each contained a piston that pushed air into the furnace hearth. They provided steadier air flow at higher pressure than did bellows. Most North American furnaces soon installed blowing tubs; nearly all U.S. furnaces had them by the early nineteenth century.[59]

It usually took at least two days to initiate a blast. Once the furnace was ready to smelt, fillers charged it with ore, flux, and charcoal by dumping them down the stack every thirty to forty-five minutes. Approximately every twelve hours, the founder and the keeper tapped the crucible, sending liquid iron spewing into trenches dug in sand on the furnace floor. Some became pig iron. Moulders ladled the rest into molds to make pots, skillets, stove plates, and other castings. At least twice a day furnace hands drained off the slag, which the banksman gathered and dumped, leaving behind one of the principal clues that tell industrial archeologists that a furnace once operated on a site.[60]

The success of a blast rested on its founder's shoulders. "A good Founder will make an extra saving of his years wages every month," Delaware Furnace manager Derick Barnard reminded his employer. In an earlier letter, Barnard defined a good founder. "In the first place," the manager explained, a founder must

> at all times be an accurate observer of the proceedings in the Furnace, especially during the Manager's absence; he should pay *particular attention* to the putting up of the stock, the raking of the coal to prevent waste, and in

Figure 4. Warwick Iron Furnace, Chester County, Pennsylvania, 1803. Watercolor on paper by Benjamin Henry Latrobe. Courtesy of the Maryland Historical Society, Baltimore, Maryland.

> fact see that all the duties of the banksman and ore wheeler and burner are properly attended to, at least during the absence of the Manager; for I am confident the correct working of the Furnace depends so materially upon a proper attention to these objects that they *must* be attended to constantly; furthermore 'tis absolutely necessary that he attend to the fillers at all hours in the day, *every thing* depends upon their attention to their business . . . and lastly his situation demands from him a constant attention to the working of the furnace, as much on one turn as on the other.

Some furnaces authorized founders to hire and fire furnace staff, especially keepers, who served as founders' assistants and supervised the furnace in their absence.[61]

It took a founder years to learn how to monitor a blast and a crew well. English ironmaster John Fuller claimed that "a Furnace is a Fickle Mistress and must be Humoured & her Favours not be depended upon," so that "the Excellency of a Founder is to Humour her Dispositions but never to Force her Inclinations." Coaxing good iron from a furnace taxed a founder's abilities and it reflected on his masculinity in the eyes of his peers and superiors. They depended on his expertise and vigilance. Many problems could surface during a blast. It fell to a founder to identify and address them, often from the clues contained in the qualities of the flame which emanated from the stack.[62]

Moulders depended heavily on their founder. Their trade experienced deeper technological change than did any other within colonial Anglo America's iron industry. Before 1740, moulders cast iron in one of two ways. The easiest and most common was to press a carved wooden pattern into damp sand on the furnace floor and allow iron to run into the imprint. Fabrication of hollowware, which included items such as skillets and kettles, was far more difficult. Moulders had to construct their molds out of clay, a time-consuming process. To make matters worse, each mold yielded just one pot; it had to be broken to free the casting.[63]

English ironmasters adopted a new way to cast hollowware and smaller flat items in the early eighteenth century. Moulders made castings by pouring iron into patterns pressed into damp sand packed in wooden flasks. This increased productivity; it demanded far less time, labor and expense than clay molds did. A moulder could prepare more molds and reuse the same pattern rather than manufacture a new one for each piece of hollowware. Flask casting of pots spread to North America: by 1743 Thomas Potts, ironmaster of Pennsylvania's Coalbrookdale Furnace, had begun experimenting with the technique. Most colonial furnaces soon followed suit.[64]

The application of flask casting to the production of stove plates has most captured the attention of curators and historians. The technique gave moulders greater control over the thickness and evenness of the plates, making stoves easier to assemble. It permitted manufacture of curved plates, as well as of plates that had finer detail than was possible from castings poured in a furnace floor. Flask casting enabled furnaces to market cast iron stoves to German-speaking immigrants. Many of their purchases, which often featured ornate floral designs or Biblical illustrations and inscriptions, now reside in museum displays of early Anglo American material culture.[65]

For moulders, flask casting proved a mixed blessing. It enhanced their ability to earn income. Growing consumption of hollowware and stove plates meant more demand for their services. The technique required considerable skill and experience; moulders had to take into account several variables which ranged from how iron behaved at different temperatures to the texture and humidity of the sand that they used. If flask casting boosted moulders' status, it also made their work more dangerous. Higher risk of serious injury accompanied greater productivity. In 1822, John G. Smith informed Delaware Furnace's owner that "John Collins has burnt himself severely (his foot) and am apprehensive he will not be enabled to mould any more during his stay here." The following year, on two consecutive days, moulders at Delaware scorched their own feet by pouring molten iron on them. In 1827, Joseph Lanning burned his eye badly, probably because a pocket of steam exploded in his face when liquid iron hit wet sand in a flask.[66]

Danger also lurked for other furnace hands. Fillers dodged smoke, singeing drafts, sparks, and flames. When the stack of Martha Furnace exploded, John Craig, then on duty as filler, "got very much burnt." Derick Barnard tried to discourage his employer from hiring Robert Downs to serve simultaneously as founder and keeper because burns so often incapacitated keepers that an injury to Downs would exacerbate the shortage of skilled hands that already plagued Delaware Furnace. All furnace workers endured searing heat and inhaled soot and charcoal dust that filled the air and covered the ground around the furnace.[67]

Most pig iron went to refinery forges. Cast iron's high carbon content and grainy microstructure rendered it too brittle to tolerate the pounding and bending necessary to make horseshoes, nails, wire, or most tools. A refinery forge, equipped with at least two hearths, plus bellows and trip hammers powered by water, converted pig iron into tough and malleable wrought or "bar" iron that would handle such treatment.

Making bar iron was a long, complex process that proceeded in two stages. First, one forgeman, often known as a "melter," gradually pushed two or three pigs into the hearth, where they collapsed into a pasty ball of glowing metal and slag. Someone had to stir the hearth constantly to expose the iron directly to blasts of air so that it would, as forgemen in Wales told Robert Erskine, "get its Nature." When the master finer determined that the iron had reached the proper temperature and consistency (about an hour after it began to melt), it was removed with hook and tongs and carried to a trip hammer weighing several hundred pounds. There the master finer, assisted by one or two underhands, carefully positioned the mass under the pounding hammer, which elongated the iron and expelled slag from it. The iron then was returned to the hearth to soften it and to reliquify the slag before it underwent more hammering. A master finer and his assistant(s) repeated this procedure at least four to five times to produce an anchony, a thick, flat bar of malleable metal with a rough knob on each end. Anchonies went to the chaffery forge, where hammermen worked them into bar iron. They reheated each anchony to soften it and then placed it under another trip hammer to flatten out the knobbed ends and draw the iron to a standard length and thickness, known in the iron trade as "merchant bar," or to order.[68]

Forgemen were strong and highly skilled. Robert Erskine marveled of some Welsh forgemen that "habit certainly increases there strength very much as a lad much more slender than I can take a Pig of about an hunder'd & half, carry it with seeming care, and throw it behind the Fire." Expertise rather than brawn made a good forgeman. A master finer had to recognize by sight and feel exactly when melting pig iron had attained the proper temperature and consistency for placement under the hammer. He had to know, based on a quick examination of what stuck to the end of a poker, when to drain slag from the hearth. Hammermen required tremendous agility and dexterity; it was hard to deform iron with a trip hammer without breaking it. Often forgemen had to work with pig iron that they called "red-short" (brittle when heated) or "cold short" (brittle at room temperature). Either took longer to forge and consumed more charcoal.[69]

Forgemen risked serious injury. Pulled muscles and bad backs plagued them. The constant pounding of hammer against anvil created a deafening din. A forgeman's hands, arms, and shoulders absorbed the impact of trip hammer against hot iron and colder anvil. The hammer's blow sent slag and flecks of hot metal into the air. Brumall and Ned, slaves who fled a Virginia forge in 1766, had "scars on their arms, from burns which they got by melted cinders flying on them when at work." In 1783, cinder burned John Collins so badly that the manager of Ridgely's Forge thought it was "very uncertain when he will be able to work." Forgemen did not wear protective eyegear; bits of metal or slag sometimes found their eyes. In 1832, William Davis noted that Sam, a slave hammerman, "objects to his eyes (which is in fact a very great objection [as he] might in all probability loose them if [he] continued in the forge.)" Sam could not have been more than forty years old. His work may have cost him his sight; an inventory of William Weaver's slaves taken in 1863 declared that he was worth nothing.[70]

Sam lacked the power to decide when he would stop being a hammerman. Most forgemen, despite the risks that they regularly confronted, practiced their trade until they could no longer perform it. A forgeman spent a significant portion of his youth in training. William Norcross, a hammerman at Union Forge, testified that it took at least two years to become a "first rate" underhand. Those who wished to become master forgemen required more years of training. Having invested so much time and energy to learn the mysteries of forging iron, few who pursued the craft willingly abandoned it. The considerable control over work that forgemen enjoyed also encouraged them to remain in the iron business. Unlike founders and moulders, who were bound to the rhythms of a furnace, they had more discretion to set the hours and pace of their work, which they oriented around the completion of tasks. They sometimes determined what they would produce and what their employer could sell. Henry Drinker alerted Atsion customers that he could not fill many orders for iron drawn thinner than was customary because his forgemen would object. The quality of their work—how it looked and how well it performed—determined the price that bar iron commanded and an adventurer's reputation in the iron market. Even one parcel of inferior bar iron might offend customers, such as the group of smiths who wrote Robert Coleman in 1801 to protest poor bar iron that Spring Forge had sold them, and make it nearly impossible to win them back.[71]

Exceptional Iron? The Dilemmas of Anglo America's Industrious Revolution

In 1786, Richard Blackledge proposed building an ironworks, and he wanted Henry Drinker's advice. The future of his venture, Drinker warned, would depend largely on recruiting and retaining men "in the different branches of the business who will in their respective departments execute the parts entrusted to their Care with fidelity and uprightness," especially "those sober managing Men who do well for themselves &

Families will upon the whole do best for the Employer." Any ironmaster on either side of the Atlantic would have agreed; Drinker's American peers knew that it was often wishful thinking. Getting experienced and conscientious workers was difficult; keeping them and ensuring that they continued to perform their duties well was even harder.[72]

That was especially truc in early Anglo America, where most believed that a man's work should make him independent. Thousands considered a job within the iron industry a path to a "competency" or a way to secure one—to "do well for themselves & Families" as Drinker put it. Many were farmers or their sons who lived near ironworks, sold produce to them, and offered their services to them, especially after the harvest. Wood cutting in particular generated income and funded purchases of consumer goods. Working for adventurers enabled them to participate more fully in the industrious revolution on their terms.[73]

But adventurers and ironworkers knew that Anglo America's industrious revolution and the needs of the iron industry often collided. Many ironworkers could never become truly independent and they knew it. An increasing number of unskilled ironworkers would never be their own bosses. Most tradesmen would always be employees; they had to redefine what they understood to be the appropriate relationship between manhood, industry, and independence to fit their lives. This kept them within the iron industry and proud of their craft. It also led them to insist on autonomy while at work and to shop their scarce expertise around. Both threatened ironmasters' control over their enterprises. Colonial and American adventurers succeeded in molding ironworkers and the industrious revolution to suit their purposes—the iron industry could not have grown so big and so quickly had they not. But to do it, they had to sever the link between industry and independence for thousands of ironworkers by enslaving them.

Adventurers believed that bound labor served their enterprises well. An ironworks needed labor year-round; there was enough work to justify the expense of purchasing and boarding servants, slaves, or both. The servants and slaves of the Baltimore Company spent 1776 "principally employed at Raising of oar & making charcoal in the cource of the summer, & at wood cutting in the winter." The iron industry encouraged slavery to take root outside the southern colonies. In colonial Pennsylvania and New Jersey, where relatively few farmers owned slaves because of religious convictions or because the crops that they grew did not require it, adventurers brought slavery to the countryside on an unprecedented scale.[74]

They did so in part because they undertook risky and expensive ventures in, as Pennsylvania's Lieutenant Governor Patrick Gordon reminded the British Empire's Board of Trade and Plantations in 1734, "a Country where Labour is so dear." Competitive pressure from outside the British Empire sharpened colonial Anglo America's need for industrial slavery. In an address to Pennsylvania's Assembly nearly four years earlier, Gordon lamented that the province's iron industry had "proceeded with Vigour, till the vast Quantities unexpectedly imported into Britain from the new Works in Russia, where the poor People labour almost for nothing, have given some Damp to that Manufacture." If Russian ironmasters exploited serfs, colonial adventurers had to have slaves to prosper. As a result, on Europe's eastern and western peripheries, more ironworkers were mastered by furnaces and men during the eighteenth century.[75]

I

IRON AND EMPIRE

The Colonial Era

2

MOLDING MEN

In 1671, Edward Vickers faced New Haven's town fathers. They accused him of "Cursing & sweareing" and of "giveing threatning speeches against" ironworks clerk Patrick Morran. The defendant had more to answer for. He had claimed to be a fugitive servant from Virginia, who, as his master pursued him, "shott at him & thought he had killed him, for he saw him fall downe." Several witnesses attested that Vickers cursed often, to which he "owned, & sd he was sorry yt he had soe done, but he sd he had beene frequently used, & he hoped he should reforme for ye future." If that mollified New Haven's magistrates, it did not stop them from showing Vickers "the greatncs of his Evill" for not minding his mouth. They found him "highly guilty of Common & frequent Curseing & sweareing in a most prophane & blasphemous manner, horrible to be hearde or uttered, & the like not formerly knowne among us, to ye great dishonor of god & danger of infection to others." They also judged Vickers "dangerous" for having threatened repeatedly to kill Morran, which confirmed that the ironworker was possessed of "a violent & furious spirit & behavior." The court ordered that he be whipped and that he post a £20 bond to ensure his good behavior until Vickers and his wife left New Haven and he had managed to "acquit himselfe or be acquitted from ye sd suspicion of the guilt of bloud."[1]

Did the leaders of New Haven expect Vickers would reclaim the money? He had no incentive to go to Virginia because he might face charges and would likely be convicted. Vickers would serve that colony's elite as an example to servants who were increasingly challenging their masters' authority. Either way, the bond would become a fine; New Haven would get some revenue; and its magistrates would have effectively banished a menace before he contaminated others or brought divine retribution upon the town.[2]

But if New Haven's magistrates simply wanted to get rid of Vickers, why didn't they whip him and exile him and his wife permanently? They did not because they hoped he would return. Perhaps then he would change. After all, there were promising signs that Vickers could be reformed. He was married; he had apologized and stated that he wanted to moderate his speech and his temper. All suggested to the court that he might change his ways. Above all, the magistrates knew, New Haven needed men like Vickers, because the town fathers believed that their community needed an ironworks. Industrial enterprise, the public good, and the community's salvation went hand-in-hand. To make iron was to serve God. Part of the price of redemption was to suffer the likes of men like Vickers and undertake the hard, frustrating, but necessary work of trying to mold them into new men.

The trial and sentencing of Edward Vickers underscore what many English entrepreneurs who planned and led the colonization of the North Atlantic during most of the seventeenth century believed: work, industrial enterprise, and institutions could together improve societies and individuals. Adventures in ironworks became a vehicle to wealth, power, and a better world by helping English colonists refashion men in their own image whom they deemed inferior. In Ireland, conquest, the establishment of ironworks, and the recruitment of Irish men to staff them proceeded together. Adventurers such as Sir William Petty aspired to anglicize Irish men by making them ironworkers. In Virginia, adventurers linked the birth of British North America's iron industry to the conversion of Algonkian boys into Christian Indian subjects. As in Ireland, the recipients of their largesse considered the ironworks to be symbols of conquest and destroyed them.[3]

No one combined industrial enterprise and attempts to mold men more explicitly than did Puritans who conquered and resettled what they called New England. The name described their aspiration—they wanted to create a new, more perfect England, one which honored God's word and did God's will. New England's fathers soon determined that they could not build their model society without furnaces and forges. But few men in the first wave of migrants who sailed across the Atlantic knew how to make iron. For that, New England adventurers had to recruit unruly and ungodly strangers, who embodied what Puritans sought to escape when they left England. Necessity prompted them to try to remake ironworkers into new Englishmen by seeing to it that they worked hard and behaved well. Entrepreneurs, neighbors, churches, and courts joined forces in that endeavor. That made what New England's ironworkers said and did a public matter. Had it been otherwise, historians could not tell the story of the region's early iron industry or of its relationship to the industrious revolution. Few accounts and little correspondence of New England's seventeenth-century iron industry survive. Its workers principally left their mark in the records of courts that sought to police, punish, and perhaps redeem them. Presenting the industrious revolution through sources that mostly portray ironworkers as troublesome distorts our vision. It also tells us how seriously New England's fathers took their duties, how seriously they took the strange sinners whom they had invited among them, and how skeptical they sometimes were that they would succeed.

Did they succeed? The institutions upon which they relied for reform served their purpose, largely because ironworkers made use of them for their own purposes. The ventures that employed them failed, though they provided a foundation for regional economic development. But New England's model for combining industrial development with early Anglo America's industrious revolution eroded quickly—and from within. Its adventurers also began to develop a new system, one founded on the coerced labor of outsiders whose reformation mattered little to them. They pointed the way forward for the colonial iron industry.

Dust and Ashes

The mission was clear to Virginia Company leaders as they met in London in July 1621: the ironworks at Falling Creek had to fire up soon. By May, the financially strapped company had already dispatched over 150 workers and spent over £4,000 on them. It had little to show for its efforts except delays, bills, and broken promises. Some workers had died; others had abandoned their posts to plant tobacco so that they could profit from the colony's only booming enterprise.[4] Developing the ironworks as the company envisioned needed the full support of the colony's leaders in Virginia. Officials in London requested that John Berkeley (the agent whom they had charged with supervising construction and operation of the ironworks) and his men "may bee Cherished by yow and supported by ye helpe of the whole Colonie if need shall requier" because of what the company had already invested and because iron was a commodity "so necessarie as few other are to be valewed in comparrison therof." The company had staked its future on the Falling Creek ironworks. Upon them "mens eyes are generally fixed" and were they to "fall to ye ground, ther were little hope that evr they would bee revived againe." The colony's fate rested on its adventure in iron. If it failed, its leaders believed, so might Virginia.[5]

How did an ironworks become so important to the Virginia Company? To be sure, its shaky finances and tarnished reputation were on the line. So were goals that it considered far more noble. Falling Creek represented, the company's leaders proclaimed, a path to deliverance for England, for Virginia, and for the colony's native neighbors.

To the Virginia Company, iron was the most significant of several "more solid" and "reall" commodities which colonists had neglected in their zeal to grow tobacco. The tobacco boom, they claimed in a 1620 broadside, "redounded to the great disgrace of the Countrey, and detriment of the Colony" and had served to "greatly deceive them which have trusted to it." A new society propelled by addictive smoke had damaged Virginia's image and it had left the colony too dependent on one commodity. Making iron promised to diversify Virginia's economy and "restore due reputation to that Land and people" by supplying something useful which demanded the mustering of capital and industry.[6]

Another way to salvage Virginia's battered image was to argue that a colonial iron industry would help to resolve England's social and ecological dilemmas. Ironworks and glassworks had decimated English forests and prompted legislation to regulate them. Two decades before colonists founded Jamestown, Thomas Hariot promoted colonization by arguing that iron would be "a good marchantable commoditie" partly because of the "infinite store of wood" in North America and "the want of wood and deerenesse thereof in England." In 1620, "His Majesties Counseil for Virginia" echoed Hariot, claiming that iron "which hath so wasted our *English* Woods, that it selfe in short time must decay together with them, is to be had in *Virginia*, (where wasting of Woods is a benefit) for all good conditions answerable to the best in the world." The venture at Falling Creek would save England's forests for building ships that would defend the realm from enemies, especially the Spanish Empire.[7]

The Virginia Company's leaders wanted to win new allies for England's global struggle with Spain, and they believed that they had found some in Algonkians. They followed a tradition established by promoters of English colonization, who had long claimed that America's native peoples wanted and needed England's help to liberate them from Spanish rule. The English, boosters insisted, were to be the good colonists who would bring prosperity, peace, and truth to the Americas by vanquishing Spanish tyrants and by replacing Catholic and Native superstition with true Christianity. Converting Natives into Anglicized Christians would make them helpful neighbors and loyal subjects and provide the moral foundation for English expansion into North America.[8]

Here too the Virginia Company had a problem. Besides the well-publicized and carefully scripted example of Powhatan's daughter Pocahontas, it could point to little success at persuading Algonkians to adopt English ways. Clashing beliefs about gender, power, and land use barred most from crossing the turbulent, and often violent, gulf between Natives and English immigrants. Some on the scene diagnosed the problem and proposed remedies. George Thorpe and John Pory blamed colonists who were "not soe Charitable . . . as Christians ought to bee" for alienating Algonkians, and they thought that gifts might "make a good entrance into their affections." Two years earlier, Governor George Yeardley, while discussing the ironworks, warned Company leader Edwin Sandys that "[t]he Spirituall vine you speake of will not so sodaynly be planted as it may be desired, the Indians being very laoth upon any tearmes to part with theire children." He thought that the Virginia Company should "draw the people in to live amongst us" by giving them "some aparel and cattell and such other nessisaryes" and setting aside land for them to plant so that colonists might "Instruct theire Children" without separating them from their families. It was an imperfect solution, but it would provide a more solid base from which to battle for the hearts and minds of the next generation of Native Virginians.[9]

The campaign to transform Algonkian boys into English Christian men required schools. Company officials envisioned an educational system to be capped by a college which would graduate young men who "may be sente to that worke of conversion" of others. To their delight, wealthy patrons backed the plan. In 1620, the Virginia Company revealed that Nicholas Ferrar had directed the executors of his estate to allocate £8 a year to rear three Native children "in the grounds of Christian religion" and willed £300 to be paid when ten children were ready to enter the college. Ferrar's bequest followed an even bigger donation from an anonymous benefactor, "Dust and Ashes," who pledged £550 in gold for "the Convertinge of Infidles to the fayth of Christe." The funds, he insisted, should maintain "a Convenyent nomber of younge Indians taken att the age of Seaven years or younger" and teach them reading and "the principalls of Xian Religion" until they were twelve years old. Some should then "be trayned and brought vpp in some lawfull Trade w[ith] all humanitie and gentleness" until they became twenty-one, after which they were "to enioye like liberties and pryveledges" with Virginia colonists. "Dust and Ashes" did not trust the Virginia Company with his money, warning that "guiftes devoted to Gods service cannot be diverted to pryvate and secular advantages without sacriledge."[10]

His suspicions were well-founded. The Falling Creek ironworks badly needed funds; Edwin Sandys diverted what "Dust and Ashes" donated to them. "Dust and Ashes" demanded a public account; Sandys provided one. He had asked the Societies of Martin's Hundred and of Southampton Hundred to use the bequest to "undertake for a certaine number of Infidell Children to be brought up by them, and amongst them in Christian Religion and some good Trade to lyve by accordinge to the Donors religious desire." Both refused. The Adventurers of Southampton Hundred offered £100 to dodge the responsibility. They backed down under pressure and directed the donation of "Dust and Ashes" plus "a farr greater Some" of their own money "toward the furnishinge out of Captaine Bluett and his Companie being 80 verie able and sufficient workmen . . . for the settinge up of an Iron worke in Virginia." Profits from the ironworks "were intended and ordered in a ratable proportion to be faithfully imployed for the educatinge of 30 of the Infidell Children in Christian Religion and otherwise," as "Dust and Ashes" had stipulated. Bluett and several key workers died shortly after reaching Virginia, but Sandys offered assurances that the setback was only temporary.[11]

Sandys praised "Dust and Ashes" for his charity and then argued that he and the company knew best how to make it do God's work. "Dust and Ashes" had offered another £450 if the Virginia Company sent some Algonkian boys to England to be educated. That, Sandys insisted, would backfire. Many Natives who visited England only seemed to become more alienated from the English. It would also be a mistake, he contended, to devote the money to a school in Virginia. Workers to build it could only be hired "at intollerable rate." Their wages might empty the "sacred Treasure" that "Dust and Ashes" had filled, leaving behind only "some smale fabricke" to show for it, whereas an ironworks would provide a "foundation, as might satisfie mens expectations." Falling Creek, Sandys exhorted, was "a worke whereon the eyes of God, Angells, and men were fixed." A vibrant iron industry was the best way to harvest Native souls, make Englishmen of Algonkian boys, and ensure Virginia's success.[12]

Company officials eagerly awaited word that Falling Creek had gone into blast, especially after hearing from Berkeley that soon they "may rely upon Iron made by him." It never came. In 1622, Native warriors stormed Virginia and killed nearly one-quarter of its colonists, including more than two dozen at Falling Creek. The Virginia Company of London was another casualty. Two years later, James I revoked its charter and Virginia became a royal colony under his direct control.[13]

Virginia's Native peoples paid far more dearly for the war. The attack, their response to accelerated encroachment on their lands and to years of accumulating grievances, seemed an inexplicable act of treachery to the English. To Edward Waterhouse, who tallied English killed in the "Barbarous Massacre," the war was a physician. It administered a painful but necessary treatment by bleeding colonists and English investors of delusions which had blinded them to the Native threat. The war justified conquest and demonstrated the futility of "civilizing" Natives "by faire meanes, for they are a rude, barbarous, and naked people, scattered in small companies, which are helps to Victorie, but hinderances to Civilitie." Vanquished Natives could "justly be compelled to servitude and drudgery," replacing "men that labour" so that they might "imploy themselves more entirely in their Arts and Occupations, . . . whilst Savages performe their inferiour workes of digging in mynes, and the like." Natives, Waterhouse implied, were evil and inferior by nature. They could never become English or even reliable allies.[14]

John Berkeley and twenty-six others died at Falling Creek at the hands of men whose sons and nephews they were working to save and transform. So did the ironworks. When word reached London, the Virginia Company's Treasurer and Council ordered that the "people remaining of the Iron works" be charged to Maurice Barkley so that they could "be imployed" until "we may againe renue that bussines, so many times unfortunately attempted, and yett so absolute necessarie as we shall have no quiett vntill we see it pfected." Their enthusiasm lagged, especially after having to acknowledge under withering criticism that the ironworks were "wasted." The Falling Creek venture soon dissolved, and Virginians erected bloomeries to make small amounts of iron for local use. A century passed before furnaces or refinery forges again arose in the Chesapeake. When they did, their owners made no pretense of doing God's work or of transforming Natives or anyone else for the better.[15]

"Under the Discipline of Your Country"

In 1653, Puritan propagandist Edward Johnson urged English readers to support New England. Its past was an inspirational story which clearly demonstrated God's "Wonder-Working Providence," most vividly "in fitting" its colonists "with all kind of Manufactures, and the bringing of them into the order of a commonwealth." Johnson listed accomplishments that owed to colonists' industry and the Lord's favor, most impressive of which was the ability to feed themselves, "their Elder Sisters" Virginia, Barbados, "and many of the Summer Islands that were prefer'd before her for fruitfulness," as well as "the Grandmother of us all, . . . the firtil isle of Great Britain," and even Iberia. Provisioning the North Atlantic rim with grain, meat, and fish spurred urban growth and stimulated enterprises which included lumber mills, food processing, and shipbuilding. All provided the visible foundation for what Johnson called "a well-ordered Commonwealth."[16]

Johnson extolled the industry of New England's colonists in part because he feared that the commonwealth's foundation was shaky and he wanted to stabilize it. The diverse economy that he celebrated arose from the region's inability to monopolize northeastern North America's lucrative fur trade, to mine precious metals, or to grow a lucrative staple crop such as tobacco or sugar cane. In 1640, Puritan victories in the English Civil War intensified the need for a wider range of activities. The chance to reform England prompted thousands of Puritans to cancel plans to migrate to North America and lured colonists back. Steady immigration had propped up the Massachusetts Bay economy. Without continued immigration, its future looked bleak unless colonists launched new enterprises. They pinned many of their hopes on ironworks built and operated at Braintree and Lynn by the Company of Undertakers, a group of mostly English investors.[17]

Edward Johnson worried that their venture might collapse. Indeed, he hoped to stiffen the resolve of English owners and to attract new investors to the ironworks. Johnson began his account of local manufactures by noting that Massachusetts's "very good iron stone" had prompted men "of good rank and quality in England to be stirred up by the provident hand of the Lord to venture their estates upon an iron work." The first furnace at Braintree sputtered out quickly "and profited the owners little, but rather wasted their stock, which caused some of them to sell away the remainder." Johnson attributed its demise primarily to "the high price of labour" and to unforeseen complications, since "the way of going on with such a work here, was not suddainly to be discerned." Management had learned its lessons, he insisted, and the cost of labor could be addressed. Johnson pleaded that "those Gentlemen who have undertaken the work" should remember "where their works are, namely in N. E. where the Lord Christ hath chosen to plant his Churches in, to hide his people under the covert of his wings, till the tyranny of Antichrist be overpassed, and any that have disbursed pence for the furthering of his work, shal be repayed with thousands." He promised that investors would have the support of "the Gentlemen that govern this Colony," who would "rather take any burthens upon themselves and the Inhabitants, that in justice they ought, then that those Gentlemen should be in any wayes damnified." Indeed, New England's track record in agriculture, industry, and commerce, all of which testified to God's favor, indicated to Johnson that an ironworks could not fail. Adventurers could reap the financial and spiritual rewards by backing the Lynn venture and remaining patient.[18]

Investors may have accrued spiritual credit from the Lynn ironworks, but they saw few worldly returns from it. In 1653, as Johnson's words appeared in print in England, authorities seized the assets of the Company of Undertakers. Litigation halted Lynn's ironworks for a decade and long outlived attempts to revive them. William Hubbard's 1682 history of New England remembered that Lynn, "instead of drawing out bars of iron, for the country's use," produced "nothing but contention and lawsuits." Many ironworkers who had staffed Lynn's works before bankruptcy stilled them moved to a new venture outside New Haven. That too expired by 1680. Artisans affiliated with Lynn, Hubbard noted and recent historians have agreed, furnished the expertise that made New England the center of British North America's iron industry into the eighteenth century. But the vision that Johnson and New England's fathers shared, that a grand iron industry would provide the material foundation for their model society, never came to pass.[19]

In blaming expensive labor for the Undertakers' problems, Johnson echoed a common refrain among New England's elite. Hired hands came far too dearly to them. The difference between what ironworkers earned in the British Isles and what they commanded in New England burdened adventurers, who imported most of their equipment and faced steep startup and maintenance costs. It was likely no coincidence that Emmanuel Downing in 1645, just before he sold off his share in the Company of Undertakers, celebrated news that Massachusetts might wage war on the Narragansetts: "If upon a Just warre the lord should deliver them into our hands," it would be "more gaynefull pilladge for us than wee conceive." Downing saw no way that the colony could "thrive" without "a stock of slaves suffitient to doe all our buisines, for our Childrens Children will hardly see this great Continent filled with people, soe that our servants will still desire freedome to plant for them selves, and not stay but for verie great wages. And I suppose you know verie well," he reminded John Winthrop, "how wee shall maynteyne 20 Moores cheaper than one Englishe servant." Though slaves never worked at Braintree or Lynn, his former partners later followed Downing's reasoning by employing Scottish prisoners seized in another English holy war.[20]

War in the British Isles may have encouraged the Undertakers to finance New England's first ironworks, but it compounded their difficulties in recruiting ironworkers. New England generally had trouble attracting servants; those who crossed the Atlantic saw better opportunities in the Chesapeake colonies or the Caribbean.[21] In 1643, Robert Child probably spoke for his partners when he complained that "these times put me to my wits Ends." In 1641, Irish rebels targeted English-owned ironworks and their English employees. They left the industry in ruins and killed many workers. Joshua Foote, an Undertaker based in London, reported in 1643 that he had "inquired and sought out For to gete a blomrie man and can here of non. I was with Sir John Clattworthie about blomry men . . . he telles me that times are so in Irland that he thinks thay are kild or ded For he can here of non, and I haue inquird much after some and can here of non."[22]

That was not all. Three ironworkers bolted before boarding ship and the Undertakers never heard of them again. Three others spent weeks in port while their ship awaited permission to leave. They complained that "thay have layne so longe here and have lost ther laber" and that their stipend for room and board did not cover their expenses. Joshua Foote directed John Winthrop, Jr., to promise them that "when thay com in new England that when you se som good proced and proffit com in of the workes that you will further Raward them." The ironworkers also had to be shown tokens of "good Raspict" such as a higher daily allowance, "good bedes bought them," and "all acomendation in the shipe that you can posibly aford them to kept them in health." The voyage was such a debacle that Winthrop petitioned Parliament for the approximately £1,000 that the "unjust" delay had cost him and the Undertakers. The ship departed and then waited six weeks for a wind to propel it across the Atlantic. It took fourteen weeks to reach New England. Most aboard fell "dangerously sick of feavors," and arrived "so weakened" that the "servants and workemen were not fitt for any labor or imployment. . . ." With winter fast approaching, Winthrop had "to keepe his workmen and servants at great wages and charge without imployment." It appears that Parliament never acted on his request.[23]

The war also forced the adventurers to accept men who had not mastered their craft. After explaining that the Irish rebellion had made forgemen even scarcer, Foote advised Winthrop to "Joyne all your workmans hedes togather and see to breed vpe blomries. [A] smith aftre a lettell taching will make a blomer man." Perhaps, but brief training would never convert a blacksmith into a good forgeman, especially without anyone to instruct him. Five years later, the Undertakers apologized to Winthrop for the men they had sent, declaring that "want of experience in the Minerals in most of our workmen hath bin loss, and charge to us: And worse qualificacions in some of them have beene a trouble to you. It is our earnest desire, and we have endeavored all wee can to be furnished with better men than some of them are."[24]

The ironworkers that the company dispatched in 1648 promised to create more problems for their supervisors and their new neighbors. They were, their recruiters sheepishly acknowledged, men "for whose civilities we cannot undertake, who yet we hope by the good example, and discipline of your Country, with your good assistance may in time be cured of their distempers." Maybe they believed that positive role models and vigilant supervision would cure the workers of what ailed their employers and mold them into fit soldiers for a godly society. The Company's reputation in Massachusetts, already suspect, rested on fulfillment of that hope. Its leaders bore responsibility for the qualities of those whom they hired. John Winthrop, Jr., was probably the best person available to defend them to colonists. Moreover, if the English Undertakers believed that "the good example and discipline" of New England would transform their employees, theirs remained mostly an untested faith. Hands of suspect character might become a liability on company ledgers, but those adventurers did not have to live with the men whom they sent across the Atlantic.[25]

The company could not blame Massachusetts officials for lack of encouragement. In 1645, the General Court, anxious that "such an oportunity for so great advantage to ye comon wealth might not be let slip," promised to notify each town to attract investors. After more than a year of negotiation, leaders conceded nearly everything that the adventurers wanted: a virtual monopoly over the colony's iron market; land grants of up to three square miles each for as many as six ironworks; and a twenty-year exemption from provincial taxes. The Undertakers owed the commonwealth relatively little: Massachusetts authorities regulated how much they could charge for iron and barred them "from selling iron, or iron worke, either to Indians or enemies."[26]

Massachusetts also excused the Company of Undertakers and its employees from most communal obligations. Company agents and ironworkers did not have to pay provincial taxes. They exempted key employees—"fin[e]rs, found[e]rs, & ham[e]rmen, &c., & oth[e]rs constantly implied"—from keeping watch or militia training, though each was to "at all times be furnished wth armes, powder, shott, &c." Most significantly, the adventurers persuaded the General Court to rescind its directive that they, wherever "any iron worke is set up, remote from a church or congregation, unto wch they [ironworkers] cannot conveniently come," provide "some good meanes wherby their families may be instructed in the knowledge of God, by such as the Cort or standing councell shall approve of."[27]

In their eagerness to promote an iron industry to enrich the commonwealth, the rulers of Massachusetts made it harder to mold ironworkers into New England men. Their concessions to the Undertakers strained the bonds between community, faith, enterprise, and manhood that characterized their holy experiment. They had released many ironworkers from military duties and rituals that tied men together for the common defense. By excusing them from personal taxation, the General Court released ironworkers from an obligation to the commonwealth that all other freemen shared. In addition, the Court left the Lynn ironworks with no direct religious supervision. The job of policing its workers fell squarely on neighbors and county courts. In 1648, soon after Lynn's fathers realized the burden that they had assumed, they petitioned the General Court to compel the company and its employees to share the bill. The Court agreed; it ruled that the tax exemptions it had granted applied solely to Massachusetts, and not to those that the town or its church might levy. If Lynn had to remake ironworkers into tolerable neighbors, then they and their employers had to pay their way.[28]

The residents and the magistrates upon whom it fell to administer the "discipline" that the Undertakers celebrated had their work cut out for them. The men who staffed New England's ironworks were emphatically not saintly material. Pugnacious, unchurched, foul-mouthed, contemptuous of authority, and inclined to imbibe, ironworkers indeed seemed at best a "distempered" lot to Puritans. Magistrates had to supervise them closely, in part because even the men charged with running the ironworks sometimes created more problems than they prevented. Like fishermen and mariners, New England's ironworkers often were ethnic strangers and outsiders to their neighbors. Many had only recently become "English" and were at most a few generations removed from French and Walloon ancestors who had introduced the indirect process to England. The thinly Anglicized surnames of many of the ironworkers betray their ethnic origins: Pinion, Vinton, Leonard. To respectable Puritans, mariners and fishermen seemed an indigestible mass and a necessary evil, but at least their work often carried them far away for months. When out of sight, they could remain mostly out of mind. The same could not be said for ironworkers; their duties kept them in Lynn.[29]

Ironworkers could not be ignored, because to ignore them, Puritans believed, would be to risk their own damnation. Magistrates and neighbors had to police, punish, and seek to reform them. Only New England's institutions—town governments, courts, and Congregational churches—could inscribe duty and deference into ironworkers' hearts and minds. Such values, though certainly applicable to the workplace, were communal. How ironworkers behaved and how they governed their households determined how well they served the communities in which they lived. With proper oversight and discipline, particularly by neighbors through courts, perhaps they would become new men who would neither disturb the peace nor provoke God to punish colonists for inviting such sinners among them.[30]

Lynn's ironworkers allowed the town and the Essex County court little peace when they arrived. They fought one another; some attacked travelers on the road; others blasphemed; many made a habit of skipping church; most drank heavily. All of this pointed to magistrates' key concern: ironworkers could not or would not honor their duty to serve as governors of orderly households.

It was the court's responsibility to hold them accountable. Essex County authorities upheld the sanctity of companionate marriage, for example, by ordering John Baily to return to his wife in England or to send for her. Sometimes ironworkers or their spouses required discipline for flaunting the fruits of their industry. In 1652, Nicholas Pinion and his wife Elizabeth, Ester Jenckes, wife of Joseph Jenckes, Jr., John Gorum, and John Parker appeared in court for violating sumptuary laws—that is, for dressing above their station by wearing silver lace. All had notorious reputations and it rankled their neighbors to see them wearing such finery, especially since their worldly success apparently did not owe to upright conduct. Nicholas Pinion and Joseph Jenckes, Jr., had done too well at providing for their wives—a key feature of New England manhood—and not well enough at following the law.[31]

The three toughest cases that the Essex County Court heard in 1647–1648 addressed domestic violence within ironworkers' households. All involved abusive men and unruly women. Magistrates tried to compel the men to behave better and intervened to help them reform by later prosecuting the women involved. Each case reveals the lengths to which the court went to remake men. In two of the three cases, it succeeded in getting the workers to obey it.[32]

Consider the case of John Turner. He allegedly stabbed his daughter-in-law Sarah Turner and swore "by the eternal God that he would kill John Gorum." The court determined that Turner should "be severely whipped at Salem," dispatched to "Boston prison until he be whole; and later to be whipped at the iron works," and then it revoked the sentence. Why? It is most likely that Sarah Turner's behavior prompted the court to excuse

John Turner's. In a defamation suit that her husband Lawrence filed two years later against Henry Leonard, a fellow ironworker, and his wife Mary, magistrates heard that Sarah had pursued other men and said that "she wished the devil would take those, body and soul," who had tampered with an outdoor tub. They ordered two men to deliver Sarah, who was "to be whipped for her many offences." John Turner perhaps tried to get her to mend her ways. Since her husband could not or would not, the court had to act.[33]

Conflict between Nicholas Pinion and his wife Elizabeth was harder to resolve. The court knew Nicholas Pinion; he had been presented the previous session for swearing and for "missing four Lord's days together, spending his time drinking, and profanely." Elizabeth had to pay a fine or face a whipping after continuing an affair with Nicholas Russell and saying that if Russell obeyed her husband's order to leave the house and stay away, "she would depart also." She tried to turn the court against her husband by accusing him of killing five children. The following month Elizabeth Pinion appeared for "fighting three times with her husband in the night since she was bound to keep the peace." He beat her and caused her to miscarry. The General Court of Massachusetts tried Elizabeth for adultery and found her "not legally guilty thereof" but ordered that she be whipped in Boston and in Lynn "for her swereinge & adulterous behavior." The Pinions' marriage thereafter was relatively peaceful, though Nicholas still attracted the attention of the court. In 1649, he was fined forty shillings—thirty for "swearing three oaths" and ten for swearing, as ironworker Quinton Prey deposed, "by God, all his pumpions were turned to squashes, and by God's blood he had but one pumpion of all."[34]

Richard Pray headed the ironworker household that perhaps most troubled Essex magistrates. In 1648, Jabish Hackett testified that Pray had often called his wife Mary Pray "jade and roundhead, and curse her, wishing a plague and a pox on her," particularly after he had returned from meeting. Hackett overheard Pray, who had already beaten his wife once that day, say that "he would beat her twenty times a day before she should be his master." Another time Hackett intervened to stop Pray from striking Mary with a stick, only to watch him kick her against the wall. Hackett also reported that Thomas Wiggins "spoke to Pray about cursing and swearing upon a Lord's day" after the Prays had skipped meeting. Pray protested "that it was a lie"; Mary challenged him; he "took his porridge dish and threw it at her, hitting her upon the hand and wrist, so that she feared her arm was broken." Someone warned Pray "that the court would not let him abuse his wife so." He retorted "that he did not care for the court and if the court hanged him for it he would do it." When told "that the court would make him care, for they had tamed as stout hearts as his," Pray remained the defiant master of his household: "If ever he had trouble about abusing his wife, he would cripple her and make her sit on a stool, and there he would keep her." The court had to try to tame Pray—he was a wife-beater, and a seditious one at that. It sentenced him to a whipping at the ironworks or a fine of eighty shillings: "10s. for swearing, 10s. for cursing, 20s. for beating his wife, and 40s. for contempt of court." A year later, Pray had not paid the fine and faced the lash at Lynn on lecture day.[35]

Richard Pray too came to respect the magistrates whom he had vowed to defy. Indeed, they sought to help him reform by fining John Hardman and binding him "to good behavior for profane swearing, for calling Mary a vile name and seeking to provoke her husband against her and for excessive drinking." Pray also let the court handle his disputes with Mary. In 1650, it convicted her of telling her mother-in-law, "get you whom yow old hogge get you whom," of throwing stones at her, and of attacking Richard after he took away a letter "shee had gotten wrighten for England." Richard Pray may not have become a saint, but he did become a better man in the eyes of his neighbors and his superiors by conforming more closely to their view of how a man should behave.[36]

Pray's deference to the Essex County court suggests that ironworkers obeyed New England's institutions partly because they benefited from them. Ironworkers were key witnesses in criminal or civil cases involving other Lynn hands. They acted as magistrates' eyes and ears. After all, it was Thomas Wiggins, who performed several jobs about the ironworks, who told Pray to watch his mouth and his drinking on Sunday. It was probably another ironworker who warned him that the court would make him stop beating Mary. The quarterly meetings of the Essex County Court acted as the principal venue for resolving disputes between ironworkers. There they, like their more upright neighbors, sought satisfaction for slights and defended their reputations. They needed the court. By using it to pursue their own interests, they became invested in its legitimacy.[37]

A change in management at the Lynn ironworks probably helped its employees respect local government more. Richard Leader's relations with colonial officials, like those of company member Robert Child before him, were strained at best. Leader expressed contempt for local authorities after he left Lynn. In 1651, the General Court heard accusations that he had "threatned, & in a high degree reproached & slaundred, the Courts, magistrats, & goverment of this common weale, & defamed the towne & church of Lin, also affronted & reproached the constable in the execution of his office." It fined Leader £50 (then the second largest fine in Massachusetts history) and forced him to acknowledge the Court's power. Leader refused to confess that he had uttered such words, violating the custom that convicts offer only penitent speech. His replacement, John Gifford, had lived in New England before he became manager. Gifford also knew the iron business far better than Leader did and so probably commanded workers' respect more easily.[38]

The collapse of the Company of Undertakers deepened ironworkers' dependence upon the Massachusetts system of justice. No one felt the firm's 1653 bankruptcy more keenly than its unpaid employees. Lynn hands followed developments in the legal wrangling over the ironworks and they tried to influence the process. Seventeen of them, including Nicholas Pinion, Joseph Armitage, and Samuel Bennett, petitioned a special court in Boston after they learned that an imminent judgment on satisfying the company's creditors had excluded them. As men who lived "by theire labors" and had "wrought some tyme for the Iron workes whereby more wages is Groune due unto them then at present they cann receive or are able to beare," they implored their "Honored Majesties" to rule that they be paid before the court determined the status of other creditors or that they be paid shortly thereafter. They pressed their claim as responsible household heads who pleaded for the Court's mercy, without which they and "theire wives & children" would "suffer greatly." The ironworkers concluded their plea with a promise to "pray for yor Honnored Happiness long to Continew." Bennett and Armitage attended the Boston hearing, perhaps to dramatize their plight before those assembled.[39]

Two weeks later, John Gifford filed suit at Essex County Court to skirt liability for ironworkers' unpaid wages. The jury found that the company, and not the former manager, owed them a sum that Gifford had calculated at nearly £1,400. Ironworkers subsequently filled the dockets of the Essex court with claims against the Company of Undertakers to pay what it owed them. Others pursued their wages through other venues. In 1658, Samuel Bennett and Joseph Armitage accused Thomas Savage of failing to uphold an agreement he had made five years earlier before the Special Court in Boston. After the Court received the ironworkers' petition, the governor advised Savage that if he should win judgment against the Company of Undertakers, Savage "would paye the workemen," to which Savage assented. Savage's failure to honor the bargain, the two workers asserted, "hath browght great damages" to the company's reputation because "seuerall Executions have bin granted against this saide estate which would haue bin preuented had this promise bin performed."[40]

The ironworkers who tried to collect their wages became more tightly bound to the colony's judicial system. Their claims and their ability to be responsible masters of their households rested on the court's power. Iron-

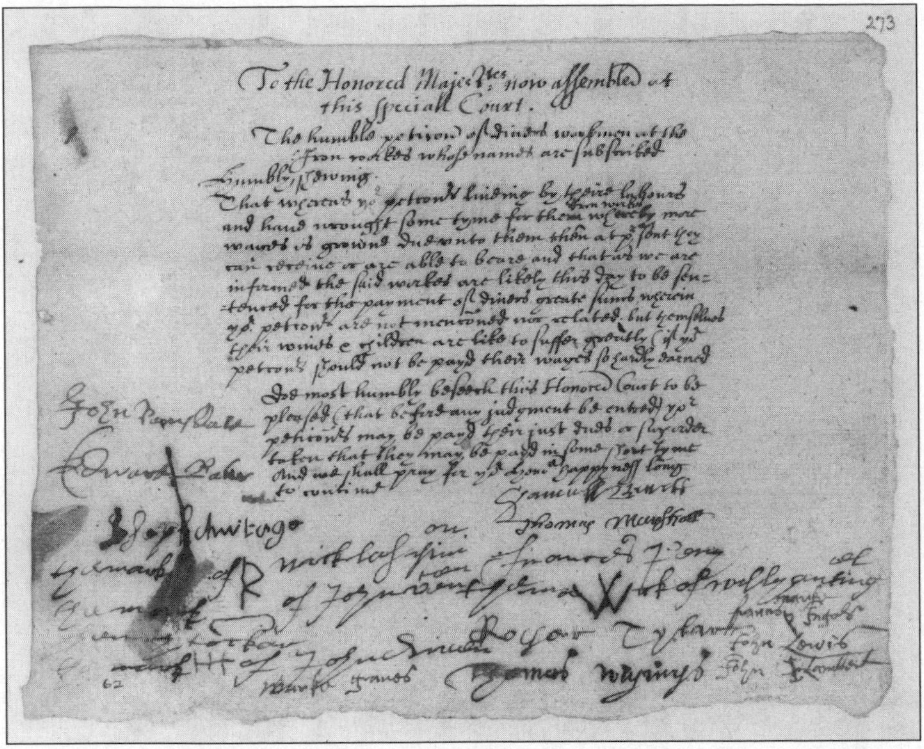
To the Honored Majestes now assembled at this speciall Court.

The humble petition of diverse workmen at the Iron workes whose names are subscribed

Humbly shewing

Figure 5. *"Petition of various workmen to special court for their wages due, 1653." Lynn Iron Works Collection. By permission of the Baker Library, Harvard Business School.*

workers had no choice but to acknowledge its legitimacy; they had to mind their behavior and stay in the good graces of magistrates. This perhaps helps to explain why few appear in Essex County's court records except as plaintiffs in cases against the Company of Undertakers. To be sure, shuttering the ironworks for several years prompted many to leave the area. Still, most who remained managed to stay out of trouble. They needed the courts and they needed neighbors' sympathy and support.

Essex County magistrates may have brought ironworkers to heel, but the household heads of New Haven, the destination of many former Lynn employees, doubted that the lessons had stuck. In 1657, at a meeting called to discuss the ironworks, they fretted that the costs of accepting ironworkers among them would far outweigh the benefits. They complained that two adventurers, John Winthrop, Jr., and Stephen Goodyear, had already leased their interests in the New Haven ironworks to outsiders. If their partners were to follow suit, "the trade may be caried to other places, and a disorderly company of worke men brought in here, wch may be much annoyanc to the Towne." A few months later the town's fathers reaffirmed that the ironworks existed "for the good of New-hauen & Brainford, for bringing and setling a trade there" and required their approval before sale or lease of any portion of them. "[A]ll seruants, workemen, and others imployed in any respects aboute this Iron-worke" were to "attend and be subject to all Orders and Lawes, allready made or wch shall be made and published by this Towne or Jurisdiction, as other men." Ironworkers were to assume the regular obligations of all men in New Haven; they would receive no exemptions from taxes or from militia duty as they had in Massachusetts. If the ironworks was to serve New Haven, then its employees had to meet its expectations of adult white men. Even though a bad reputation still followed ironworkers, the most self-consciously Puritan community in New England believed that it could digest and perhaps reform them.[41]

For a while such faith seemed well placed. New Haven's ironworkers either behaved reasonably well or authorities and neighbors turned a blind eye to them. By 1665, however, the town could no longer ignore them. Authorities heard three cases between December 1664 and April 1665 which concerned people directly connected to the ironworks.[42] Two involved members of the Pinion family. Robert Pinion, son of Nicholas and Elizabeth, faced charges that he had made "some contemptuous speeches in reference to ye Authority" after magistrates had questioned him. He allegedly boasted that he had told the town's leaders that "he had as good be bitt with a mad dog as snapt by a company of fooles." When asked if he had anything further to say, Pinion admitted that he had committed an evil deed. His contrition mollified his prosecutors. Though "his evill deserved sharpe Corporall punishmt," Pinion was to them "a stranger" and this was "ye first time yt he was brought in Publike." In other words, he did not know the rules in New Haven and he had not done anything else to merit their attention. Pinion escaped with two days in the stocks and a fine.[43]

Then again, Robert Pinion had an excuse for his behavior: he came from a family whose head, Nicholas Pinion, shirked his duties. Magistrates had heard an accusation filed by Elizabeth Pinion and her daughters Ruth Moore and Hannah Pinion against ironworks clerk Patrick Morran. They charged that Morran had tried to "Violate the Chastity" of Ruth and Hannah. Hannah, then around fifteen years old, testified that Morran had promised that "if she would come to ye furnace with him & let him ly with her he would bring her a payre of gloves." She spurned him. Morran persisted, telling her that he would signal when he was expecting her by placing "a bush . . . in ye furnace Bridge." He continued to pursue Hannah by upping his offer to "a payre of gloves and a shilling in silver." After showing her the coin, Morran "went to take her up in his armes & fling her on the bed, & she sd to him yt if he would not be quiet she would call out to ye folke below, & so he set her downe againe." According to Hannah, Morran still refused to take "no" for an answer. He propositioned her when she purchased goods at her mother's request and refused to sell again to Hannah unless she agreed to meet him at the furnace. Her difficulties in filling her mother's orders brought Morran's behavior to Elizabeth's attention. She largely seconded what Hannah had said, as did Ruth Moore, who claimed that Morran also propositioned her.[44]

Where was Nicholas Pinion? As the household's governor, he should have addressed the matter privately with Morran or, if necessary, before authorities. When asked what he knew, Pinion stated that he was merely a witness. He dissuaded his wife from filing a complaint, learned from her of Morran's advances on his daughters, and still did nothing. Morran protested his innocence and charged that the Pinion women had falsely accused him. Ruth Moore was, he asserted, a woman of "such an ill report" that he dreaded being alone with her lest it become "a scandall to the gospell & a Blemish to his name." Morran denied that he had offered anything to Hannah or that he had propositioned her. As for the "signs" that he had allegedly used to beckon Hannah, the clerk claimed that the plaintiffs had concocted the story because "he would not let the old woman have soe much Commodities as she desired." Once, Morran recalled, he visited the Pinions and Elizabeth "fell out with him, because he would not let her have soe much blue Linnen as she would have had, & abused him wth her tongue & tooke up an axe & calld him Scotch dog & Scotch Rogue & sd she would knocke him downe." Elizabeth Pinion admitted that she had threatened the clerk, though not until he followed her into another room and raised a fist at her.

The court refused to convict Patrick Morran, though it thought him "suspitious of something of the like nature." He refused to let the matter rest. His accusers had damaged his reputation and he intended to restore it. Morran sued the women for £200 for slander and defamation. The Court found for him, but reduced the penalty to £5 in damages, plus 50 shillings and court costs. Morran did not escape entirely unscathed. The Court admonished him to "carry it more prudently for the future then he hath in ye former business . . . that it may be more for his owne advantage & the advantage of his owners."[45]

The ironworks and its employees had become the "annoyanc" that the town had feared they would. Theft was bad. Sedition was worse, especially when New Haven and Connecticut authorities disputed who had the right to govern New Haven. The case of the Pinions and Morran was probably the most disturbing. The women were found to have borne false witness against an important man and they had acted loosely and aggressively. Morran certainly looked like a sexual predator who abused his authority; his conduct reflected badly on the ironworks, its adventurers, and the town that hosted them. Nicholas Pinion had abdicated his responsibility to his family and to the community by refusing to police his putative subordinates. The case indicated to New Haven's leaders that the ironworks could not govern its employees and that some ironworkers would not govern their households.

If adventurers could not or would not impose order, the town would by regulating whom they could employ. The ironworks could not engage anyone without first providing "a Certificate or Certificates from some persons of knowne reputation, & good Judgemt of his or their Civill life, & blameles Conversation" to magistates and obtaining their approval. Agents faced a forty-shilling fine for each violation. Anyone hired illegally was to be dismissed at once, and he could not return to work without permission. Adventurers were to assume financial responsibility for maintaining dismissed workers or risk having them sent out of town until authorities had cleared them for return.[46]

Benjamin Graves was likely a victim of more vigilant oversight. In December 1665, he stood accused of "frequent suspicious & offensive society" with Ruth Briggs (formerly Ruth Moore, daughter of Elizabeth and Nicholas Pinion) when she was a widow, of "kissing & embraceing" her after she remarried, of defying her husband's orders, and of saying that she was a whore and that he had "had Carnall knowledge" of her. None of Graves' alibis rescued him. "The Court laboured much wth him, to bring him to a sight of his sin, but little prvayled" and then fined him and demanded that he "make acknowledgemt of his evill to ye Court, or Else be severely whipt, & being noe allowed inhabitant here, that he speedily depart ye place." The magistrates also made an example of Briggs, who had, they charged, disobeyed her husband, and while a widow had ensnared and deluded several men with promises of marriage "to countenance & cover unlawfull familiarity with them." She confessed that she had promised to marry another man besides Graves and had misled John Luddington by "calling goodm Moulthrop unkle in open Court upon prtence of marriage" with him. She was as guilty as Graves and she, in the Court's eyes, merited a heftier fine, "or else be whipt, alsoe shee speedily depart ye place." If Briggs left New Haven, she later returned and moved into her father's house.[47]

A week after New Haven demanded control over hiring at the ironworks, Giles Blach and Robert Pinion appeared before the court. Blach, Patrick Morran's servant, faced charges of stealing from his master and of doing it on a Sunday. He claimed that he was alone when Pinion boasted that he could break into Morran's cellar, and then Pinion proved it. He stole rum, took it to his brother Thomas's house, and looked for sugar to sweeten it. Robert Pinion found none; his sister Mary and his mother gave him keys to Morran's residence. He snatched some sugar and returned to his brother's. The brothers reentered Morran's that afternoon to steal more rum and sugar, a pair of stockings, some cloth, and gunpowder. They, Blach added, also entered the clerk's chamber and picked up his account book, which he told them to set down. Morran could not calculate damages because he could not find his account book. For his "miscarriages as a Treacherous & unfaithfull servant, stealeing & embezling his masters goods, &c," the Court sentenced Blach to the lash and fined him.[48]

Robert Pinion insisted that he knew nothing. Circumstantial evidence corroborated Blach's story: broken locks, spilled sugar at Thomas Pinion's, and word that Robert Pinion had met an Indian and tried to trade liquor for wampum. Elizabeth Pinion's key opened Morran's door. The clerk was missing an account book; Robert Pinion had threatened to burn Morran's account book. Somebody had torn only the pages that contained Pinion's accounts from the ironworks ledger. Because the Court could not determine the value of what Pinion allegedly stole, it dismissed the charge, even though it believed he was guilty.

Pinion faced four other charges: breach of Sabbath, "lieing & slaundering the Authority & people here, . . . Lascivious & Corrupt speeches & Carriages," and "Threatening the lives of some against ye peace of his maj[es]ties good subjects in this plantation & of ye governmt of this jurisdiccion & c." He confessed to the first because he had no good alibi to counter Blach's account and "his too frequent absenting from ye ordinances" led some to claim that they "were afrayd to leave their houses" unoccupied on Sunday. The other charges involved what Pinion had allegedly said, and they mattered more to New Haven authorities. Some witnesses testified that they had overheard him slander the court, which Pinion largely admitted, and that he had boasted of "makeing maids loveing of him & kissing him in ye stockes, & yt he sd to his sister yt if had but halfe an houres speech wth her he could make her come to him (if he were in the stockes) & kisse him, to which she sd, doe you goe & sit in ye stockes & see if I will come to you." Henry Morrill and his wife asserted that Pinion "vowed to be ye death of ym yt punished him though it was seven yeares hence."

Robert Pinion denied little. He admitted that "he had been apt to speake very vilely . . . & for his threatening speeches, he spake them in his wrath." He would have fared better had he rested his case. Instead, he challenged the magistrates' authority to judge him. He could not have slandered the court in many of his speeches—at the time he gave them; New Haven's government had no right to operate because it had no royal charter. "There was noe Authority nor Law here," he asserted. This sat poorly with the magistrates. "There was ye same law & ye same Authority," they corrected him, "onely they had not tooke oath." They then told Pinion "of his evill & wt a sad accot he had to give to god for ye fame" and ordered that he be severely whipped "for a future warneing & terror to himselfe & others against such miscarriages." To ensure that he would behave well, the court directed Pinion to post a £100 bond and fined him to cover the costs of prosecuting him and for selling liquor to an Indian.

To New Haven officials, such cases exhausted what little faith they had in the ironworks' management. In 1666, two months after the town fined Patrick Morran "for selling Liquors contrary to the law whereby some young persons did much abuse themselves," the town put the ironworks under its direct supervision by appointing Matthew Moulthrop to "see yt persons doe attend ye ordinances on Lords dayes; and alsoe to looke after other disorders there, & prsent ym to Authority." For nine of the next twelve years New Haven charged a constable to watch over the ironworks.[49]

The workers kept Moulthrop busy over the next two years. His successors though, to judge by town records, had relatively little to do. Their oversight apparently worked. Moreover, of the cases presented before New Haven's town court between 1666 and 1668, all but one directly involved a Pinion, particularly Ruth Briggs or Hannah Pinion. Richard Nicolls was the exception, and he was linked to Ruth Briggs. In November 1667, the town court ordered Nicolls "to attend his duty & returne" to his wife in New York within eight days. He did not comply because, he claimed, he could not settle accounts with the ironworks. The court fined him and again ordered him to go—a married man who had abandoned his wife set a bad example and might be tempted to commit adultery. Nicolls picked a propitious moment to leave. In February 1668, Briggs stood accused of adultery and infanticide. She named Nicolls the likely father. He never faced questioning. The Court of Assistants in Hartford convicted Briggs and had her executed.[50]

As at Lynn, New Haven's campaign to discipline ironworkers, to mold them into men fit to live among the godly, seems to have largely succeeded. It was communal work, the responsibility of the commonwealth for the commonwealth, to promote economic development and to avert divine wrath for having invited the ungodly among them. Puritans did not expect work alone to reform them. A work ethic could not redeem ironworkers without a concerted campaign that tied the newcomers to civic life. In the process, ironworkers became members of their communities—men who at least partially assumed the roles and norms of New England manhood. The ironworks that they staffed did not survive long, but they became valued if unequal contributors to the society they had been brought in to enrich and to perpetuate.

The Way Forward

If New England Puritans associated the redemption of ironworkers with that of their commonwealths, they also indicated another way forward, the exploitation of unfree outsiders to sustain industrial enterprise and economic development. Nowhere in the seventeenth-century Atlantic world were Puritans averse to relying on coerced labor to enrich themselves and avoid arduous work. Puritans who settled on Providence Island quickly turned to plantation slavery before Spanish forces terminated their venture. Emmanuel Downing saw Native captives and African slaves as vital to the future of a society plagued by men who demanded high wages and then quit to strike out on their own. New England colonists seldom bought unfree labor mainly because most could not afford it, not because they had moral qualms against it.[51]

In 1650, the English Civil War delivered hundreds of Scottish prisoners to the Company of Undertakers. One of the principal partners, John Becx, sent over two hundred to Boston. Most were sold; over sixty arrived at Lynn in 1651. The ironworks provisioned nearly thirty Scottish servants more than a year later.[52] Most were woodcutters and farmhands, saving the company what it had paid to men who lived nearby and worked seasonally. The adventurers leased several Scots to colliers, who saw to their room and board. This allowed the company to recoup some of the colliers' wages that Nicholas Bond considered far too high. It exasperated him that he and his partners paid men so well just for turning wood into carbon. "I wonder," Bond complained, "all the Workemen doth not Turne cauliers."[53]

The Scottish prisoners also promised the Undertakers relief from the high wages that their tradesmen commanded. The adventurers urged John Gifford to "gett workemen Under those we have to learne ther trade or else we shall alwayes be to seeke for workemen espetially if we should be hearafter incoraged to sett Up more workes." Gifford complied. He featured his efforts to train some of the Scots to bolster his case that he had been a good steward of Lynn. Among the Scots, Gifford attested, there might be "6 Colliors yt may be able to save the Comp.a all Chargis in Coaleing except one m[aste]r Collior to over looke them." John Clark had begun to learn smithing. James Mackall, John Mackshawne, and Thomas Tower were learning to become forgemen; each boarded with one. Two "Scotts Carpenters Comeing on," Gifford hoped, "may save all Carpentors Chargis except in makeing some worke Extraordinary." According to Gifford, many of the Scots had "attained trades" and, had the company not collapsed, would "neare have mannaged the Comp.a business themselves and haue Saved them many hundreds of pounds in a yeere." He had incentive to exaggerate the training that the Scottish servants had received under his watch, but his larger point was clear. Fettered labor would serve the company well.[54]

The Scots did not save the Lynn ironworks, but they did represent the colonial iron industry's future. The Company of Undertakers had resorted to a group of ethnic outsiders, unpaid prisoners of war who stood out among their new neighbors even more than other Lynn employees did. They signified the growing importance of empire to colonial Anglo America's industrial development; the expansion of England into the British Isles, North America, the Caribbean basin, and Africa would provide the labor that adventurers believed they needed to make their industrious revolution.

Nowhere was this clearer than in the first ironworks founded in New Jersey. James Grover, part of a Connecticut group that resettled lands that England had just conquered from the Dutch Empire, built an ironworks in Monmouth County. In 1675, Colonel Lewis Morris, a Barbadian merchant, bought a half share of the Tinton Iron Works. Within two years Morris had secured rights over "the Mannor of Tinton" from the Proprietors of East Jersey which included mining concessions, guarantees that ironworkers could not be arrested for debt, a special court to try suits against employees, and a provision which exempted them from "pressing, mustering, watching & trayning, but amongst themselves, & by such officers of their owne, whome the Governor shall appoint and commissionate." Over the next eight years Morris hired some two dozen white workers to erect a furnace and expand the forge.[55]

They were a minority at Tinton. Africans, the first slave ironworkers in colonial Anglo America, accompanied Morris from Barbados. They helped to build the ironworks and probably supplied it with ore, wood, and provisions. George Scot visited Tinton in the early 1680s and estimated that "60 or 70 Negroes" lived and worked there. In 1691, Morris willed Tinton and its sixty-six slaves to his nephew (and future New Jersey governor) Lewis Morris. He inherited twenty-two men, eleven women, and thirty-three children. By then the ironworks had probably closed.[56]

For Lewis Morris, industrial slavery seemed a natural extension of the system that English planters in the Caribbean had adopted to grow and process sugar cane just thirty years earlier. Morris personified connections to a newly reorganized English empire. He depended on it for political advancement; and he profited from it as a sugar planter. Morris brought a new order to Anglo American industry, one that saw its workers as civically dead and their work as incapable of effecting their redemption or full acculturation. It would be an industrious revolution rooted in empire, geared to the belief that economic development must rest on the coercion and exclusion of those who worked for others.

3

PASSAGES THROUGH THE LEDGERS

In May 1769, Kingsbury Furnace's clerk recorded that the Principio Company had recently spent nearly £9 cash to catch "Negro Mingo," a slave who ran away from Lancashire Furnace. Mingo, who according to the itemized list of the costs of retrieving him "pretends to be a freeman," had fled the previous October to petition the Baltimore County Court to emancipate him. John Murray, overseer at one of Principio's mine banks, pursued Mingo, seized him, and returned him to Lancashire in handcuffs, but not before the court heard his motion and referred it to the Provincial Court in Annapolis. A year later, company managers James Baxter and Francis Phillips spent seven days attending Mingo's hearing in Annapolis.[1]

"Mingo" had a story to tell, which he began by reclaiming his name. He stood before the assembled justices as Juan Domingo López, shoemaker, "native born of the Island of Hispaniola in the West Indies and a Subject of the King of Spain," under whom he had "enjoyed his natural right of being a freeman which he humbly apprehends cannot be forfeited or taken from him without his own consent." The plaintiff had told the Baltimore County Court that he left Curaçao aboard a privateer in 1761. English privateers seized the ship, took him captive, and transported him to Jamaica. In 1764, the Court of Vice Admiralty ordered his sale. López was taken to Maryland, where the Principio Company bought him. López argued that the court should set him free because he had never been allowed to defend himself during the Admiralty sale and because none of the company's members lived in Maryland. He beseeched the Court to "grant him that Liberty which is his Natural Right and save him from that Slavery which is odious to the British Constitution, and from whose Laws your Petitioner hopes and expects relief." The Court dismissed his petition. López and his attorney, Samuel Chase, requested an appeal, which the Court granted. López then vanishes from our view. A record of his appeal no longer exists. By 1781, when the State of Maryland confiscated its holdings, the Principio Company no longer owned him. Between his final Provincial Court appearance and 1781, López likely met one of four fates: he won his appeal; he found another way to escape slavery; the firm sold him; or he died a slave to faceless masters who lived across the ocean.[2]

Juan Domingo López's passage to and through Maryland in many ways parallels those of his peers and encapsulates the history of the colonial South's iron industry. War delivered him into bondage. Lancashire Furnace's voracious appetite for labor made him a slave ironworker named "Mingo." As property of the Principio Company, López confronted Anglo American slavery and the harsh routines of colonial iron production. He also learned how he might escape both—he invoked British liberty and he retained an attorney to plead his case.[3]

Passages shaped and characterized the industrious revolution at the ironworks of Maryland and Virginia from 1715 to the Revolution. Chesapeake adventurers assembled work forces which were among the most racially and ethnically diverse in colonial British North America. They employed a stunning variety of workers: African and American-born slaves whom they owned, hired slaves, transported British felons, unpaid indentured servants, contract indentured servants who earned wages or piece rates, salaried free workers, and casual laborers. Ironmasters and their agents constantly weighed the advantages and disadvantages of different labor systems and adjusted staffing in response to shifts in demand for iron or in Atlantic labor markets. Most who staffed the region's first long-lived ironworks chose to sail to the Chesapeake—principally as indentured servants. Within one generation, men and women who arrived in shackles had largely replaced them. Some were British and Irish convicts sentenced to years of servitude for their crimes. Enslaved Africans who survived the Middle Passage joined them. As colonial ironworkers, they encountered new ideas about time, work, gender, and even the meaning of iron itself. Their children experienced their own passages—they entered the world as slaves and made their way in it accordingly. All underwent passages of another sort as they struggled to define where they stood with masters, peers, and neighbors. Some tried to flee their bondage; more did what they could to make the best of it.[4]

Anglo America's industrious revolution and industrial enterprise were nearly incompatible in the eyes of Chesapeake adventurers. They were sure that most people would not work hard or well for someone else unless they confronted the threat of having work beaten out of them. Nor did Maryland and Virginia ironmasters aspire to transform ironworkers into neighbors or valued members of their society. Indeed, their power and, they believed, the survival and success of their ventures depended on their right to own and coerce workers whose status as slaves or transported felons defined them as outsiders.

But slave ironworkers could not remain complete outsiders if the region's iron industry was to grow and flourish. They had to be acculturated—but not assimilated. Africans had to learn some English, the rhythms of colonial life, and the harsh demands that making iron and American slavery would impose upon them. Their children, born into slavery, posed a different challenge for adventurers. By relying so heavily on unfree labor, ironmasters put their enterprises in the hands of people who had little incentive to perform their duties well. The demands of their business required that they not rely primarily or too openly on coercion and exclusion—slave ironworkers needed incentives to work harder and better. Adventurers' success would rest largely on their ability to make slaves who were in but not of Anglo America: who would be industrious but would never attain the independence that most Anglo American men expected to achieve as a result of their work. Slave ironworkers, especially those born in America, created better lives for themselves by participating more fully in the industrious revolution than plantation slaves did. They also helped adventurers forge industrial slavery and so strengthened the shackles that bound them.

"More Subservient and observant of orders"

In December 1773, Charles Carroll of Carrollton wrote his partners to argue that the Baltimore Company should replace all its hired hands with unfree workers, particularly slaves, as soon as possible. To bolster his case, Carroll cited a letter from manager Clement Brooke which estimated that the ironworks would need at least forty-five more hands by the following spring and which warned that "'the company has but few hands of their own, and hirelings are so scarce, and their wages so high, that it will be impossible to carry out the business with such but at a very great loss.'"

Carroll had closely examined the ironworks' payroll years earlier and calculated that the Baltimore Company could save at least two-thirds of what it had spent on wages in 1768 by purchasing "35 or 40 young, healthy, and stout Country born Negroes." At those wages the slaves could pay for themselves within three years. Even better, "the business would be carried on with more alacrity and fewer disappointments." Brooke agreed; his superiors needed more slaves because "the sooner you get rid of hirelings the more it will be to your Interest."[5]

To Carroll, born into Maryland's planter elite, using slave labor was the best way to run an ironworks or virtually any other enterprise. The partners whom he joined in 1767 had long thought so. In 1734, just three years after it formed, the Baltimore Company, the region's first ironmaking firm owned entirely by colonists, employed at least eighty-one workers, forty-four of whom were slaves. In 1764, Carroll's father bragged to his son that he stood to inherit part of a company that owned at least 150 slaves. By then slaves served nearly every Chesapeake ironworks, often in far greater numbers than on neighboring farms and plantations. Ironworks owned 12 percent of adult slaves in Baltimore County in 1773, and ironmasters were the six largest slaveholders there. Well over one thousand slaves worked in the region's iron industry by the Revolution. In 1781, Maryland authorities confiscated 153 slaves from the Nottingham Company and 137 from the British-owned Principio Company.[6]

Industrial slavery on such a scale was natural to Carroll, his partners, and their peers. It was not always so. The Chesapeake's first adventurers of the eighteenth century, the Principio Company and Alexander Spotswood, tried to make iron primarily with people who chose to work for them. They soon concluded that they could not because white indentured servants and hired hands were too expensive and too unreliable. By 1730, the aspirations and actions of white ironworkers convinced Principio shareholders and Spotswood that they must chain the future of their ventures to as much unfree labor as they could afford.

In 1723, John England arrived at Principio Iron Works to find its workers "allmost Ready to Muteney" and

> so rude & Drunken yt for my part I Dispare of doing any good amongst ym they have gon on in such a Loose way so long yt they tell me to my face they will be wors instead of better & tell me they have not have there bargains pformd to ym & they will Do ye Compny no good so long yt they stay if they can help it. [T]his they tell me to my face so what course to take with ym I know not for they hav had ye Rains laid in there neck so long yt now they come to be checkt a little they wont bare it at all, & out of 60

> men I Cant find above 12 or 14 many times & when I find or any more appear they are so Drunken yt there is no such thing as speaking to ym.

So hopeless did the situation seem to Principio's new manager that a company partner had to dissuade him from boarding a ship back to England the next day. Over the following months, as he confronted sullen and rebellious ironworkers, John England probably wished that he had left when he had the chance.[7]

If nothing else, England could take cold comfort that his predicament had caught the attention of his superiors across the ocean. They agreed that Principio's problems stemmed from lax discipline rather than their failure to honor agreements with employees. The firm's partners met and decided to order England to identify the most troublesome workers, "take them before the proper magistrates and reclaim them by such Laws as are there in force but if [you] should find that ineffectual to discharge the Ring leaders of them wholly from our Service." Recalcitrant servants were to be "turnd over to such planters as will keep them to the stow upon the best terms [we] can make with them." The company's hands did not know how good they had it, the adventurers believed. Perhaps they would after they had served less indulgent masters who would put them in the fields and ride them harder. Two weeks later, the company repeated that anyone who "will not be reform'd may be turned off according to our former letters."[8]

What perhaps most bothered England and his partners was that they and their workers had reached conflicting understandings of the proper relationship between employers and employees. This situation was hardly unique to the iron industry; Chesapeake masters' complaints of English servants who refused to conform to colonial work routines seem to have spiked during the 1720s and early 1730s. They and the Principio Company believed, in accord with centuries of English thought and law, that employers reserved the right to demand loyal service of workers, even when they failed to compensate them in a timely fashion or meet other obligations to them. John England conceded that many of the workers' grievances had merit. In July 1723, he noted that "ye workemen complain verey much for want of there wages, & I fear Other Debts beside." But that in no way justified their decision to withhold their services, which the firm considered its property.[9]

The workmen viewed the matter differently. By failing to pay them or to honor its other agreements, the Principio Company had effectively forfeited its right to demand labor from them, at least until it upheld its side of the bargain. Some workers probably left the ironworks. Those who stayed saw no reason to exert themselves to benefit the company so long as it failed to deliver. Nor, since Principio owed them money, did it suit them to depart without settling accounts to their satisfaction. Besides, Maryland law barred many from leaving. Under the 1715 "Act Relating to Servants and Slaves," any servant, "whether by Indenture, or according to the Custom of the Country, or hired for Wages" who traveled more than ten miles from the ironworks without written permission from the company's agent risked being considered a runaway. If caught, fugitives had to serve ten days for every day that they were gone and assume the cost of their capture and return. Finances and the law made staying about the ironworks and doing "ye Compny no good" until they won what they wanted the best option for most Principio workers.[10]

The ironworkers held fast in the face of the firm's punitive measures. As John England directed construction of a race to power the forge at Principio Iron Works, he complained that "ye men are ye Cheefest Discouridging object at psant, for sometimes they will worke very well for 3 or 4 days & then for as long time they will be as bad agane." Nine days later, England, protesting that "I was forst to Do it or stand still," sheepishly requested funds to pay off the workers. Nothing could go forward "untill ye Company is upon Even hands with ym."[11]

The standoff at the wheel race was not all John England had to contend with; Robert Durham was again making his life difficult. The manager believed that Durham, Principio Iron Works's founder, personified the contagion that plagued the company. His superiors agreed; they suspected that Durham was the principal "ring leader" of their disgruntled employees. That worried them because they needed him to supervise the furnace closely—wishful thinking, England believed, so long as Durham, a "verey loose Careless Drinking man," was in charge. To set an example, the manager fired the founder, and then had to rehire him because the firm could find no one better to replace him. The incident chastened Durham briefly, but soon England deemed the founder "as bad as Ever, & what to do with him I know not." The lesson was clear to Durham and his peers—the Principio Company could do little to discipline them, especially if they possessed a skill that few colonists had mastered.[12]

If threat of dismissal did not render workers more obedient, two other tools, both solely in the possession of the company, might. The firm, not ironworkers, kept time and account books. Both enabled the company to quantify and track workers' activities. England owned a watch, and he probably used it to follow the progress of the furnace blast and the comings and goings of furnace hands during the night shift. On occasion, Principio agents consulted clocks to calculate labor costs more precisely. In 1732, the company withheld over £72 from Richard Snowden's bill for the hire of fifteen slaves because they had accumulated "44mon:25ds:5hrs lost time." Still, clock time had little direct impact on the daily routines of most Principio hands.[13]

Account books did. Because the Principio Company compensated workers primarily by extending them credit, it could discipline them accordingly. In 1724, John England and his partners imposed a clearer distinction between workers' time and Principio's by debiting their accounts. Over the next year, the Principio Company fined eight hands for missing between two and one-half and eighteen weeks of work. Robert Durham paid £9 for eighteen weeks "Lost time" and John England forced him to sign a statement acknowledging his debt to the company. Eighteen weeks of "Willfull Neglect" of his duties cost Basil Tyler nearly £4. In 1725, forgeman Benjamin Turley tried to flee Principio and the more than £17 that he owed it after England charged him for missing nearly four days of work. The firm wanted to teach workers a clear lesson: absences would translate directly into money off their side of its ledgers.[14]

John England tried to ensure that forgemen in Britain knew that their time would equal money before they sailed to Maryland. In January 1724, England advised a partner that contracts should stipulate that if forgemen lost "any time Excepting by sickness they shall Pay Dammage & not to have it set to make good there Lost time at Last." He did not get his way. In October 1725, Joseph Reading bound himself to serve the Principio Company for seven years. He promised that he would not "either in unreasonable manner or at unusual times neglect or absent himself" and that he would "make good" when his term ended "what time he shall Lose by Sickness."[15]

Perhaps the partners declined the manager's advice because they knew that few forgemen would consent to such terms. Most of the adventurers lived in England and had close ties to both the British iron industry and transatlantic commercial networks, which enabled them to keep abreast of demand for ironworkers' services. Under John England's proposal, a forgeman might have ceded control over when he worked to employers and he might have had to shoulder liability for events, such as bad weather, that were beyond his control. That was no way to get proud English forgemen to board a boat headed for Chesapeake Bay.[16]

Charging employees for lost time or persuading more forgemen to set sail for Maryland did not satisfy the company's desire for more dependable and obedient workers. More slaves, it believed, would. The firm employed slaves almost from its inception. It owned many before John England arrived—the company purchased eighteen slaves in 1722. Within a year the partners sold two and three died of the yaws. The rest probably dug ore. By 1723, Principio's adventurers considered mining a task most suitable for slaves. When England complained of little ore near the furnace, he cautioned a partner that "phaps thee mayst Expect it should be gott by Negros; but thee mayst ashur thy Selfe thy victuals will be Eat & thy worke undon."[17]

England soon argued that slaves should assume more responsibility. In the same letter in which he complained of Durham and the men constructing the mill race, the manager asserted that "if we had more Negros [it] would be an advantage . . . for they would be brought in a little time to fill & wheel as well as any white man & Cutt wood allso." His partners in Britain agreed and William Chetwynd authorized England "to buy at any time 40 or 50 Blacks or what number you think proper."[18]

England followed Chetwynd's advice, with disastrous results. Several of the Africans England acquired soon died, victims of the same diseases that killed many recent arrivals to the Chesapeake. Chetwynd castigated his manager for being penny-wise and pound-foolish; he advised England to buy only "seasoned blacks in ye country" at £40 each rather than risk "loosing ym as we must ever do when we buy ym raw ones"—especially since the five who had survived "cost us above fforty pound per head." That said, he tried to set England at ease: "we must expect such accidents but [I] would not have you be att all discouraged by ym." To Chetwynd, Principio needed slaves, because "no works can be well & steadyly carryed on without ym."[19]

Chetwynd's partners stayed the course. Within a few years lobbyists invoked their commitment to slavery to justify why the British Empire should encourage economic development in its North American colonies. In 1731, Fayrer Hall, who argued that Parliament should abolish duties on colonial bar and pig iron which entered Britain, projected that slave labor would soon enable Principio "to work full as cheap, if not cheaper than here." The firm had proven that slaves were "as useful as any white Men, when they are instructed, in cutting Wood and making Charcoal, and stocking it near the Furnace," and "in all Cases where Labour is principally required."[20]

But the company's commitment to slavery was already far deeper than Hall stated. Confining slaves only to those positions "where Labour is principally required" did not guarantee Principio's adventurers enough power to organize work. They thought that it left them too dependent on white tradesmen, whose recruitment proved risky, whose expertise gave them excessive influence over making iron, and whose scarcity promised them high wages and ready employment elsewhere. In September 1725, the company informed England that Stephen Onion, whom they had just hired to clerk the Principio Iron Works, would "take care yt all such blacks as you and in your absence ye Clerk of ye Stores shall appoint be taught in all things necessary to ye makeing of Iron." William Chetwynd reinforced the message, exhorting England to "encourage ye blacks & all such workmen as teach them any thing."[21]

The partners especially wanted slaves to learn how to forge iron. For that they needed white forgemen to cooperate. Some would not. In April 1725, William Russell reported that two Gloucestershire finers, James Jarrett and John Hughes, had refused to train slaves. "All ye Arguments yt Cou'd be used Cou'd not prevail," he lamented, "to admit them of a Clause to teach Negroes; they said they were murdering Rogues they wou'd have nothing to doe wth them."[22]

What made Jarrett and Hughes so hostile to the idea of training slaves? Why did they insist that Africans were "murdering Rogues?" The second question is harder to answer than the first. Their view of Africans probably did not stem from personal experience. Few Africans lived in rural Gloucestershire or Wales, the center of western Britain's iron industry, although Jarrett and Hughes may have had contact with the sizable community of Africans and Afro-Britons in Bristol or some other port. Most slaves in Britain served as domestics or personal retainers. Few were artisans, because British tradesmen often sought to exclude them. London artisans discouraged slaves from entering their trades and in 1731 won legislation barring slaves from apprenticeships. British forgemen were probably even more opposed to slaves in their ranks than were other tradesmen. They were a tightly knit group which relied on kin networks to gain entry to their craft, to master it, to find work, and to keep it. They jealously guarded their knowledge from outsiders, be they ironmasters or potential competitors.[23]

If other white forgemen at Principio agreed with Jarrett and Hughes, they did little to show it. In 1726, William Coslett earned £10 for "Teaching [a] Negro." The following year he regularly made anchonies with Patrick and drew over half a ton of bar iron with Andrew. Principio probably apprenticed Patrick to Coslett. In 1727, founder George Williams paid Coslett "for Negroe Patrick Working 6 Days upon the Bank" and twice credited him for work that Patrick had done with Joseph Reading in the forge's upper finery in 1728.[24] Nor was Coslett the only forgeman who took a slave apprentice. In 1727, John Hughes, who William Russell had not previously been able to persuade to train slaves, pocketed £5 for "Taking Negro Prince apren." James Jarrett also dropped his opposition to teaching slaves; he and Hughes together received nine shillings for allowing Prince to work at the furnace.[25]

Why did Jarrett and Hughes change their minds? Why did so many of Principio's white forgemen choose to work with slaves? The company may have concealed who they would work with until they arrived. Joseph Reading signed his indenture months after Jarrett and Hughes rebuffed Russell. He pledged "to Teach & Instruct" whoever the company directed "in ye said business of a finer or Bloom maker" and "to find him or them Cloaths & provisions in consideration of his haveing ye service of ye said person or persons for ye Term of five years if he shall continue so long in ye said partners service after takeing Such person or persons."[26]

Financial necessity surely spurred some white forgemen to take on slave underhands and train them. In September 1728, Joseph Reading and James Jarrett owed Principio approximately £86 and £28 respectively. By the following March, Reading's debt had decreased to nearly £82. Jarrett's had increased to over £49, which compelled him to bind himself as a servant to the firm. Reading had already promised to instruct whoever the partners directed; his debt made it even harder for him to renege. Fear of sinking into arrears may have encouraged Hughes and Coslett to instruct slave forgemen. By March 1729, Principio owed Hughes over £7, £5 of which he earned by agreeing to apprentice Prince.[27]

Hughes and the other white Principio forgemen likely saw instruction of slaves as an opportunity rather than a burden. Some may not have cared what the Principio Company might do in the future because they had journeyed to the Chesapeake to earn all they could before they returned to Britain. Coslett and Hughes could use the extra money to improve their standard of living. Joseph Reading obtained help that would enable him to turn out more anchonies. He and his peers also gained something else—a chance to wield power over slaves, even if they only oversaw them for their employers.

The introduction of slaves to Principio's forges proceeded fitfully. In 1730, the partners, dissatisfied with their ability to find enough reliable hands in Britain, proposed engaging a group of German forgemen and handing one entire forge over to them, but they never did. The company slowly expanded its ranks of slave artisans; by the Revolution they nearly monopolized the hearths and hammers of the firm's forge.[28]

Alexander Spotswood turned to industrial slavery less quickly but more completely than did the Principio Company. He began to operate the Tuball Iron Works in piedmont Virginia around 1715. With Christopher von Graffenried's help, Spotswood recruited a group of German-speaking servants which included several skilled miners. Over the next four years, he bound sixty more German families. Few stayed longer than required. In 1719, several left as soon as their terms expired to establish a settlement nearby. Most others departed Tuball after they had served out their agreements. Spotswood's servants, like most men who migrated voluntarily across the Atlantic, had come to North America hoping to obtain enough land so that they would no longer have to work for others. The same aspirations that had lured ironworkers to the Tuball Iron Works drew them away once their servitude ended.[29]

Their pursuit of independence meant perpetual bondage for their replacements. When William Byrd toured Spotswood's furnaces in 1732, he saw few white employees. Scarcely a dozen worked at the Fredericksville furnace that Charles Chiswell managed: a founder, a collier, a miner, a clerk, a carpenter, a wheelwright, a blacksmith, and several carters.[30] Chiswell told Byrd that he supervised 80 slaves and that 120, "including Women, were necessary to carry on all the Business of an Iron Work." After meeting with Chiswell, Byrd proceeded to Tuball, where Spotswood had "contriv'd to do every thing with his own People, except raising the Mine and running the Iron, by which he had contracted his Expence very much" and "believ'd that by his directions he cou'd bring sensible Negroes to perform those parts of the Work tolerably well."[31]

Employing so few white hands, Chiswell and Spotswood stressed, saved them money. Chiswell noted that the founder, the collier, and the miner at his furnace supervised many slaves for piece rates. The annual wages of the other hired hands totalled only about £100. Spotswood began to explain to Byrd "the whole charge of an Iron-work" by stating that "there ought at least to be an Hundred Negroes employ'd in it, and those upon good Land would make Corn, and raise Provisions" to feed themselves and cattle, "and do every other part of the Business." In 1739, Spotswood advertised Tuball for rent. He boasted to prospective tenants how slave labor had enabled him to economize. He promised them the services of "Sixty able Working Slaves," along "with 12 or 15 of their children," who had "all been trained up and Employed in the Iron Works." They had brought Tuball "to the present frugal state which is such as I believe no other Iron Work in the World can be Carryed on at so little Expense to the owner." Not only did slaves spare Spotswood the cost of hiring anyone except "one Cheif ffounder and one Generall Overseer," they also, he claimed, made his ironworks nearly self-sufficient. Slaves grew enough grain and fodder to provision themselves and his cattle; they built "their own Cabins or Habitations in the Woods without one farthing of Expense to me"; and they made "all the Coal Basketts and Mine Baskets that I use." Because slaves cut wood, Spotswood had "no Occasion to be at the Trouble and Expense of Cording and Examining the same." Hired choppers, paid by the cord, would try to exaggerate how much wood they cut.

Slave woodcutters, he implied, did not or could not deceive so easily, partly because they had less incentive to try.[32]

The savings that slaves afforded Spotswood perhaps mattered less to him than his conviction that they were simply less trouble than white hands. Slaves could not depart his ironworks without leave. White ironworkers moved easily between employers, and that worried Spotswood greatly. He confided to Byrd that he hoped for "many more Iron works in the Country" as long as "the partys concerned wou'd preserve a constant Harmony among themselves and meet and consult frequently, what might be for their common Advantage. By this they might be better able to manage the Workmen, and reduce their Wages to what was just and reasonable." Spotswood expected free workers to seize any opportunity to play ironmasters against one another. Only collusion could empower adventurers to suppress wages and discipline free labor.[33]

The harmony between adventurers that Spotswood wished for seldom materialized. During the 1730s the Baltimore Company and the owners of the Patuxent Iron Works squabbled over white tradesmen. When Baltimore partner Charles Carroll of Annapolis heard that Patuxent ironmaster Richard Snowden had "very Dishonestly & Dishonnorably inveigled away One of our founders, & has Entered into an Agreement with him," he lectured Snowden's partner Edmund Jenings on the dangers of such behavior. "[W]ee may hurt Each other by such a Conduct in raising Wages [&] making workmen insolent," Carroll scolded: "I hope you will not imploy this man or any other from Our Works without a formall Discharge & wee shall act in the same manner in regard to you."[34]

Jenings pleaded ignorance; Snowden insisted that there was more to the story. He blamed the founder, Michael Hodgkiss. Snowden asserted that Hodgkiss had approached him and told him that he would leave the Baltimore Company unless its owners ceded complete authority over its furnace to him. Snowden tried, to little avail, to mollify Carroll by offering him someone to replace Hodgkiss. The conflict provoked discussion within the Baltimore Company. Dr. Charles Carroll refused to believe that Snowden had not poached the founder, though he thought that Hodgkiss was no innocent either. What annoyed him most was that the founder had violated his bargain with the company. Hodgkiss had posted a bond that carried a £200 sterling penalty should he engage with another employer while under contract with the Baltimore Company. The partners had promised Hodgkiss "not to turn him off from our work without cause or six months warning." Neither secured the founder's loyalty. He had mastered a skill that he could market, and perhaps Snowden might give him what the Baltimore Company would not.[35]

Industrial slavery permitted adventurers to deal with men such as Michael Hodgkiss as little as possible. Slaves could not negotiate behind their backs with competitors, nor could they run up debts and then flee before making good on them. The memory of making iron primarily with white workers haunted Spotswood as he prepared to lease Tuball. He considered his slave ironworkers so valuable that the £10,000 that he demanded as security from a lessee "would scarse make me amends if they should be made away with" and "no others trained up in the Business" to replace them. He dreaded the thought of again operating an ironworks without slaves. Not only did slaves earn no wages, they "will always be more Subservient and observant of orders than ffree men in this Country will be." Anglo America's industrious revolution, Spotswood concluded and his peers and successors agreed, had made free men so free that industrial enterprise was nearly impossible without slavery.[36]

Bound for the Ironworks

By chaining the progress of their ironworks to unfree labor, colonial Chesapeake adventurers needed a steady supply of convict servants from Britain and of slaves from Africa and the region's plantations. Many could supply them personally—either because they had imported servants or slaves themselves or because they owned plantations from which they transferred their own slaves.[37]

Between 1717 and 1775 royal authorities sentenced approximately 50,000 felons to servitude-in-exile in North America. The overwhelming majority sailed to Maryland and Virginia, where masters eagerly bought them as an inexpensive alternative to slave labor. Most male convicts cost less than £13 sterling—about one-third what an adult male slave sold for—and served at least seven years.[38] Such low prices were attractive to adventurers, who often traded iron for servants. The Chesapeake iron industry probably employed hundreds of convicts at a time, mostly to dig ore and char wood into charcoal. The Bristol firm Stevenson, Randolph & Cheston sold at least sixty-six convicts to Chesapeake adventurers between August 1767 and August 1770, including fifty-six to Northampton Furnace owners Captain Charles Ridgely and Benjamin Howard.[39]

Many of the slaves who worked alongside convict servants also were shipped across the Atlantic. Charles Chiswell informed William Byrd that few of the slaves whom he supervised were "Virginia born." The names of five of the thirteen slaves listed on a 1720s Principio Company roster of workers suggest African birth: four men—Cujo, Quash, Quamini, Tantaro—and one woman, Quatheba. In 1732, Charles and Daniel Carroll sent five African slaves to the Baltimore Iron Works; some of whom may have been on a 1734 list of workers bearing names such as Coffee (of whom there were two), Qua, Caleboy, and Quame. By the 1760s, some ironworks identified African slaves by what their owners believed was their ethnicity. In 1767, clerks at Kingsbury Furnace recorded the deaths of "Mandingo Sam" and "Negro Capt. Ibo," while Elk Ridge Furnace journals and ledgers tracked the activities of "Gola Jack," who may have hailed from Angola or from the interior of what is today Liberia.[40]

By 1750, slave ironworkers born in North America outnumbered those born in Africa. Most ironmasters, like Chesapeake planters generally, preferred "country-born" slaves to Africans. When Chiswell told Byrd how many slaves he would need to staff an ironworks, he advised "the more Virginians amongst them, the better." American-born slaves were less susceptible, as William Chetwynd reminded John England, to many diseases that killed up to one-quarter of Africans within a year after they landed. They spoke English and had less trouble understanding directions. Ironmasters and managers also felt more comfortable with American-born slaves, who were already acculturated to white norms and thus seemed far less "outlandish" and dangerous to them than Africans did.[41]

Demographic changes brought more American-born slaves into the iron industry. During the 1720s, the Chesapeake's slave population began to grow through reproduction as well as immigration. By 1750, American-born slaves outnumbered Africans everywhere except in Virginia's Piedmont. By the Revolution the vast majority of slaves were American-born, and many could claim American nativity for at least three generations.[42]

The diversification of activities on the region's farms and plantations turned even more slaves into ironworkers. Low tobacco prices prompted many farmers, especially on Maryland's Eastern Shore, to grow less labor-intensive crops such as wheat. They sold slaves whom they no longer wanted or needed, some to adventurers. Jack, who fled the Baltimore Iron Works twice in 1748, was one. So was Dick, who ran away from the Northampton Iron Works in 1765 months after his master in Dorchester County sold him.[43] Many Chesapeake adventurers who were also prominent planters had already begun to train slaves in trades that their plantations needed. Some skills that they learned, such as blacksmithing and carpentry, transferred readily to the iron industry. The growing population of "country-born" slaves encouraged adventurers to introduce slave labor into skilled positions because they, like many Chesapeake planters, questioned the ability of people born in Africa to master a trade.[44]

Carolina planters tried to acquire West Africans who had expertise in cultivating rice. Many peoples of West and West Central Africa had ancient metalworking traditions, but there is no evidence to indicate that Chesapeake ironmasters specifically recruited Africans who they thought might know how to forge iron. Brumall, "a Gold Coast Negro" and a forge hand who fled the service of Archibald Cary in 1766, was apparently one of few skilled Africans in the Chesapeake iron industry.[45]

Adventurers may have been especially reluctant to teach African-born slaves a trade because they had such a large stake in the outcome. Several trades within the iron industry, especially forging iron, took years to master under close supervision. Ironmasters could defray training costs by requiring white artisans to board slaves who served under them, as did the Principio Company of Joseph Reading. The Baltimore Company attempted to lure hammerman John Cary away from the Patuxent Iron Works in 1737 by proposing to "allow him a Negro Boy for Two years to work with him whose diet he should find." In 1753, Dr. Charles Carroll pledged to direct the agent of a proposed ironworks to "get Young Negro Lads to put under the Smiths Carpenters Founders Finers & Fillers" as soon as possible. Training slaves when they were so young improved the odds that they would master a trade by the time they reached their most productive years. In addition, as Joshua Gee of the Principio Company contended before a committee of Parliament in 1738, a young slave apprentice would cost his master considerably less than what his competitors in England paid a laborer.[46]

Nothing testifies more powerfully to the investment that adventurers made in training their slaves than the prices that skilled ironworkers commanded on the auction block. Most valuable were forgemen, who took longest to master their trade and whose labor and knowledge had the most impact on the quality and price of bar iron. In 1768, Bernard Moore advertised Isaac, a twenty-year-old finer and hammerman, for £230, and asked £220 each for Joe, age twenty-seven, a forgeman, and Mingo, age twenty-four, a finer and hammerman. Abraham, a twenty-six-year-old forge carpenter, cooper, and clapboard carpenter, also cost £220. Moore requested £150 for Bob, age twenty-seven and a master collier. Moore's slaves promised to bring far more than did other slaves for sale whose skills were more suitable for plantations, such as George, a twenty-two-year-old sawyer and carter, whose asking price was £80, or Ben, age twenty-five, a house servant and carter, who was advertised for £75.[47]

Few adventurers could afford to buy and support as many slaves as they wanted, so most leased some. The Principio Company credited Richard Snowden £150 for the "hire of 15 Negroes" for one year each in 1732. Sometimes agents or partners in an ironworks leased slaves to the company, as the Snowdens did to the Patuxent Iron Works. White ironworkers who owned slaves occasionally leased them to their employers. In 1746, Kingsbury Furnace compensated founder Thomas Finley for "2 weeks & 3 days of his Negroe in the Furnace as Keeper" and for over three months' work that his slave completed. A few years later John Honey, a miner, hired out "his Negro Man" to Accokeek Furnace for nearly eleven months.[48]

Hiring slaves offered ironmasters several advantages. Above all, it took less capital than did outright purchase. Ironmasters could engage slaves as they needed them. To Chesapeake adventurers, who often sold iron in an uncertain British market, leasing slave labor was a good deal. It gave them access to workers whom they could discipline with more impunity than hired white hands without having to assume the obligations and risks associated with ownership.[49]

Ironmasters sometimes leased slaves to white employees. Charles Chiswell employed a miner who worked with a crew of slaves whom he hired for twenty-five shillings a month each. In 1749, founder George Williams leased four Principio Company slaves to help him run Accokeek Furnace. During the 1750s and 1760s master colliers frequently hired slaves from the Principio Company to assist them. In 1768, for example, Francis Turner and William Bannister each leased four slaves and one servant from Lancashire Furnace. They obtained help, while adventurers gained overseers, someone to train slaves, and a way to recover some of the expense of purchasing and provisioning them.[50]

Slaves leased to white overseers had little say in the matter and often benefited little from it. But others saw opportunities in the iron industry that their masters could not or would not provide. In 1768, John Mercer's foreman Scipio heard that an iron foundry gave slaves six pounds of meat weekly. He threatened to run away unless Mercer, who recruited hired slaves for the ironworks, agreed to let him work there. Scipio might get to eat better as an ironworker; he would certainly get a chance to travel and expand his social network. More significantly, the iron industry offered him a rare chance to assess the value of his labor and enjoy some of the fruits of it.[51]

Passages through the Mines

For Scipio and for many American-born slave men, becoming an ironworker promised a chance at a better life. Convicts and African-born slaves found the experience an ordeal—a passage that some perhaps found even harder to endure than the one which carried them across the ocean. Convicts and African slaves shared the stigma attached to fettered outsiders who came to North America against their will. Together, they comprised much of an industrial work force which was hard to discipline and which posed a threat to adventurers and to other white colonists.

Maryland's 1732 "Act to Encourage Adventurers in Iron-Works" reveals how leery ironmasters and Chesapeake elites were of ironworkers. The legislation, sponsored by Daniel Dulany, a Baltimore Company partner, mandated that an "Ordinary-keeper, Victualler, or Publick House-keeper" could not "harbour or entertain any Person who shall be hired or employed in any Manner about any Iron-Work, or give them Credit for any Liquor, or other Accomodations" worth more than five shillings "Current Money in any One year" without first securing written permission from the ironmaster(s) or the ironworks' principal manager. Dulany and his colleagues in the Assembly and in the iron business did not simply wish to make workers more industrious by regulating their access to alcohol. Taverns and public houses offered refuge to ironworkers, largely because they were spaces where ironworkers could socialize with other working people away from the prying eyes of adventurers and their agents. It was no coincidence that violators faced the same penalties as those which applied for "entertaining, harbouring, and trusting Sailors" under the 1712 "Act Restraining Victuallers and Keepers of Publick Houses from entertaining Sailors to the Prejudice of Trade and Commerce"—ironworkers were as fearsome as mariners and had to be kept apart from them if possible. Sailors had the means to spirit servants and slaves away. They also were a bad influence as far as adventurers and Maryland's elite were concerned—notorious throughout the British Atlantic world for their rowdiness, their willingness to transgress racial boundaries, and their ability to spread news and ideas that undermined the authority of masters and employers.[52]

Ironworkers seemed especially threatening to colonists in part because so many were convicted felons who endured their sentences sullenly. In 1754, *Pennsylvania Gazette* readers learned that John Oulton, a runaway convict servant belonging to the Baltimore Iron Works, stabbed his captor John Orrick after he unbound Oulton's hands to let him eat. Adventurers agreed that it was unwise to own too many convict servants. In 1774, Clement Brooke acknowledged a Baltimore Company directive to buy convicts by warning that purchasing ten of them over two years "will be as many of that sort as will be necessary or proper to have at a time." His superiors agreed; ten would give them "the full Complement of Convicts."[53]

Too many convicts often meant too many fugitives to chase down. They were notorious for trying to outrun their bondage, so much so that

Charles Ridgely directed his agents in 1772 to prepare detailed physical descriptions, each a paragraph long, for about forty white servants at Northampton. The information that they assembled provided handy copy for handbills or newspaper ads to retrieve them should they flee. Some convicts succeeded in liberating themselves, as did James Hall and Henry Jones, who departed Elk Ridge Furnace together in October 1761. The following April their masters gave up hope of catching them and wrote off over £43, the balance that remained on what they had spent to acquire them. Most were caught and returned, often to run away again.[54]

Unfree workers especially wanted to escape Ridgely and Northampton Furnace. At least twelve servants and two slaves fled Northampton between 1775 and 1778, some repeatedly. Among them was Phillip Beall, who ran off in 1775 "for fear of being Whiped and pilloried" for stealing sugar and rum. Northampton charged Roland Bates more than £10 for prison fees and "expens. when Whip'd & piller'd." Thomas Gray had to ride to Lancaster, Pennsylvania, to collect Robert Brown and William Orton and bring them home. A hired slave identified in furnace records only as "Abraham Patterns Man" escaped Northampton's mines in 1774, not by running away, but by cutting his own throat after he had tried and failed to flee. Abraham Pattern, unwilling to put his property at greater risk, arrived a few weeks later to take him home.[55]

Some fled Northampton because Ridgely was literally working them to death. Five of the furnace's servants died between November 1775 and December 1777, the demise of each dutifully recorded in company accounts as a debt incurred by "Profit & Loss." They included William Moses, who expired in March 1776, just two months after someone dragged him back to Northampton. The situation sickened the furnace's doctor, Randolph Hulse, who resigned in protest in 1777. Serving as doctor to "the Iron Manufactorys in Maryland," he asserted, "is an Imployment adapted for those only whose Ignorance Poverty or trivial private Practice will induce them to submit to the meanest Indignity or the haughty mandate of some imperious Task Master a Manager or Overseer." "Such acts of Cruilty," the doctor insisted, "have prevailed at some Iron Manufactorys as would extort a Blush from a Turkish Bashaw and he must possess a heart of stone and be deaf to every Sentiment of Humanity."

Hulse punctuated his protest by reciting a story which, though perhaps apocryphal, resembled life at Northampton. An ironworks punished a fugitive servant by forcing him to work with a fifty-pound ball chained to his leg. After the servant somehow managed to run off again, his manager beat him so badly that the servant died the next morning. "What a pity it is," Hulse chided Ridgely, a reluctant revolutionary, "that one Quarther of

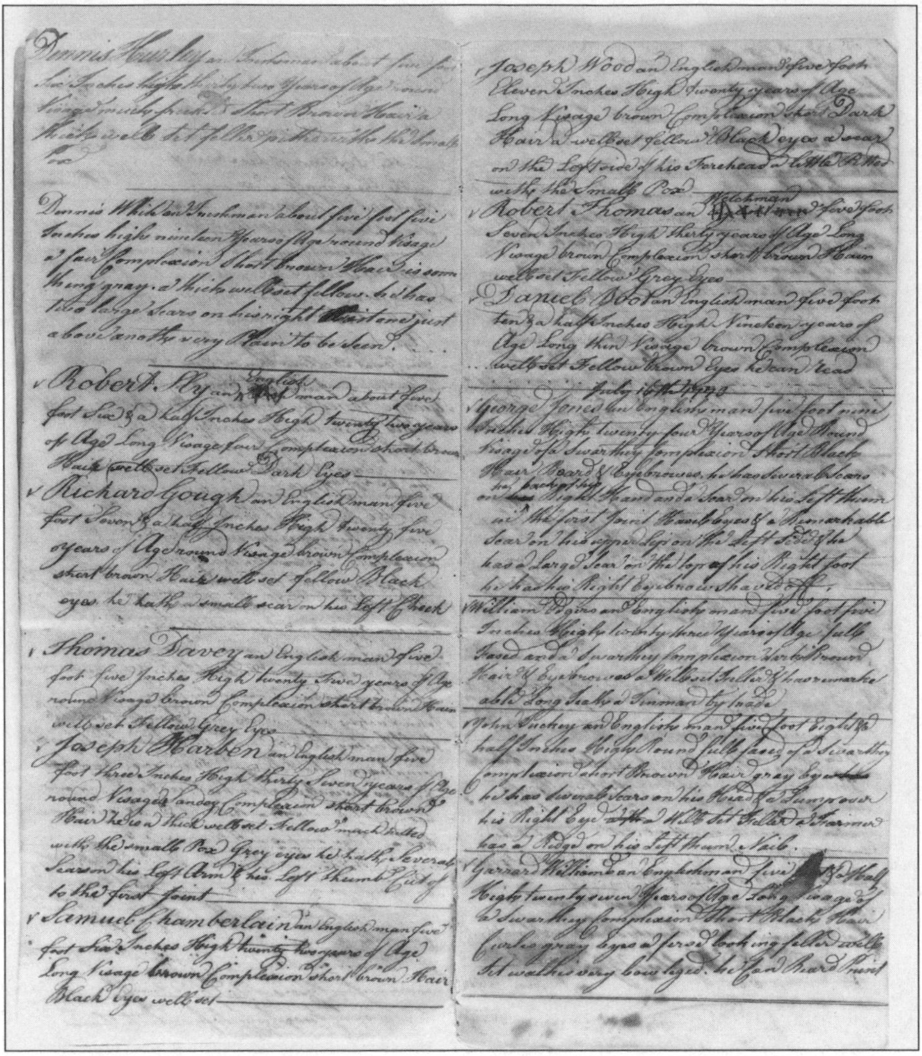

Dennis Hourley an Irishman about five feet Six Inches high thirty two years of Age round Visage much freckled Short Brown Hair a thick well set fellow pitted with the Small Pox

Dennis White an Irishman about five feet five Inches high nineteen years of Age round Visage a fair Complexion Short brown Hair is some thing gray. a thick well set fellow he has two large Scars on his right [illegible] just above one another very Plain to be Seen.

Robert Fly an English man about five foot Six & a half Inches High twenty two years of Age Long Visage fair Complexion short brown Hair well set Fellow Dark Eyes

Richard Gough an English man five foot Seven & a half Inches High twenty five years of Age round Visage brown Complexion short brown Hair well set fellow Black eyes he hath a small scar on his Left Cheek

Thomas Davey an English man five foot five Inches High twenty Two years of Age round Visage brown Complexion short brown Hair well set Fellow Grey Eyes

Joseph Harben an English man five foot three Inches High thirty Seven years of Age round Visage Sandy Complexion short brown Hair he is a thick well set Fellow marked with the small Pox Grey eyes he hath Several Scars on his Left Arm & his Left Thumb Cut of to the first Joint

Samuel Chamberlain an English man five foot Six Inches High twenty two years of Age Long Visage brown Complexion short brown Hair Black Eyes well set

Joseph Wood an English man five foot Eleven Inches High twenty years of Age Long Visage brown Complexion short Dark Hair a well set fellow Black eyes a scar on the Left side of his Forehead a little pitted with the Small Pox

Robert Thomas an Welchman five foot Seven Inches High thirty years of Age Long Visage brown Complexion short brown Hair well set Fellow Grey Eyes

Daniel Boston an English man five foot ten & a half Inches High Nineteen years of Age Long thin Visage brown Complexion well set Fellow brown Eyes he can read

July 16th 1772

George Jones an English man five foot nine Inches High twenty four Years of Age Round Visage of a Swarthy Complexion Short Black Hair Beard & Eyebrows, he has several Scars on his Right Hand and a Scar on his Left thumb at the first Joint Hazle eyes & a Remarkable Scar on his upper Lip on the Left Side of the [illegible] has a Large Scar on the top of his Right foot he has his Right Eyebrow Shaved off.

William Evans an English man five foot five Inches High twenty three years of Age full Faced and a Swarthy Complexion Dark Brown Hair & Eyebrows a Well set Fellow & has remarkable Long Teeth a Tinman by trade

John Suckey an English man five foot Eight & a half Inches High Round full face of a Swarthy Complexion short Brown Hair gray Eyes he has Several Scars on his Head & a Gunpowder [illegible] his Right Eye a Well set Fellow & has a Ridge on his Left thumb Nail.

Garrard Williams an Englishman five foot [illegible] & half High twenty seven years of Age Long Visage of a Swarthy Complexion Short Black Hair Curles gray Eyes a fierce looking fellow well Set walks very bow leged. he can Read and Write

Figure 6. Description of White Servants, 1772. Ridgely Account Books, MS 691. Courtesy of the Maryland Historical Society, Baltimore, Maryland.

the Glove contending for Liberty should tolerate a wanton abuse of power."[56]

Convict servants tried to escape nasty and dirty work, whips and chains, and an early death, all of which invited others to associate them with slaves. Like African slaves, law and social custom had rendered convicts pariahs in their native lands; they were captives sold into bondage to distant masters across the ocean. They entered societies that viewed them with contempt and fear. Adventurers disproportionately assigned convicts and African slaves to the same gang labor—breaking rock and smoldering wood into charcoal. Others thought that they could pass for black. In 1767, the owners of the Patuxent Iron Works informed *Maryland Gazette* readers that they might recognize runaway convict Joseph Smith, whom they labeled a "Gypsie," because he "very much" resembled "a swarthy Mulatto in Colour."[57]

Convict servitude, in short, blurred the connections between place of birth, skin color, and freedom. Convicts were cheaper "white" slaves who made condition matter nearly as much as color in sorting out the Chesapeake's social hierarchy. John Adams visited Baltimore in 1777 and concluded that reliance on unfree labor had led "Planters and Farmers to assume the title of Gentlemen, and they hold their Negroes and Convicts, that is all labouring People and Tradesmen, in such Contempt, that they think themselves a distinct order of Beings." Sometimes the similarities of their predicament caused convict and slave ironworkers to make common cause: on at least five occasions they fled ironworks together. Sometimes the differences mattered more—especially to their masters. Convicts represented smaller and more temporary investments than slaves did, and that sometimes translated into harsher bondage. English traveler William Eddis determined from a visit to Maryland that because masters owned slaves "for life," they were "almost in every instance, under more comfortable circumstances than the miserable European, over whom the rigid planter exercises an inflexible severity." Randolph Hulse and the Northampton servants whose wounds he tended to might have agreed. But most convicts had reason to believe that they would outlive their bondage. Slaves did not.[58]

For African slave ironworkers, industrial bondage in the Chesapeake was a greater nightmare than it was for convicts. They too faced dirt, drudgery, and the threat of the lash. Worse, they confronted technologies and beliefs that often conflicted with those they knew. For African ironworkers, slavery was perhaps an even more disorienting ordeal than it was for the other Africans who accompanied them through the Middle Passage. Many plantation slaves could draw upon skills and knowledge learned in Africa to adjust to and even shape the work routines that planters demanded of them. Many Sierra Leonian slaves of rice plantations in South Carolina or Senegambians who grew tobacco in the Chesapeake colonies cultivated crops that they had grown in Africa and they used familiar methods to do it. Indeed, the expertise of African slave rice growers empowered them to claim more autonomy while doing their masters' work and more leeway to do as they would with the rest of their time. African legacies underwrote their participation in the industrious revolution as slaves and as builders of informal economies that they made mostly by themselves and for themselves.[59]

It was not so for African ironworkers. Few brought work experiences with them that might have eased the physical and psychological burdens that adventurers and ironworks imposed on them. Few could have been smiths or smelters; those with such skill and status were too valuable to send out of Africa. The men and women dispatched in their stead discovered that Anglo American routines for making iron and the cultural assumptions that underlay them often clashed violently with those of sub-Saharan Africa. Colonial ironmaking stripped away many of the social and cultural meanings that West and West Central Africans almost universally attached to iron's manufacture.

Despite the stunning cultural diversity of West and West Central Africa, some beliefs and practices surrounding iron and its production seem to have been nearly universal. The division of labor in ironmaking was highly gendered, often caste-specific, and inextricably connected to ritual and spirituality. During the era of the Atlantic slave trade, iron conferred power on those who owned it, those who wielded it, and those who knew how to make and shape it. Bars of iron signified wealth and status; they also served as a common circulating medium which Africans often accepted in exchange for slaves. Iron implements enabled West Africans to improve agricultural productivity, making it possible to eke out a living in often harsh environments. According to Ian Fowler, the ideas which underpinned iron production in Africa "appear to be very closely associated with core beliefs about the nature of things and the role of human, mystical, and other material agencies in providing for the satisfactory reproduction of human society."[60]

West and West Central Africans commonly likened smelting to procreation, in which male smelters attended to the needs of a furnace so that she would "give birth" to good iron. Most peoples who made iron barred all females except girls and postmenopausal women from participating directly in smelting or smithing for fear of angering the spirits and ancestors who oversaw the process and determined its success. Women could help to dig ore or prepare charcoal. During blasts, smelters were enjoined from sexual intercourse lest they make the furnace jealous and give her cause to ruin their efforts. In Central Africa, smelters often demonstrated their understanding of furnaces as "female" by constructing them with features such as breasts.[61]

In many West African societies, smelters and smiths belonged to some of the most powerful castes. Ironworking castes were generally endogamous, within which men handed down knowledge to their sons. In some cultures, mastery of the ability to shape metal conferred upon smiths the power, and sometimes the duty, to practice sorcery, both for their work and to protect themselves from harm. The rituals involved in smelting and forging iron also often accorded their practitioners the power and the responsibility to heal others.[62]

The association of men who smelted ore or shaped metal with sorcery and healing indicates the degree to which ritual and spirituality infused all aspects of making iron in West and West Central Africa. Success rested on how well smelters and smiths achieved and maintained the proper balance between the living and the dead, between the physical and spiritual worlds, between the masculine and the feminine. A good blast required invocations to ancestors and the spirits, strict adherence to sexual proscriptions, and the offering of medicines, herbs, and sometimes animal sacrifices, to ward off evil and guard against violation of taboos.[63]

The world that Africans encountered at Chesapeake ironworks shared features with the one they had to leave behind. Like Africans, Anglo Americans viewed some aspects of making iron as gendered. Founders and adventurers sometimes characterized furnaces as "fickle mistresses." As in Africa, men dominated production of iron, particularly its handling. Anglo Americans esteemed and rewarded men who knew how to oversee a furnace, shape molten metal, or hammer out wrought iron. Fathers often passed down their trades to sons. Ideas and practices common to colonial and African ironmaking, reinforced by how industrial slavery developed, helped to preserve some African ideas about metalworking and metalworkers among African Americans for generations.[64]

If such similarities helped African ironworkers adjust to their new lives, most aspects of making iron in Anglo America contradicted or violated their beliefs and practices. The balance of spirit and matter, male and female, the dead and the living—so central to Africa's iron industry—mattered little to masters or white coworkers. No taboos explicitly barred women from entering a furnace shed or forge. No elaborate rituals initiated the furnace blast. No one offered medicines or sacrifices to ensure production of good iron. To African eyes, colonial ironworks largely operated absent the need for ritual, taboo, or spirituality.

Making iron also seemed to proceed with less human intervention than in Africa. The scale of production that mill races and waterwheels enabled dwarfed anything that Africans had ever seen. Water-powered bellows or blowing tubs, not natural drafts or hand-operated bellows, kept furnaces in blast. The constant flow of air that water power provided enabled

Anglo American furnaces to be bigger, to maintain higher temperatures, and to run far longer (months instead of days) than their African counterparts. Hotter furnaces could use a fluxing agent, which allowed them to produce cast iron—something that African technology did not permit. Refinery forges seemed large and noisy to Africans accustomed to smiths using hand bellows, hammers, and anvils. Colonial adventurers and the European ironworkers whom they engaged sought not balance, but to harness nature in the hopes of mastering it. A furnace was to be humored, not propitiated. To Africans, such a system, which functioned with no obvious effort to seek the protection and mercy of the supernatural, as well as with machines, must have been at least bewildering and at worst have prompted some to question, if only for a moment, their understandings of the physical, the supernatural, and the relationship between them.[65]

The work that most Africans did on ironworks made their adjustment to life in the Chesapeake colonies even harder. Probably a few were lucky enough to ferry people and supplies. Watermen had more opportunities to travel, to form wider social networks, and to draw upon African seafaring traditions and religious beliefs.[66] Chopping wood, a common assignment for slave ironworkers, resembled the clearing of land that West Africans usually reserved to men. A sixteen-inch statue made of iron and found under an eighteenth-century Virginia blacksmith forge suggests the survival of some African cosmology concerning ironmaking.[67] But few Africans had a chance to become smiths, and adventurers generally relegated African slaves to the dirtiest and most backbreaking tasks. Pit mining was the most disagreeable, perhaps especially for Akan-speaking slaves from the Gold Coast. During the seventeenth century their ancestors had enslaved others to mine gold. In the eighteenth century, the expanding Asante empire took Akan speakers captive and traded them to European slavers. Many of those who were sold to ironworks were likely dispatched to mines, condemned to labor that they considered abhorrent and servile. Other African slave men may have viewed mining as emasculating, as work fit only for women and children.[68] A passage to the iron mines of the colonial Chesapeake offered Africans little but sweat, tears, and toil. It did not allow them to buy freedom—as slave gold miners in Brazil often did—nor did it allow them to create an economy that they mostly controlled, as female slaves in Cuban copper mines did.[69]

African ironworkers suffered the physical, psychological, and cultural toll that new, or familiar but intensified, work routines exacted from them. So did nearly all Africans who became slaves in British North America between the late seventeenth century and the Revolution. Yet those sentenced to the mines and coaling pits of Chesapeake ironworks gained an even more harrowing, humiliating, and destructive introduction to colonial slavery and to the industrious revolution. Their children who grew up to become ironworkers adjusted better to both. But they paid a high price for it.[70]

The Wages of Industrial Slavery

Chesapeake adventurers soon discovered that, although they could introduce slaves and convicts to the routines of ironworks, getting them to work as they wished posed a greater challenge. Often relegated to tasks considered oppressive and even humiliating, unfree ironworkers had little reason to exert themselves fully or perform their duties carefully. Adventurers could not simply whip work out of them. Careful coordination of interdependent tasks and strict attention to detail often meant the difference between success and failure for their enterprises. Expensive equipment, wooden buildings, and live charcoal made ironworks particularly vulnerable to sabotage. Ironmasters had little choice but to find ways to motivate slaves. Their need to invent effective incentives assumed greater urgency once they placed slaves in positions in which workmanship and judgment directly affected iron's quality and quantity. Slave colliers, furnace hands, and forgemen—by the nature of the tasks they performed—held the power to cripple their masters' ventures. Adventurers who punished slave ironworkers excessively ran the risk of demoralizing or angering them, either of which could result in slowdowns, feigned illness, acts of vandalism, or fugitives.[71]

Rewards for exceptional service promised one way to make slaves more industrious. In 1728, the Principio Company credited Prince, a forgeman, more than five shillings for "being carefull." Three years later, the company gave the "forge Negroes & Carters" over nine shillings cash at Christmas. In 1734, Mingo and a hired slave earned ten shillings from Principio "for diving to retrieve 8 tons pig iron which was sunk." Alcohol might also motivate slaves. In 1741, the Principio Company gave rum to "Negroes & all sorts of Wrkmen in Winter when twas very difficult to get Coles or other work."[72]

These were hardly ways to make industrial slavery all it could be for adventurers. Passing out alcohol was problematic. It might encourage productivity; it might also make whoever consumed it too drunk to work. Handing over small sums of cash at holidays or paying slaves for completing extraordinary tasks did not address adventurers' needs either. Rewards distributed occasionally, and solely at masters' discretion, failed to establish a clear link for slaves between their productivity and what resulted from it.

Planters in the Chesapeake and Lowcountry South Carolina and Georgia often bought slave-made handicrafts and produce from their slaves' gardens. Sometimes ironmasters did as well. In winter 1759–60, "Boy Hercules" sold two bushels of peas and Charles sold two bushels of peas and cabbages worth one shilling, six pence to Elk Ridge Furnace. The furnace bought a total of twenty-four coal baskets made by six slaves in December 1760. Four years later, Elk Ridge Furnace bought even more produce from slaves' gardens. Scarborro earned more than eighteen shillings for peas and potatoes; Phoebe and "Old Kate" earned eight and seventeen shillings respectively by selling their peas to the company. Henry Dorsey's slaves focused on their cucumber patches. Isaac and Peter each prepared pickles. Bobb sold some three-dozen cucumbers to his employers.[73]

Slave gardens and handicrafts were central features of the industrious revolution throughout the British colonies of southern North America and the Caribbean, but adventurers could hardly consider them the best use of their slaves' time, energy, and expertise. They wanted and needed slave ironworkers, especially those in whose training they had invested heavily, to apply themselves to producing better iron more cheaply. From the 1730s to the 1750s the Principio Company paid slave forgemen such as Prince and Dick one shilling to one shilling, six pence for every ton of anchonies that they made. Jack earned four pence for each ton of bar iron that he helped to draw in 1755. To the Principio Company, this system hitched slaves' earnings directly to their productivity. Dick, Prince, and Jack could take a different lesson—who did the work mattered far more than how much work was done. White forgemen earned at least fifteen to twenty times more than slaves for making the same amount of wrought iron.[74]

Adventurers and slave ironworkers eventually created a system, one that historians have come to call "overwork," which rewarded industry and invited slaves to strive for more comfortable lives. By the Revolution, nearly all Chesapeake adventurers offered overwork to slaves, whether they owned them or leased them. Overwork was essentially a modified task system in which slaves earned cash or credit for any work that they did which exceeded the quota for their job, usually at the same rates as free workers. Slaves who amassed credit could use it to purchase goods from ironworks or withdraw cash to spend largely as they wished.[75]

Adventurers embraced overwork because it seemed the best way to harness slavery to industrial enterprise. Sugar plantations with mills, the principal consumers of slaves, blended agriculture and industry, which created many opportunities for slaves to earn income. So did ironworks. Overwork helped to keep slaves occupied throughout the year; most slave ironworkers earned credit by cutting wood—a winter activity.[76] Paying slaves for additional service in a systematic way guaranteed that those who participated would complete their designated tasks. If they exceeded their quotas, so much the better for ironmasters, who obtained more labor from the same number of hands. Compensation for "extra work" gave slaves a financial stake in how well the ironworks fared. It also channeled their aspirations through those of their masters and employers. Here overwork differed from plantation task systems in telling ways. Tasking permitted slaves more freedom to do as they would once they finished their assignments—they could cultivate large gardens, sell produce, or hire themselves out. A task system, in short, allowed slaves time and leeway to create their own economy. Often that served masters' interests; sometimes it seemed to threaten slaveowners, especially when slaves chose to market their labor or the fruits of it in a more independent fashion.[77]

Overwork afforded slaves fewer outlets for their industry. It enabled adventurers to keep closer tabs on their economic activities, ensuring that owners would capture as much of the value of slaves' labor as possible. Overwork credit, recorded meticulously in account books, allowed ironmasters to track slave productivity and document it for prospective buyers—ambitious and industrious slaves left a paper trail that only made them more valuable to masters. Overwork took some of the sting out of industrial bondage; adventurers could discipline slaves without having to resort to force or the threat of it so quickly. By charging "Johnson's Bobb" and Pompey two shillings each "for staying a Day longer than he was allowed in the Hollidays," Caleb Dorsey underlined and enforced the distinction between his enterprise's time and theirs while dodging the risks that confronting them more publicly might entail.[78]

Adventurers also liked overwork because so many slaves embraced it. Payments to slaves for extra work, according to Ronald L. Lewis, comprised 45 of 316 entries on a 1766 list of the Elk Ridge Company's debts. Why did so many slave ironworkers undertake overwork? They obtained something for their time and labor and got a more concrete sense of the value of both. Overwork allowed slaves to distinguish between *their time*, when they chose to work and for which they deserved whatever compensation the market would bear, and that which adventurers claimed. Some ironmasters acknowledged and accepted that interpretation. In 1767, the Patuxent Iron Works credited Osburn £1 for ten days' work "in his own time," and in 1768 several other slaves together earned a total of more than £1 for twelve days' work "in their own time." More significantly, overwork permitted slave ironworkers to claim a degree of independence unavailable to most other Chesapeake slaves, the vast majority of whom worked in gangs on inflexible schedules. It was not just more meat that attracted slaves for hire such as John Mercer's Scipio to the iron industry. With cash and credit that they had earned through their own industry, slave ironworkers could decide for themselves which goods and services to purchase, and improve life for themselves and those they loved.[79]

Account books speak most eloquently about what slaves bought with their overwork earnings. Doing extra work and getting paid for it enabled slave ironworkers to accumulate personal property, so much so that Cato and Coffy, credited for weaving coal baskets at Elk Ridge Furnace, purchased padlocks to safeguard their possessions. Many, particularly at the Patuxent Iron Works, spent what they earned on rum. Others, such as the Principio Company's Dublin, bought trappings of gentility such as wine glasses. In 1763, "Boy Jack," a moulder at Elk Ridge Furnace, spent one shilling for "a Knot Perch Line & Doz. Hooks," perhaps to enjoy his time away from the furnace shed, more likely to get more food and relieve a monotonous diet. Most slave ironworkers spent overwork credits on cloth or apparel, such as hats, shoes, or stockings. Some hired tailors to cut and sew garments for them. In 1757, Prince paid "for making his fine Shirt"; Dick and London each ordered a coat and breeches from Joseph Elliott. Slaves' desire to clothe themselves was one of the most powerful motors that propelled colonial Anglo America's industrious revolution on plantations as well as in ironworks. Clothing allotments dramatized masters' power over slaves. What slaves wore was one of the few ways in which they could influence how others perceived them. If clothes made the man, then overwork was a way for male slave ironworkers to be their own men, at least within the chafing constraints of bondage.[80]

Overwork also may have given African slaves a chance to be more African. Igbo beliefs in reincarnation, for example, encouraged men to be more industrious so that they would become wealthier and more powerful. The Igbo honored and prized those who climbed the social ladder through hard work and business acumen above those who inherited their wealth and status. Akan speakers also celebrated socioeconomic advancement and saw personal industry as the best way to achieve it.[81] For some African ironworkers, overwork might have provided a unique opportunity, albeit within the narrow confines of slavery, to advance themselves and, for the Igbo, acquit themselves better at their next incarnation. Adventurers hoped that overwork would lead slaves to conclude that they shared a common interest with their masters in the success of ironworks. Some ironworkers may have viewed the system as a means to express and hold fast to beliefs and values that their enslavement otherwise besieged.[82]

But overwork also served as a means to acculturate all slaves to mainstream ways of life and thought in colonial society. It encouraged slaves to develop, in a limited fashion, the acquisitive individualism that characterized the behavior of most colonial men. Whether most did is almost impossible to say. Sometimes slaves pooled their labor to generate income. "London & his partners" and "Emanuel & his partners" cut wood at North East Forge in 1755; Will and Cato chopped over forty cords together and sold them to Kingsbury Furnace in 1768. It is clear that slave ironworkers' earnings enabled them to participate directly in the mushrooming trade in consumer goods that increasingly knit the Anglo American world together during the eighteenth century. Overwork promoted slave acculturation in other ways. It increasingly distinguished slave ironworkers from plantation slaves, and may have served to alienate them from slaves who lacked such opportunities.[83]

Nor could slave ironworkers participate equally in overwork. Few women ever collected it. For them, the system was an extreme example of how economic diversification and occupational specialization had almost exclusively benefited slave men. Skilled slaves who performed individually oriented tasks, such as forgemen or moulders, were best positioned to claim rewards from overwork. Ironworkers relegated to time-oriented jobs, such as fillers, or to work undertaken in gangs, such as mining or coaling, had to work on off-days, avail themselves of unusual opportunities as they arose, or shoulder other tasks, usually in the off-season, such as cutting wood. Nearly all slave tradesmen in the Chesapeake iron industry were American-born; a disproportionate share of gang laborers were African-born. Overwork may have widened the gap between African and African American slave ironworkers.[84]

Despite the opportunities and "privileges" that overwork offered slaves, it served above all to remind them of their bondage. Overwork depended on the voluntary participation of slaves, who used it to establish that adventurers could not compel them, at least not openly, to do more work once they had completed their shifts or finished their assignments. The power to set limits on what masters demanded of them was a tangible accomplishment—one which slaves defended whenever they saw that it was threatened.[85] But overwork proved to be a gilded cage. That slaves might command market rates for their "extra" labor only underscored that they had little to show for most of the service that they rendered. The wages of industrial slavery certainly almost never bought slaves, no matter how ambitious or industrious, their freedom.

Slavery within the iron industry imposed other burdens. Most jobs were dirtier, more demanding, and more dangerous than those commonly assigned to plantation hands. Slave ironworkers had more trouble building and maintaining stable families than did plantation slaves. Slave men vastly outnumbered slave women at nearly every Chesapeake ironworks. In 1781, seventy-two of the ninety-nine adults enslaved by the Principio Company were men, as were seventy-five of the ninety-five adult slaves of the Nottingham Company. Charles Steffen's analysis of the 1773 Baltimore County tax records tells a similar story; the county's seven ironworks owned a total of 171 males and 41 females. Such skewed sex ratios made it hard for men to have families where they worked and hard for them to honor their obligations to partners and children, who often lived far away.[86]

The lives of slave ironworkers differed from those of many plantation slaves in another important way. By the Revolution, prominent Chesapeake planters had begun to reimagine and sentimentalize their relationships with their slaves in familial terms that stressed reciprocity. Slaves owed them obedience, trust, and faithful service; planters owed them protection and guidance. Such masters told themselves that they were obligated to care for the physical, spiritual, and emotional welfare of those who had to endure bondage under them. They designed these principles mainly to serve themselves, but such ideals could operate to ameliorate bondage by casting slaves as household members as well as tools of production. Slaves thought so; they exploited such beliefs by acting, individually or collectively, to cajole masters into honoring them.[87]

Slave ironworkers had no such standard to which they could hold adventurers. Colonial Chesapeake ironmasters, among them Charles Carroll of Carrollton, even though they figured among the region's most prominent planters, did not consider their slaves family; they owed slaves nothing save what might make them work harder or better. Geography and business organization partly help to explain why. Sentimental patriarchs resided among, and interacted constantly with, their slaves; most Chesapeake adventurers lived miles or an ocean away and either contributed some of their own slaves to the firm or owned slaves jointly through it. They learned of slaves' activities through an agent's reports or a review of accounts. Not only distance made ironmasters' hearts less fond of their slaves. They valued slave ironworkers for the energy, knowledge, and skills they possessed, but they never pretended that slaves' spiritual or emotional well-being mattered to them. What slave ironworkers did that could not be recorded and quantified in timebooks or ledgers was of little consequence to colonial adventurers. If slaves were human in ironmasters' eyes, they were human principally because they were economic actors as well as assets. These too were the wages of industrial slavery.[88]

If slave ironworkers enjoyed more opportunities to improve their material lives and more freedom from open coercion than did plantation slaves, they and adventurers knew that industrial bondage rested ultimately on force. Ironmasters reminded slaves that they possessed the power to beat work out of them. In 1735, Jemy ran away from the mine bank at the Baltimore Iron Works "for an Easier Berth." Manager Stephen Onion urged his superiors to deny his bid lest they set a disastrous precedent. "Humoring" slaves "by removall," Onion observed, "makes them Lazy and of little Value." The next day the manager insisted that Jemy had received "no correction" to justify flight and warned that should slaves "be punished by removing them upon every complaint you would have but little service from them, and consequently no Occation of an Industrious Overseer." The Baltimore Company's partners followed his advice. They ordered a collar placed about Jemy's neck and sent him back to the mines. Four months later, he ran again.[89]

Jemy was not the only runaway slave whom adventurers bound with iron. In July 1767, Guy was captured in Joppa, Maryland, returned to Kingsbury Furnace, and fled again two days later. In early August, Kingsbury's clerk noted: "Guy retaken and Chained at nights." In 1775, Will and two convict servants stole away from the Marlboro Iron Works. Their owner Isaac Zane warned *Virginia Gazette* readers that they should beware because Will "had about his neck an iron collar, with the horns cut off, being a notorious runaway, and much given to drinking," and one of the convicts "has been heard to say some atrocious things in respect to the dispute between Great Britain and the colonies, from which it is suspected they may part, and some of them make to seaports." If Will boarded a ship, it did not carry him to freedom. Six years later he again tried to escape Zane and Marlboro.[90]

Slaves seldom outran their bondage for long. American-born slaves generally fled alone; Africans often absconded in groups, which sometimes included American-born slaves. Tom left Elk Ridge Furnace in 1762 with four African slaves who could "speak but very little English." The four had come to Elk Ridge as part of a group of nine African slaves a few months earlier. Tom undoubtedly organized and planned the escape. In 1759, soon after Elk Ridge bought him, Tom ran away, making good use of his knowledge of Chesapeake Bay and of his ability to speak "pretty good English, and a little French," while probably trying to conceal "a remarkable large Scar proceeding from each of his Temples down his Cheeks" which he received in Africa.[91] His skills did little to help his companions. The four "New Negroes" were returned to Elk Ridge two days after the ad seeking the return of the runaways appeared in print. Tom remained at large and likely stayed that way. In December 1762, the furnace incurred a debt of £20 from "Account of Negroes for 1 free."[92]

Others were far less lucky. In the summer of 1763, three slaves stole away from Elk Ridge. No one bothered to place a newspaper ad seeking their return, perhaps because Thomas Hooker apprehended them so soon after they were discovered missing. The matter did not end there. The next line in the Elk Ridge Journal records a debt owed "Acct of Negroes" for "one dead, viz Scipio, who was kill'd in taking him up home run away, he cost in Sterling Money £40." The following December witnessed a flurry of activity surrounding Scipio's death and his masters' efforts to recover some of their investment in his body: compensation for someone dispatched "to get Scipio valued by the Men who saw him killed" and the expenses of sending someone to attend the Provincial Court in Annapolis to plead their case.[93]

Scipio's story testifies powerfully to the meanings of the colonial Chesapeake's industrious and industrial revolutions. All available evidence suggests that he was one of thousands of Africans who entered British North America during the eighteenth century; slavery stripped away even his name. He went to Elk Ridge because Caleb Dorsey and his partners had determined that it was nearly impossible to make iron without people like him. Only unfree men and women, be they enslaved Africans, those who inherited bondage from them, or British felons bound for America as punishment for their crimes, permitted adventurers the control that they believed they required to stay in business in British North America. Convict servants were cheap and unruly outcasts whom ironmasters could discipline with impunity. Slave ironworkers faced an even greater ordeal. Africans confronted the shock of enslavement within a foreign culture—a shock amplified by gang labor and their exposure to new technologies and to different assumptions about the relationship between humans, nature, and spirit. American-born slaves had a different but perhaps no less trying experience. They knew that the iron industry afforded them opportunities for autonomous economic activity and recognition of the worth of their labor that few who toiled in tobacco and wheat fields enjoyed. They also knew that such opportunities carried a high price tag: greater difficulty in establishing and maintaining a family and exclusion from a patriarchal ideal that plantation slaves could sometimes exploit for their own benefit. Even overwork proved a double-edged sword. In undertaking it, slaves strengthened the shackles that bound them by making them more supple.[94]

Nowhere did this prove more true than in the stories that ironmasters and their agents have left us of the industrious revolution that they and their slaves made together. We historians have come to understand slavery and teach about it most vividly through the examples of those who resisted it. Tales of sullen, deceptive, and rebellious slaves fill the diaries and letters of great planters and the writing of most scholars who have recently studied slavery. The transformation of the slave "Negro Mingo" into fugitive, plaintiff, and aspiring freeman Juan Domingo López fits those scripts well. So does Scipio, who met his death while trying to flee his bondage. The attention paid to him thereafter symbolizes the degree to which Chesapeake adventurers, in their zeal to make iron for profit, reduced the value of work and of human life to a matter of pounds, shillings, and pence. But so did the overwork system. The same journal that recorded Scipio's death remembers most of the black men and women whose names appear in it as people who earned and spent the wages of slavery.

4

THE BEST POOR MAN'S COUNTRY

WAR. To William Allen and Joseph Turner, Philadelphia merchants and owners of New Jersey's Union Iron Works, it meant opportunity. As European empires battled to claim dominion over the St. Lawrence and Mississippi Valleys, the market for wrought iron exploded in British North America. Allen and Turner needed capable forgemen to seize the moment. Finding and engaging them was a tall order. Forgemen were always hard to find locally or in the Atlantic world; war only complicated matters. Heightened demand for iron kept forgemen busy on both sides of the North Atlantic. Hostile navies and privateers would dissuade many from setting sail for America.

Still, the adventurers persisted in trying to lure forgemen to Union. They promised steady work. Union's furnace made pig iron that was easy to work with; the ironworks had enough timber to provide charcoal for years and enough water to run the forge even during the dry season. Allen and Turner had more to offer, they claimed. To John Griffiths, who they hoped would help them recruit several British finers and hammermen, they observed:

> What encouragement it must be for any of them that are Sober men to come over wee know its four times more than whats given in England as wee have been often there & no Strangers to each workmens wages or what the employer expects from them. . . . Wee have houses for each fam-

> ily to Live in & thats all wee allow them fireing they may have for Cutting. Wee must be known to many in England & no doubt your father has made you acquainted with our Caracters so that no doubt will arrive but all that comes will meett with honesty & good Ussuage. . . . The Articles of overyeald wee leave to the workmen to do as they please. Wee dare not dispute with workmen for burning or waisting while at work nor choose to differ with them about triffles for Should Such Leave our business it may be a week ere wee get [news].[1]

More money than they could hope to make at home, housing, and pledges of fair treatment surely would appeal to British forgemen. So would the ironmasters' promise to cut them a wider berth lest they offend them and lose their services. In short, New Jersey was, they assured—echoing former indentured servant and ironworker William Moraley—"one of the best Poor Man's Country in the World," which "agrees very well with English Constitutions." But their pitches persuaded few English forgemen, forcing Allen and Turner to seek hands in the Palatinate and in Sweden.[2]

If New Jersey and Pennsylvania together comprised the Anglo Atlantic world's "best Poor Man's Country," they did not sit so well with everyone's constitution. The forgemen Allen and Turner sought were to have served alongside "Some negroes of our own." When Turner turned his attention to the Palatinate, he insisted that prospective forgemen must consent to "take any Negroes under them to be Instructed while they Live with us." The slaves received the training that their masters wanted. In 1770, Allen and Turner advertised for hire or sale "six Negroe Slaves" who were "good Forgemen, and understand the making and drawing of Iron well."[3]

A decade before William Allen and Joseph Turner began their fitful search for forgemen, the mid-Atlantic colonies had become the center of British North America's iron industry. Pennsylvania remained so for more than a century. Enterprising men and abundant supplies of ore, wood (and later, coal), and water enabled such an impressive achievement. So did adventurers' ability to harness workers' muscles and minds to the demands of making iron. They invited men to cross the North Atlantic to a place where they might enjoy material comfort and personal autonomy unattainable at home. When enough Britons did not answer their call, they turned to German-speaking Europe. Those who came learned to attach new meanings to work, meanings that largely signified their acceptance of many of the principal values of their Anglo American neighbors. That proved a mixed blessing for adventurers, who sometimes found that recent arrivals soon expected more from their work than their employers could or would give.

Confrontations with white ironworkers' expectations of work led adventurers in colonial Pennsylvania and New Jersey to conclude that they needed slavery to forge an industrial revolution out of the industrious revolution. By 1776, ironmasters figured among the largest slaveholders in the Middle Colonies. But industrial slavery operated differently in the mid-Atlantic than in the Chesapeake colonies. Slaves were a small minority of ironworkers in New Jersey and Pennsylvania. A disproportionate number of them learned a trade, but they and the scattered population of African Americans who lived nearby struggled to create and sustain an autonomous culture like that which southern blacks enjoyed.

Industrial slavery was far more significant to ironmasters and white ironworkers than the relatively small number of slaves would suggest. For some white tradesmen, slavery provided an opportunity to enhance their earnings and social status. For others, slavery represented a powerful tool that adventurers used to discipline their workers by limiting the latter's ability to negotiate terms of employment. Slave labor, along with the concentration of ownership of ironworks in a few hands, enabled Pennsylvania and New Jersey ironmasters to manage a market for free labor that they helped to create. As in the Chesapeake colonies, slavery served to undergird a system in which entrepreneurs acted on the belief that men would not labor on others' behalf unless coerced. Mid-Atlantic adventurers made little pretense that they bore any responsibility for transforming those whose toil sustained their enterprises. They promised many ironworkers the chance to experience their part of Anglo America as the best poor white man's country. They cared little whether it agreed with or fortified their constitutions.

Forging Slavery

In 1728, the best-laid plans of James Logan seemed about to collapse. Word that "great quantities" of American pig iron sat unsold in England surprised and disheartened Logan and his partners in the Durham Iron Works. It especially worried him because Durham had proven a far more expensive venture than "all our first Calculations" had projected. Without "better Encouragemt," Logan warned, "tis to be apprehended you will not long be troubled wth much from this Province, divers of those concern'd" in ironworks "seeming already resolved to Drop them."[4]

Thanks to astute politicking by Pennsylvania ironmasters, encouragement was on its way. What principally upset Durham's owners was that they had underestimated the cost of ironworkers' services. To that they had already found a likely solution. In 1727, several "Persons concerned in the Iron-Works," claiming "that the Difficulty of getting Labourers, and their excessive Wages, are a great Discouragement and Hindrance to their Undertakings," petitioned the Pennsylvania Assembly to allow ironmasters to import or purchase slaves duty free and asked leave "to bring in a Bill for that Purpose." The Assembly considered the bill and passed it. But after a motion surfaced to allow anyone to import slaves duty free, the body revisited the issue and the bill failed. Two years later, the Assembly reduced the duty on slaves from £5 to £2. Between 1731 and 1761 there were no duties collected on slaves brought into Pennsylvania.[5]

Many adventurers were well-positioned to staff their ironworks with servants or slaves. Those who were merchants, based mostly in Philadelphia, had direct access to commercial networks that transported bound laborers to North America. The firm of Alexander and Charles Stedman, shareholders in Pennsylvania's Elizabeth Furnace and Charming Forge, brought more than 11,000 Germans to Philadelphia between 1736 and 1753. Several ironmasters, including William Allen and Joseph Turner, Alexander Wooddrop, Charles Read, and George and Samuel McCall, imported slaves.[6]

Adventurers wanted slaves because they had encountered many of the same problems with indentured servitude that plagued Chesapeake ironmasters. Even though relatively few men from England and Wales were willing to journey to North America as servants, the tens of thousands of German-speaking men and women who immigrated through Philadelphia between 1727 and 1775 offered replacements. So did thousands of migrants from Ireland, particularly after 1763. For ironmasters the supply of servants was too sporadic. They had little influence over the political, economic, and social conditions that shaped potential servants' decisions whether and where to emigrate. Nor could they predict or control the course of the wars that wracked the North Atlantic during the eighteenth century, all of which disrupted the servant trade.[7]

Nor did indentured servitude meet ironmasters' need for steady and industrious hands, especially before 1750. Most servants left the industry after their terms expired. Samuel Nutt, part owner of Coventry Forge and the probable chief sponsor of the campaign to eliminate duties on slaves whom adventurers purchased, saw nine servants complete their terms between 1727 and 1730. Most left weeks after claiming their freedom dues. None returned. A few perhaps became landowners nearby. Others may have found establishing the region's iron industry so burdensome that they eagerly left it behind. William Moraley recalled that he sometimes "got Drunk for Joy that my Work was ended" when he was helping Isaac

Pearson to launch his ironworks, and he was even happier when his servitude under Pearson expired.[8]

Many servants capitalized on wars to cut their bondage short. In 1740, Pennsylvania Lieutenant Governor George Thomas issued a proclamation seeking volunteers to fight the Spanish Empire. Ten servants of Anna Nutt and Company answered it. Nutt, proprietress of Coventry Forge and Warwick Furnace, petitioned Pennsylvania's Assembly to recover what their departure had cost her and her firm. Many of the enlistees were colliers "who had been instructed in their Business at a considerable Expence" and on whom her firm "chiefly depended for supplying the Furnace, then in Blast." After they left, Nutt claimed, production halted, causing "several Hundred Pounds Damage" to her and her partners. The assembly accepted her petition and volunteered to make restitution. The Seven Years' War was as troublesome for masters. In 1756, William Allen and Joseph Turner claimed that a new law passed by Parliament had allowed hundreds of servants, "both Pallatines and English," to end their terms prematurely. The following year, several Pennsylvania ironmasters filed claims for servants who had joined the Royal Army. Allen and Turner complained in 1760 that "many Inlist in the Soldiary & that causes a great Scarsity of hands at all our works."[9]

Replacing them was difficult. In 1740, the Pennsylvania Assembly protested enlistment of servants and cautioned that demand for them would plummet "if the Property of the Master is so precarious as to depend on the Will of his Servant and the Pleasure of an Officer." Their warning proved prophetic during the Seven Years' War. The threat that servants might leave with the blessing of imperial authorities scared off buyers. In 1756, Allen and Turner declined an invitation to reenter the trade in Palatine servants because they feared that they would not be able to sell them.[10]

Nor could ironmasters be sure that their indentured servants were safe from competitors. Artisans were especially susceptible to poaching. In 1761, Allen and Turner insisted that their agent in the Palatinate require forgemen to serve them for four years. Such contracts, they warned, would be worthless without "tyes to the Agreemts by bonds" which would exact a heavy financial penalty on whoever violated them. Otherwise, the forgemen "may leave our employ, by means of under hand practices that may be made by some [of] the Iron masters."[11]

Suspicion that adventurers lured others' servants away persisted well after the Seven Years' War. In 1771, Charles Read, Jr., emphatically denied rumors that he had knowingly employed another ironmaster's runaway servant at Etna Furnace. "We have always made it an invariable rule at our

Works," he insisted, "never to be assistant in robing a Person of his Property by Secreating his Servant. The Contrary Conduct is base and unjust as well as ruinus to the Interest of Iron Masters." Read's protest of innocence likely rang hollow to fellow ironmasters, largely because many of them had done what they accused Read of. They expected that other adventurers, if pressed to find workers to keep their enterprises running, would quickly overrule any qualms about violating another's property rights, even when doing so might ultimately harm their collective interests.[12]

Adventurers were also leery and weary of indentured servants because they sometimes deceived them. Allen and Turner wanted to replace the manager of their slitting mill, who they considered "a Very Ingenious man but a Little disorderd in his head and will imploy the mill just as he pleases." As owners of one of few colonial slitting mills, they were sure that they needed only the right man to reap large profits. After almost four years of looking in Britain, Allen and Turner thought that they had finally found him. They paid John Hughes's passage twice; he jilted them twice. Hughes then sailed to New York "without acquainting or telling us how he proposed to reimburse us." He never set out for the Union Iron Works and the partners abandoned hope that he ever would after hearing that he had met up with William Davis. Davis had been a convict servant in Maryland before Allen and Turner bought his contract "& eased him of Some years service & then paid him wages." Davis married a widow and assumed her dead husband's debts. When authorities came to collect, Davis fled to New York, where Allen and Turner believed that he would "prevail on Hughes to Stay away." It only added insult to injury when they discovered that Hughes had misrepresented himself. He was a blacksmith who knew nothing of slitting or rolling iron. In the end, the partners could do little except perhaps sue Hughes for what they had spent on his passage.[13]

The hassles and costs of dealing with white labor led adventurers in Pennsylvania and New Jersey to practice industrial slavery. Without slaves, they believed, their enterprises would never succeed. Slaves lacked legal authority to negotiate and the ability to deceive from afar. Slavery offered ironmasters relief from "the Difficulty of getting Labourers, and their excessive Wages," because slaves had little say in where they went, and they earned for their labor whatever masters chose to give them. Slaves afforded them greater control over turnover in a labor market in which employers often competed fiercely for workers, especially for tradesmen. Slaves, in effect, helped to save the region's ironmasters from themselves. They were the most secure form of human property; adventurers would not try to lure away another's slave.[14]

Account books, wills, tax lists, and county registers all testify to the degree to which the region's iron industry depended on slave labor. Historian Michael Kennedy has estimated that up to half of the workers at some Pennsylvania ironworks between 1725 and 1750 were slaves. Thomas Potts owned eleven slaves when he died in 1752. Sixteen years later, his son John's inventory included thirteen. Ironmasters figured among the largest slaveowners in rural Pennsylvania by the Revolution. According to Carl Oblinger, adventurers owned nearly one-quarter of the 824 slaves who lived in York, Chester, and Lancaster Counties. In 1780, the four largest slaveholders in Berks County were ironmasters; together they owned nearly half (57) of the 119 slaves registered under the "Act for the Gradual Abolition of Slavery."[15]

To be sure, slave ironworkers in the Chesapeake colonies vastly outnumbered those of Pennsylvania and New Jersey. Slaves never constituted a majority at any ironworks in either province after 1700. Adventurers continued to buy large numbers of British, Irish, and German-speaking indentured servants. Free men who earned wages always were a significant part of any ironworks' labor force. Indeed, adventurers in the 1720s and thereafter complained so strenuously of having to pay "excessive Wages" partly because they had so many people on their payrolls. Slave labor gradually diminished in importance at several ironworks, especially after 1750 in areas where adventurers could find enough free workers on terms that both parties found acceptable. Free ironworkers consumed a wide range of goods from their employers' stores. Their families supplied ironworks with provisions and with children whose work often preserved or led to economic independence. From its inception, the iron industry was the largest employer in the rural Middle Colonies, and as such it promoted the development of both free labor and capitalist values in the countryside.[16]

But that hardly meant that slavery melted away once free labor became available to adventurers on terms that they found acceptable. Such an explanation oversimplifies and implicitly whitewashes the key role that slaves and slavery played in making the mid-Atlantic colonies the center of early Anglo America's heavy industry. It leans too heavily on assumptions embedded in classical and Marxist economic thought—those who earn wages must and inevitably will replace inefficient, unmotivated slaves. Adventurers in Pennsylvania and New Jersey did not see things that way. To be sure, many did replace slaves with free workers, particularly when the population near their enterprises provided enough hands. But many of the same men owned many slaves and continued to purchase more. The region's ironworks were simultaneously the largest rural employers and the largest slaveowners. Adventurers did not merely deploy slave labor when they found free labor too expensive and scarce. They exploited slavery to mold free labor to their purposes.

Like Chesapeake adventurers, mid-Atlantic ironmasters leased slaves. Ironworks located in the Susquehanna Valley, which often had commercial ties to Baltimore merchants, contracted for temporary slave hands most often. In 1766, Curtis Grubb of Cornwall Furnace spent £275 to hire eleven slaves for one year from Benjamin Welsh. Marylanders Nathaniel Giles, who had previously owned part of Cornwall, and Caleb Dorsey each leased several slaves to the furnace between 1770 and 1775. Pennsylvania adventurers sometimes hired slave artisans, as did Peter Grubb when he leased Tob from Ferguson McElwaine to work in the finery at Hopewell Forge.[17]

The region's ironmasters leased relatively few slave tradesmen, partly because there were few of them available and partly because ironmasters preferred to own them. A disproportionate number of their slaves were forgemen. Ironmasters began to employ slaves in forges soon after they launched their ventures. In 1732, slaves began to work in Coventry Forge. Between 1756 and 1759, six of the nineteen hands who made anchonies or drew bar iron at Coventry were slaves. By the late 1760s, several forges in the region, like Allen and Turner's Andover Iron Works, owned at least six slave forgemen. In 1768, the owners of Greenwich Forge offered to rent the forge and "seven negro men, who have been employed for many years past in the Forge, and understand the making of Iron." A year later Glasgow Forge's proprietors marketed "FIVE FORGE NEGROES, that have been Ten Years at the Business, and are Master Workmen; three of which are HAMMERMEN, and two FINERS."[18]

Why did so many slaves become forgemen? Slave forgemen offered ironmasters respite from the high wages that white finers and hammermen commanded in North America and the effort and expense of recruiting and retaining them. In 1760, Allen and Turner noted that they paid "45/ Currency p Tunn to the finer & 35/ to the Chaffery for a Tunn of Iron but then we have Some negroes of our own good work men who Draws a part [of] that monstrous price." The partners had complained two years earlier that the "Extravigant wages wee give work men of all kind thats Employed in Carrying on Iron work is what Discorages the owners such as Colliers Carters Stock takers finers hammermen & c." and added that "wee have Lately run into the method of teaching Negros to make Anchonys & some are good hammer men." White forgemen could go ply their trade for someone else; slaves could not depart legally without their master's consent. Concerns over turnover help to explain the large numbers of slave forgemen and of slave artisans in many northern cities during the colonial era.[19]

For adventurers, slave forgemen promised more than lower costs—they represented a way to put white forgemen in their place. To some ironmasters, their dependence on the skills and expertise of free forgemen turned upside down what they believed should be the proper relationship between entrepreneurs and workers. In 1754, John Taylor appeared before the Lancaster County Court to seek damages from Caesar Andrew. Six years earlier, Taylor had hired Andrew. In 1751, the hammerman agreed to work for Taylor for one more year, during which he would earn twenty shillings per ton of bar iron. When Taylor learned that Andrew had independently struck a bargain with another forgeman, he slashed what he paid him. According to Taylor, the hammerman "then absconded, not having settled accots with me. Caesar neglected my business, destroyed my hammers & geers and wasted my anconies & coals so that . . . I am damaged by his ill conduct above £100." Andrew, Taylor anticipated, "will say we want water at ye forge & he cannot be fully Employed but must go to other work. This is not so for the works being rebuilt go with less water than ever."[20]

In his last claim, Taylor was probably right. Free forgemen wanted steady work, and they preferred to go where someone guaranteed it. When adventurers recruited forgemen, whether privately or through newspaper ads, they often claimed that their forge had access to enough water to keep operating until winter. Taylor was also correct to suspect that Caesar Andrew would seek another employer. In November 1754, Lynford Lardner paid Andrew's "prison Fees at Lancaster" and his travel expenses to Windsor Forge. Lardner and his associates probably knew something of Andrew's dispute with Taylor, but it did not dissuade them from engaging him.[21]

William Allen and Joseph Turner would have identified with Taylor's frustration. In 1759, they informed a correspondent in Britain that they needed more forgemen because their new forge would require "a number of that Sort of Gentry." Their language was a sour comment which described how they viewed their predicament. It echoed British ironmasters' mounting frustration with forgemen's ability to monopolize technical knowledge and demand concessions of them. It also reflected social insecurities more peculiar to colonial elites. Truly genteel men did not engage in manual labor and they certainly did not defer to those who did it. Circumstances mostly beyond the adventurers' control had conspired to force them to accord tradesmen who beat and flung about metal for a living far more respect and honor than their station merited. This was not what Allen and Turner had bargained for.[22]

Sometimes ironmasters looked to limit forgemen's power by taking control over work out of their hands. In 1765, the owners of the Carlisle Iron Works hired Peter Dicks to use "every manner of working that his Ingenuity & Art may suggest" to convert gray pig iron into "tough Bar Iron." Should his experiments succeed, they directed manager Robert Thornburgh to "oblige the Forgemen to pursue the same Method." Carlisle's proprietors expected resistance—some forgemen would not countenance such meddling. They had a response ready. "If John Goucher or any other refuse to do it," they instructed, "he shall be discharg'd from the Works." How would they replace those who balked? Carlisle's partners offered one option. They authorized Thornburgh to hire, if necessary, slave forgemen from Nathaniel Giles "when their Contracted time is out at Paradise Forge."[23]

Did white forgemen oppose efforts to replace them with slaves? Did they resent slave forgemen? It seems that they did not. No records indicate that white forgemen in New Jersey and Pennsylvania objected to working with slaves or to teaching them their craft. Some may have trained slaves because they had little say in the matter. In 1732, servant Joseph Tucker made anchonies at Coventry Forge with Caesar, a slave owned by William Branson and Rebecca Nutt. Some accepted financial inducements. John Goucher received "an allowance for the Negros working in his Fire" in 1743 from Coventry Forge. Others negotiated contracts which stipulated the terms under which they would apprentice slaves. In 1760, Samuel Barford agreed with John Patton, owner of New Pine Forge, to train Tom for one year. Barford pledged "to use his utmost Endavors" to teach Tom "to draw a good Bar & c." in return for twenty-four shillings for each ton of "good barr Iron" that he made with Tom's help. Patton also guaranteed Barford "a Sufficient House to live in and also ye usual customs of other forgemen," while Barford was to provide Tom with food. The following year Barford worked with Tom and Caesar and boarded both men.[24]

Such arrangements may have appealed to white forgemen because they afforded them a sense of power. Barford had the authority to supervise Tom at the hammer and at home. Many slaves whom white forgemen agreed to apprentice were probably adolescents who were considerably younger than they were. In 1762, Allen and Turner noted that their forgemen had complained of a recent shipment of forge hammers as "too heavy & a Little too Long in the bitt" and agreed because "wee have young Negro fellows often a Learning the business that require a Lighter hammer then for an Old experienced hand." Having white forgemen board slaves resembled the way masters and apprentices lived. In addition, instructing slaves saved Samuel Barford from having to find underhands to assist him, a responsibility that white forgemen and other tradesmen within Pennsylvania's iron industry often assumed.[25]

Desire for surrogate mastery and need of assistants may have prompted many white forgemen and some founders to hire slaves, often from their employers. In 1745, John Briggs credited John Potts for "ye work of his Negroes" at Pine Forge. Four white forgemen at New Pine Forge leased slaves in 1761. Eberhart Geisweid paid John Patton for Sharper's help in making five tons of anchonies at Charming Forge in 1774. Founders sometimes hired slaves to work as keepers, who may have supervised the furnace in their absence. In 1774, John Jameson leased the services of Mike for several months to help him monitor Cornwall Furnace. Artisans occasionally hired slaves for small jobs. Samuel Jones reimbursed Hopewell Forge in 1771 "for 3 days Work Negro Sam making fence." What adventurers earned from leasing slaves defrayed the expense of purchasing and supporting them. White tradesmen who leased slaves usually trained and supervised them. They benefited from slavery directly and perhaps became eager to own slaves themselves.[26]

They had examples to inspire them. Some forgemen and founders owned slaves and worked alongside them or leased them to their employers. Pine Forge frequently credited John Hanson for iron that "his Negro York" made between 1745 and 1751 and in 1754 it paid Joseph Thomas for making anchonies "with your own Negro." In 1774, Hopewell Forge compensated Thomas Mayberry for making over two tons of anchonies "with his Negro." Some founders owned slaves whom they occasionally hired out. David Short brought Cuge from Berkshire Furnace to Cornwall Furnace in 1770. While his master supervised work at the furnace, Cuge served as a laborer who cut wood or helped with the harvest. Four years later, Andover Furnace hired Samuel Patrick's slave for nearly a month.[27]

Ownership of slaves gave ironworkers some of the same headaches that troubled adventurers. Proximity to their masters perhaps whetted some slaves' desire to escape them. In 1751, Cross, James, and Dick stole away from Cornwall Furnace and their master, founder William James. Seven years earlier, Bryan Murry, a collier at Reading Furnace, placed an ad seeking Isaac's return. In 1746, Mark fled John Hanson, a forgeman at Pine Forge, twice within six months. Such incidents did not dissuade some white ironworkers, who had decided that the path to greater prosperity and social advancement lay in following their employers and hundreds of urban artisans by becoming masters of slaves.[28]

Established white forgemen had another reason to view slaves as little threat; at no time did slave forgemen play a more significant role in production than they did. Between 1734 and 1759 slave labor accounted for a growing share of the anchonies that Coventry Forge turned out. By the late 1750s, four slaves—Tom, Ben, Guinea, and Sampson—had a hand in making nearly half the anchonies. Coventry's slaves played a less prominent role in converting anchonies into bar iron. During the 1730s none were hammermen. Between March 1742 and February 1744, Lambeth and Sambo helped John Mills draw over 125 tons of bar iron. Over a two-year period some fifteen years later, Coventry slaves assisted in drawing just 27 of 163 tons of bar iron. Hopewell Forge's records also tell of a supporting role for slave forgemen. White finers who worked without any assistance from slaves accounted for more than two-thirds of the anchonies made at Hopewell between 1768 and 1775. Slaves helped to produce the rest, almost entirely with whites or under their direct supervision. Slave forgemen who worked independently produced fewer than 7 of 1,216 tons of anchonies made at Hopewell. Whenever more than one slave worked at a hearth, ironmaster Peter Grubb generally required that a white forgeman handle the metal for them under the trip hammer.[29]

Adventurers had the authority to turn over more berths to slaves. Why didn't they? They did not say explicitly. Slave underhands were expensive; master forgemen were even costlier. The youth and inexperience of many slave forgemen probably discouraged ironmasters from relying on them as much as they might have liked. Some also may have concluded that they should limit how many slave forgemen they had. In 1763, William Maybury advertised two slave forgemen and emphasized that he was selling them "for no Fault, but having more of that Calling than I have Occasion for." Financial worries shaped his decision. Two years later, the administrators of Maybury's estate auctioned off eight slaves to satisfy his creditors. But Maybury might also have determined that he had too many slaves working in his forge and sought to rid himself of some.[30]

White forgemen probably influenced how adventurers organized the work of slave forgemen. As long as their services were in demand, white forgemen could choose which offers to accept. If they perceived that a forge employed too many slaves, they might go elsewhere. For white master forgemen, though, restricting slave labor to particular tasks had appeal. They could delegate many of the most arduous tasks, such as slinging glowing balls of iron between hearth and hammer, to slave underhands while supervising the process and putting the finishing touches on the wrought metal. Slave helpers did not threaten their job security or reduce their wages. Nor did slave forgemen block their sons' path should they wish to follow in their fathers' footsteps. But slavery enabled ironmasters to pay less heed to many white forgemen and to their successors. Slave forgemen narrowed the opportunities available to white apprentices and underhands who aspired to join the "gentry" that Allen and Turner derided. Slavery gave ironmasters a valuable tool, but only one tool among many, in their quest to direct an industrious revolution in the mid-Atlantic countryside.

The Best Black Man's Country?

In 1750, the owners of Windsor Forge—William Branson, Samuel Flower, and Richard Hockley struck a deal with four of their slaves: Adam, Black Boson, Arche, and Yellow Boson. If they made one hundred tons of anchonies within one year, Adam would earn £39 and his assistants, Arche and the two Bosons, would each make two pounds, twelve shillings. That was good money to a slave, but that mattered less to Adam than the rest of the agreement. Branson and his partners pledged to find "one New Negro" who Adam would teach "the Art and mistery of being a finer." Once Adam had "Learned Two of our Negroes to be Compleat Workmen at the fire that is to Say from the puting up their fires to the making good and sound Merchantable Ankoneys," Branson, Flower, and Hockley pledged "to lett his Boy Solomon go free—And nott be Obliged To Serve us or any of our heirs Executors or Administrators or assigns."[31]

Windsor's partners stood to lose little. One hundred tons of anchonies from the hands of free forgemen would cost them more than double what they had promised their slaves. They might get three slaves trained to staff their finery forge. They would determine what "good and sound Merchantable Ankoneys" were and when Adam had taught their forgemen well enough. Even if they had to manumit Solomon, Windsor's adventurers would probably have come out ahead financially—freeing him would spare them from having to feed, clothe, and shelter him. It is hard to imagine a better way for them to have motivated Adam. He could use the money to give Solomon a more comfortable life. If Adam's industry and his craft would not end his own enslavement, they might at least earn his son's freedom.

Adam might have suspected that his masters never expected to have to honor the agreement. Many obstacles blocked Solomon's manumission. Only under the best of circumstances could Adam hope to meet his masters' terms. The money was no sure bet, but it was a reachable goal. Training others to do his job well was a taller order. It took months of constant practice to become a good forge underhand, let alone a "Compleat" workman. The "New Negro" that Branson and his partners promised to find for Adam would have to learn "the Art and mistery of being a finer." What that meant was unclear and the task could take years. If Adam ever thought that his masters would not free Solomon, he was likely right. In 1762, James Speary certified that he had written down the agreement that Adam had reached with Branson, Flower, and Hockley. This could signify many things, among them that Adam's bargain had not bought his son's freedom.

Nine years after the adventurers of Windsor Forge and their slave forgemen struck their deal, Israel Acrelius reported to the Queen of Sweden that "the negroes are better treated in Pennsylvania than anywhere else in America." Aside from brief statements on what a master might pay to buy a "good negro" and on how much one might spend to clothe and feed a slave, he had little more to say. Acrelius confined his discussion of slavery to one terse paragraph in a short chapter which assessed the iron industry in and around Pennsylvania, a topic that he knew would interest Swedish royal officials who saw it as competition for British markets. Conversations with William Allen, who estimated for Acrelius what it cost to produce iron and what price colonial iron might fetch, probably shaped his impressions of slavery.[32]

Still, Acrelius's terse report on industrial slavery demands consideration. On the "treatment" that slave ironworkers received in Pennsylvania and New Jersey, adventurers, their account books, and the advertisements that they placed when slaves ran away tell a story more confusing and in some ways more complex than that of the Chesapeake region. They speak less of whips, yet also less of opportunities for slaves to seek compensation for their industry. As for how slaves experienced and tried to shape the industrious revolution in which their masters cast them as minor but key players, the documents show less acquiescence to the slavery that their masters tried to impose and less resistance to it. Above all, the stories that adventurers have left us portray slaves as solitary and ghostly figures who rarely emerge as individuals. Some struggled to build families and struggled even harder to keep them together; others labored to forge new identities for themselves.

The account books in which ironmasters and their agents inscribed the value of slaves' labor reveal little ambition to establish an overwork system. Coventry Forge's owners occasionally doled out small sums of cash, usually less than two shillings at a time, to slave forgemen. They seldom explicitly linked those payments to work, except when slaves completed tasks on holidays. In 1736, Ben earned 1 shilling, 6 pence cash—the going rate for a day's labor—for "working on good friday." Three years later, Ben helped Caesar and Lambeth make some anchonies "in The Crismas time." A 1750 payment to Cudgo and Sambo "for drawing Iron more than their Days Work" marked the only time when Coventry's ironmasters rewarded slaves for exceeding a daily quota. The situation was scarcely different at most other ironworks. In 1760, New Pine Forge rewarded Peter, Tom, and Ned for drawing a half a ton of bar iron. The three men normally worked in the finery; John Patton hoped to make them more versatile and more valuable. Sometimes tradesmen tried to encourage their slaves to be more industrious. In 1751, Pine Forge charged John Hanson more than £2 for paying "your own Negro York one Shilling in Each Tonn Ankonies made at the Fire pr Agreement."[33]

How little overwork shaped industrial slavery becomes clearer on the rare occasions when ironmasters discussed overwork or employed it systematically. When Ferguson McElwaine leased Tob to Peter Grubb, he informed Grubb that Tob "Desires to work close & constant, & if he makes any thing Over his weeks work for his Better Encouragement he's to have it to himselfe." Tob carried a small account book in which McElwaine asked that Grubb record weekly how much work he did. The request may have struck Grubb as unusual; he did not offer overwork to his own slaves. Peter's brother Curtis, who owned and managed Cornwall Furnace, soon became far better acquainted with the practice. Between 1769 and 1772 Cornwall paid nearly twenty slaves, usually because they had chopped wood "overtask." Why? Grubb likely had decided to honor the wishes and expectations of masters and slaves. He had hired most of the slaves who earned money from two Maryland masters: Nathaniel Giles and Caleb Dorsey. Dorsey had long offered overwork to slaves at his Maryland ironworks and many performed it regularly. Dorsey and Giles may have recommended that Curtis Grubb do the same, especially if Fullo, Pompey, Nero, and Polydore had come to expect it. Grubb reserved the right to withhold such payments. In October 1772, they stopped abruptly.[34]

Though mid-Atlantic adventurers seldom used overwork, some sought to make slaves more industrious by crediting them regularly for their labor. In 1764, Charming Forge's journal recorded "Cyrus the Negroe" as indebted to Henry William Stiegel for £90, what Stiegel "paid for him." Over the next two years, as white finers paid him for his assistance, Cyrus whittled away at that sum, his progress slowed by charges for shoes, clothing, and board. By treating Cyrus, at least on the books, as an indentured servant, Stiegel had modified slavery to suit his purposes. He motivated Cyrus to work harder and he forced him to subsidize his own bondage.[35]

Peter Grubb took a slightly different approach with some slaves at

Hopewell Forge. In 1770, he paid salaries to two—Sam and Amy. Hopewell's clerk kept attendance for both, deducting absences from their annual salaries of £25 and £12. Sam and Amy also paid for cloth out of their own pockets. Regular income gave them some power to improve their lives. Amy especially needed all she earned; she had a growing family to support. For Sam and Amy, salaries were less a taste of freedom than tokens of their enslavement. Their time and energy was money, but both—and whatever they produced—ultimately belonged to Grubb.[36]

Amy knew it better than anyone. In October 1768, Grubb sent her "yellow Child" to live with Andrew Messersmith for one year, at the end of which Grubb was to give him £8 worth of store goods. They had to haggle to reach an agreement. Messersmith worried that the child might die before the year ended; Grubb promised to adjust the payment to match "what time it Lives." Amy's child spared Grubb from having to do the math. Hopewell charged her for thirteen months' nursing that Messersmith's household provided. Two weeks later, she paid for "a Coffin for Child."[37]

What prompted Grubb to separate mother and child? In 1766, he used Hopewell funds to buy Amy, then about twenty-six years old, and her three-year old daughter Nance from John Hart. The child who Grubb sent to Messersmith's had probably just been born. Amy missed two consecutive weeks of work in September, perhaps to give birth and recuperate from labor. Grubb undoubtedly viewed the baby as a distraction to Amy. He doubted that the child would survive, and Amy would pay for nursing anyway. Grubb lost nothing. If Amy were to run away to be with the child, she would not go far. Messersmith worked for Grubb and he lived nearby. Besides, Amy had Nance to care for. Grubb died in 1786. By then, Amy had become a grandmother.[38]

Amy managed to create a family and keep it together under slavery for more than two decades. Relatively few slave ironworkers in colonial Pennsylvania and New Jersey could say the same. Those who did sometimes owed their good fortune to sympathetic whites. For example, in 1748 Moravians purchased Hannah and her son. Around the same time she married Joseph, who had worked at the Durham Iron Works. Four years later Moravians baptized Joseph, which signified that they recognized the spiritual validity of their marriage. In 1760, John Hackett, manager of the Union Iron Works, sold Joseph to the Moravians. What God had joined, his new masters believed, they should not tear asunder. Joseph and Hannah probably enjoyed as much security as any slave couple could, provided that they behaved as their new masters wished.[39]

Long distances badly strained other slaves' marriages. In 1764, Betty fled the Falls of Schuylkill and Jane Blackwell noted that Betty "has a hus-

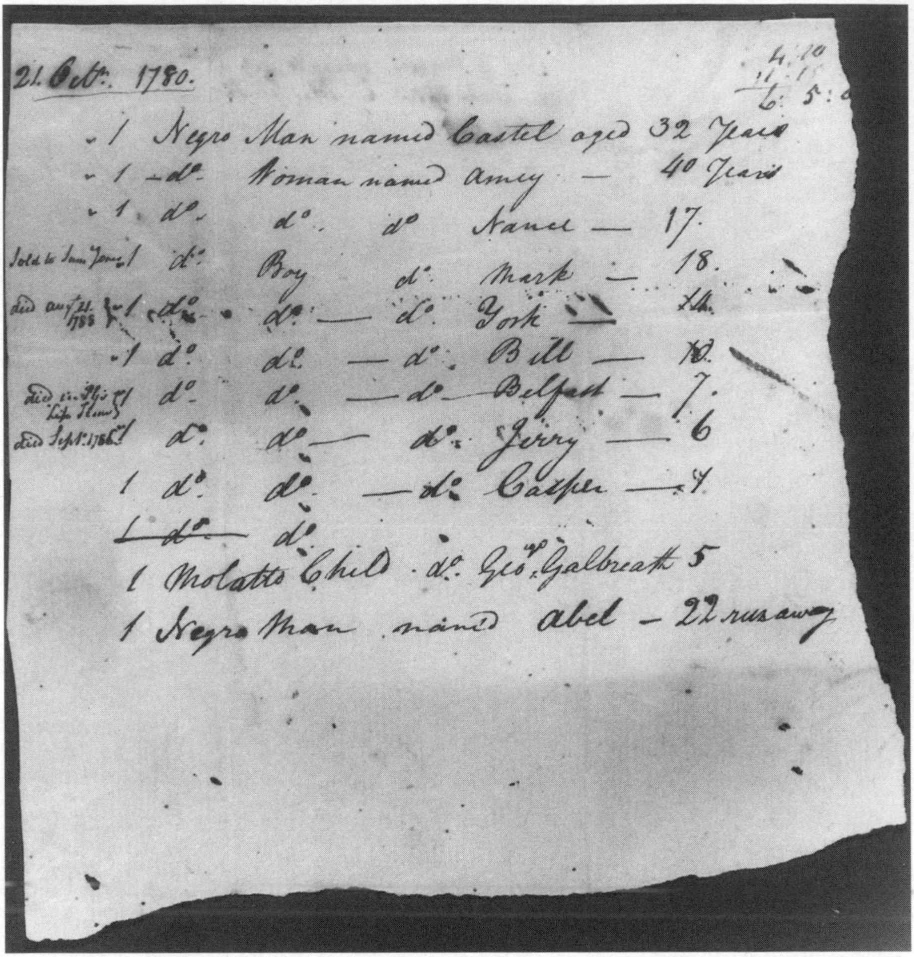
21. Octr. 1780.

1 Negro Man named Castel aged 32 Years
1 do. Woman named Amey — 40 Years
1 do. do. do. Nance — 17.
Sold to Saml Jones 1 do. Boy do. Mark — 18.
1 do. do. — do. York — 22
1 do. do. — do. Bill — 10.
1 do. do. — do. Belfast — 7.
died Sept 1786 1 do. do. — do. Jerry — 6
1 do. do. — do. Casper — 4.
~~1 do.~~ do.
1 Molatto Child do. Geo. Galbreath 5
1 Negro Man named abel — 22 runaway

Figure 7. "List of Negroes Entered into the Clerk's Office by Peter Grubb, Oct. 21, 1780," [copy with additional marginalia]. The Historical Society of Pennsylvania, Grubb Family Papers, Acc. #1967. Amy and her daughter Nance are the second and third people listed. Three people named in the original registration died within eight years after it was filed: York, 22 years old, and two children, Belfast and Jerry, neither of whom saw adolescence. One man, Mark, was sold to Hopewell Forge employee Samuel Jones.

band at Mr. Bard's Ironworks in Mount-holly, and it is thought she keeps thereabouts." The hazards of working about an iron furnace ended another marriage tragically. In 1760, Reading Furnace gave "Nokes Widow" some cash "to carry her & her two children back to Reading (her husband was the Companys Founder & Slave & killed by a Fall from the top of the Stack of the Furnace.)"[40]

She would likely have trouble establishing another long-term relationship. At least the odds of finding a man about an ironworks stood in her favor. As in the Chesapeake colonies, the vast majority of slave ironworkers were men. No large communities of slaves or free blacks existed in close proximity to the Middle Colonies' ironworks. Most masters who lived nearby had one or at most two slaves, so it was difficult, especially for slave men, to establish friendships or heterosexual relationships. The dispersed black population of rural southeastern Pennsylvania and most of New Jersey hampered the development of an autonomous African American culture, leaving those who endured slavery there to weather it without the support or comfort that such a world of shared experience and meaning provided for their peers in cities and in the South. By the Revolution, the predicament of slave ironworkers had eased a bit. Growing numbers of slaveowners in southeastern Pennsylvania increased the local black population. Slave women and children became more common at ironworks, which testified to slaves' growing success at starting families and to adventurers' deepening commitment to human bondage. Yet despite these developments, industrial slavery in the mid-Atlantic remained a lonely sentence for most who endured it.[41]

It certainly was for Sambo. On New Year's Day, 1753, he arrived at Coventry Forge lame and sick. Three weeks later, he began showing symptoms of smallpox. His masters dispatched Sambo to Mary Richards, who nursed him. He soon died; she organized his funeral and dug his grave. Sambo probably died without the company and consolation of family and friends. He also died, perhaps to his horror, alienated from African and African American traditions which considered burial one of the most significant stages of life.[42]

Sambo's demise underscores slave ironworkers' struggle to develop and maintain a distinct and coherent cultural identity. For African ironworkers in the Chesapeake colonies, the routines that ironworks followed may have been psychologically jarring, but they often arrived together and could adjust to such shocks together. Relatively few in the mid-Atlantic region hailed directly from Africa, and so a minority of slave ironworkers would have suffered a similar fate. Pennsylvania and New Jersey ironmasters, like the region's other slaveholders, usually bought slaves one at a time. The men, women, and children whom they bought came from throughout the Atlantic world. Communication among them must have been difficult. It was not just that they spoke different African languages; they often did not speak the same European tongues. Mona was "a Spanish Negroe man" who to his master's ears talked "a broken English." Joe, who his masters called "ye Portuguise" and "the Portuguese Indian," could "speak but little English, and no Dutch" when he fled Charming Forge.[43]

The diversity of backgrounds from which they came and their difficulty in creating families or a culture may have fostered a sense of individualism in many slave ironworkers which emerged when they resisted their enslavement. Slaves who ran away from mid-Atlantic ironworks usually did so alone; they fled in groups of two or more far less often than did industrial slaves in the Chesapeake region. This should hardly surprise; there were fewer slave ironworkers and relatively fewer Africans there. But slaves in Pennsylvania and New Jersey may have socialized less with white ironworkers than did slaves in the Chesapeake. Advertisements placed by Maryland and Virginia ironmasters frequently claimed that slaves and white indentured servants ran off together. To judge from the notices that Pennsylvania and New Jersey adventurers posted, slaves and servants never absconded together. Many white Chesapeake ironworkers who fled with slaves were convict servants who, like slaves, were involuntary labor. Provincial law severely discouraged Pennsylvanians from knowingly importing or purchasing convicts. Mid-Atlantic slaves encountered few white servants who shared or could have identified with their bondage. Their skin and their legal status set them utterly apart; both marked them as uniquely condemned to perpetual bondage and could only have sharpened many slaves' sense of isolation from whites and a sense that sometimes they had to stick together.[44]

As James Old learned, slave ironworkers could unite against adventurers. Thomas Cope recalled that, sometime before the Revolution, Old stopped distributing corduroy to slaves after they threatened "an insurrection. Their discontent compelled him to relinquish the idea of clothing them with the rest of the obnoxious article." For Cope, the reception that Old met anchored and punctuated a commentary on the changing winds of fashion—corduroy would not then sell and Old no doubt had paid a bargain price to Philadelphia merchants eager to dump their stocks—and on the fickle sartorial tastes of African Americans.[45]

What Cope trivialized was vital to Old's slaves. Clothing mattered to them because it was one of few ways in which they could choose to present themselves to others. If Old claimed to own them, then his slaves could at least have some say over how they covered their bodies. Chesapeake slave ironworkers devoted most of what they earned to buy cloth or garments; their peers to the north often did the same. Such purchases should remind us that what Cope described was for slaves as much an individual issue as it was a collective one. Everyone who challenged Old had a personal stake in the outcome. Dressing in corduroy, or anything else that they found too objectionable, would have only reinforced their bondage by marking them as forced to wear what others would not.[46]

By the time Old's slaves threatened rebellion, taking on new identities and resisting slavery went together more often. For some slave ironworkers, bondage in a strange land encouraged them to reinvent themselves. Joe, "the Portuguese Indian," fled Charming Forge with several articles of clothing and "a Gun, Tomahawk, and a Pair of Boots." His masters, unsure where to peg him within Anglo America's racial order, described Joe as a "Mulattoe Slave" with "long black Hair" who perhaps had "gone to join the Indians beyond the Mountain." With the help of a large cash reward, Henry Smith returned Joe to Charming Forge in a "Neck Collar." Others sought to escape slavery by renaming themselves. John Wilkinson supposed that London would "change his Name to Cuff, and pass for a free Negroe" while a fugitive. Wetheridge, who called himself Jacob, fled Pine Forge, perhaps because Thomas May demanded that he do work that he considered beneath him. Jacob was "brought up to cooking and waiting in a gentleman's family, which business he understands very well, as a gentleman in Philadelphia, from whom he was lately bought, brought him up to that business only." He packed an assortment of goods, many of which he could pawn:

> 1 fine white shirt, 2 ozenbrigs ditto, an old pair of ozenbrigs trowsers, old leather breeches, one red and white striped linen jacket, one white linen ditto, with sleeves, 2 brown cloth ditto, without sleeves, lined with shalloon, a snuff coloured broadcloth coat, almost new, with yellow metal buttons, a coarse brown great coat, with white metal buttons, good strong shoes, with brass buckles, a half worn beaver hat, which he generally wears cocked; he likewise took with him, a very old silver watch, without a chrystal, silver faced, the hour and minute hands both brass, the maker's name Moore, London, number forgot, and on the outside of the inner case is badly engraved I I, and some figures.

Jacob had put his training to good use.[47]

Few were more diligent at resisting slavery or more adept at self-reinvention than Cuff Dix. Mark Bird called him Cuff, but acknowledged that he went by Cuff Dix and advertised for him accordingly. In May 1775, he left Pennsylvania's Birdsboro Forge for several months, which galled Bird because Cuff Dix was, as the ironmaster put it the following year, "a most excellent hammerman." Nor did it please Bird that Cuff Dix "always changes his name, and denies his master," though he respected the forgeman's intelligence, warning that anyone "that takes him up must be careful in examining of him." Not long after Cuff Dix was dragged back to work, he bolted again. With exasperation, Bird noted that Cuff Dix "has often run away, changed his name, denied that the subscriber was his master, and been confined in several gaols in this province; he was employed the greatest part of last summer by a person near Dilworth's town, in Chester county." Hoping to cut off that avenue of escape, Bird threatened that anyone "who shall harbour said Negroe shall be dealt with as the law directs, and his name not omitted in a future advertisement." Bird feared that the runaway sought refuge elsewhere: "As Negroes in general think that Lord Dunmore is contending for their liberty, it is not improbable that said Negroe is on his march to join his Lordship's own black regiment, but it is hoped he will be prevented by some honest Whig from effecting it." Whatever happened to Cuff Dix, he at least got away from Bird. In 1780, he registered eighteen slaves. Cuff Dix was not among them.[48]

Cuff Dix stood virtually alone against his master. His prowess at drawing bar iron fed his self-confidence and his hatred of bondage. The chaos of war gave him a better opportunity to deliver himself. Other developments surely encouraged his dreams of freedom. By the 1760s, growing numbers of Quakers had decided their consciences could no longer permit them to own slaves. In 1769, Joseph Potts, part of the most prominent family of ironmasters in Pennsylvania, manumitted his eight slaves. They probably lived at Mount Joy Forge, a short journey from Birdsboro. In 1776, the year that Cuff Dix fled Birdsboro Forge for perhaps the last time, the Philadelphia Meeting of the Society of Friends resolved to disown members who refused to manumit slaves. Such news must have reached Cuff Dix's ears. His master was no Quaker; Mark Bird's conscience on slavery was clear. There was no sense waiting for him to see the light. Cuff Dix saw his chances and he took them. Most of the slaves he sought to leave behind bided their time as their masters clung to them tightly through war and revolution.[49]

An Industrious Revolution in German

Iron stoves and stove plates fill museums and historical societies from New Jersey, south through Pennsylvania, to North Carolina. Their distribution follows the paths that German-speaking immigrants trod most heavily. They were the principal consumers of the stoves, whose manufacture drew heavily on techniques and patterns that German moulders developed and introduced to America. The stoves testify to the impact of Germans on the iron industry. They also bear silent witness to the role that ironmasters played, sometimes unwillingly, in acculturating Germans to the norms and values of Anglo America's industrious revolution.[50]

German-speaking immigrants suffered little of the cultural dislocation or social stigmatization that African slaves endured. Most had received news, usually from kin or former neighbors, of what to expect before they emigrated. The ability to settle in or near large enclaves of German speakers eased their acclimation. Thousands emigrated as families and served as redemptioner servants together, as did the five families who Mark Bird bought in 1772 to cut wood until they had paid off their debts. Being white and Christian offered German immigrants to Pennsylvania freedom of worship and the chance to become naturalized subjects, which gave them political influence and helped to mute what prejudice they faced.[51]

Such advantages helped German ironworkers. At Pennsylvania ironworks such as Elizabeth Furnace, Charming Forge, and Oley Forge, they encountered employers who spoke their language and who were themselves immigrants. Clerks kept accounts of Tulpehocken Eisenhammer, Charming Forge's predecessor, in German. German surnames dominate lists of workers at Tulpehocken, Elizabeth, and Charming. The owners of those ironworks targeted fellow Germans for recruitment. Peter Hasenclever, agent for the American Company, engaged 535 German ironworkers and transported them, along with their wives and children, to the three furnaces and seven forges that the company would operate in New Jersey by 1768. In 1851, Elizabeth Doland recalled that Hasenclever had built houses for workers, "especially if they were his countrymen, very nice and more costly than were necessary." Finally, the techniques that Germans encountered and used differed only slightly from those they had known in Europe.[52]

Adventurers often sought German ironworkers for their technical expertise. German moulders helped to introduce flask casting to Pennsylvania, which enabled efficient production of stove plates. They included "2 dutchmen" who moulded at Colebrookdale Furnace in 1741. The "German" method of drawing bar iron, in which the same forgeman made anchonies and bar iron (rather than specializing, as forgemen did in the "English" or "Walloon" method), permitted ironmasters more flexibility. Outside of bloomeries, German forgemen found few mid-Atlantic forges which organized work as they considered customary. Those who did, such as Ludwig Hayer, a Charming Forge employee during the mid-1760s, often worked for other German immigrants.[53]

Some German ironworkers capitalized on their industry and expertise to rise socially. Several probably scraped together the few hundred pounds that it took to open one of the dozens of small bloomeries that dotted New Jersey's forests and mountains. In 1761, William Allen and Joseph Turner celebrated the durability and efficiency of forge bellows made of wood rather than of leather, to which "A German" introduced them. By "constructing them at ye different Iron Works," he had managed to amass "a pretty good Fortune—We may call it such, for he has now a Forge of his own and rents a Furnace." Through his oversight of the American Company's ironworks, his business acumen, and his close ties to networks that funneled German migrants into and through North America, John Jacob Faesch became one of New Jersey's most successful industrialists of the late eighteenth century.[54]

Many ironmasters wished that their German workers might find inspiration in Faesch's example of the diligent immigrant who made good. They discovered that German employees sometimes took a different path to acculturation. Some adapted old ways of dealing with employers to their new environment. In 1783, Johann David Schoepf reported that Pennsylvania adventurer Daniel Udree used to deal "with his workmen as is customary in Germany; that is, he furnished them with necessaries on account." He stopped doing so after many hands "made use of the opportunity to run up their accounts, and not being trammeled with families got out of the way." Udree had helped to finance the journey of many to North America. He was not about to enable them to strike off whenever it suited them.[55]

Such difficulties paled beside those of Peter Hasenclever. The hundreds of men whom he had recruited cost his superiors much of the £54,000 that they spent before they began to see any return on their money. They also cost Hasenclever his job and his reputation. He published an impassioned defense of his conduct as the American Company's agent and he blamed his workers for a large share of what went wrong. "The refractory disposition of the people," Hasenclever complained,

> was also a troublesome affair; they had engaged in Germany to be found in provisions; they were not to be satisfied; the Country People put many chimeras in their heads, and made them believe that they were not obliged to stand to the contract and agreements, made with them in Germany; they pretended to have their wages raised, which I refused. They made bad work; I complained and reprimanded them; they told me, they could not make better work at such low wages; and, if they did not please me, I might dismiss them. I was therefore, obliged to submit, for it had cost a prodigious expence to transport them from Germany; and, had I dismissed them, I must have lost these disbursements, and could get no good workmen in their stead. The desertion, sickness, and death of many of the people, and of two of my first managers, was not only a great loss and trouble, but also a terrible disappointment, which occasioned many

> things to be neglected; but, particularly, made me behind-hand with my accounts.

Hasenclever had expected docile, obedient, and grateful hands; he found what he perceived to be a collection of grasping and ornery men who tested him until he gave them what they wanted.[56]

Hasenclever's story echoes what John England told the Principio Company—uppity servants who behaved badly had brought the enterprise and its manager to the brink. Hasenclever's motive also resembled England's—to dodge blame, albeit in a more public forum, for what had gone wrong on his watch. But in England's diagnosis the contagion came from within, from rebellious underlings whom his predecessors had neglected to put in their place. Hasenclever lacked such an excuse; these were his people in at least two senses. He had engaged them and his English backers probably expected that he, like them "German," would keep them in line. The problem, in Hasenclever's eyes, stemmed from the environment to which he had brought the ironworkers. Anglo American aspirations had poisoned their minds. Their neighbors, and perhaps other ironmasters, had infected their heads with ridiculous fantasies and persuaded them to disregard their old contracts. They stuck together, first to demand more money, then to shirk their responsibilities, and then to defy Hasenclever to fire them. He folded, helpless against the ironworkers' zeal to get all they could in their new land.

The American Company replaced Hasenclever with Robert Erskine, which introduced new complications to its attempts to organize German workers. Erskine soon found himself dependent on John Jacob Faesch. The manager and his superiors believed that ethnic ties bound Faesch and the company's German workers together. The firm's adventurers worried over rumors that Faesch would leave, and that several hands would go with him. Erskine tried to reassure them. Faesch, he insisted, was not going anywhere. Even if he did, the ironworkers would not follow. They knew what a good deal they had, for "there are no iron works anywhere else where they can get so good or better wages as here. Nor anywhere else are they so sure of their money, nay, some who were represented as the most ticklish have been making their court to me in case of accidents." Steady work and steady pay, Erskine asserted, mattered to the workers far more than did working under Faesch. He was in charge and they knew it.[57]

Erskine doubted his own story. Language united Faesch and the workers, and divided the manager from them. One day, when Faesch was absent, Erskine "regretted my want of ability to scold some of the forgemen most heartily in their own language, who, through mere carelessness and hurry to get a quantity of iron worked off, had drawn a good many bars unfit for market." He assured his employers that the forgemen would "be more careful in future, as I have threatened to stop payment for this, and shall certainly do it for any such iron in future" and expressed confidence that "the Germans" would prove "tractable enough for me with proper looking-after." But Erskine still depended on Faesch's authority. Whenever Faesch was present, Erskine pledged to give him "command" of "the forgemen and founders, and when he is absent I shall endeavor to keep up the idea of his being their master, and then I have no doubt they will obey me still more readily." Many of the Germans accompanied Faesch when he became an adventurer. Some may have considered Faesch their "master." Most probably regarded him as their broker—someone who helped them to negotiate their way in a new and unfamiliar society. They, in turn, accelerated Faesch's rise.[58]

"It is time a day Not to be Impos'd upon"

By 1775, a struggle to forge an industrious revolution in the mid-Atlantic colonies had developed between adventurers and ironworkers who were mostly white and mostly free. To adventurers' delight and relief, they held the upper hand. Within one generation they had established their region as the center of British North America's iron industry. Within two they had largely succeeded at creating and maintaining a mixed and complex labor system which suited their needs.

How? In part ironmasters adapted nimbly to circumstances that were largely beyond their control. They benefited from a tidal wave of immigration that landed at Philadelphia and New York. As German immigration tailed off after the Seven Years' War, a newly organized trade from Ireland filled demand for indentured servants. A flood of migrants from England, Wales, Ulster, and Scotland crammed ships headed for the mid-Atlantic colonies. Many found their way to the region's iron industry. Thickening commercial ties between the region's merchants and the Atlantic world also brought other Europeans. In 1775, John Cox of southern New Jersey's Batsto Furnace advertised that four servants had run away: Francis Lawrence Pidginett, "a Portuguese," Matthew Serrone and Joseph Lovett, each a "Frenchman," and Francis Berrara, "a Spaniard."[59]

Adventurers also benefited as more people settled near their enterprises. Agricultural and industrial calendars dovetailed well. Local farmers wanted to earn some cash or store credit during the winter; ironworks needed them to lay in wood for coaling. The growth of nearby settlements also encouraged ironworkers to subcontract tasks such as making charcoal or hauling, often to the sons of farmers who wanted to earn additional income to smooth their path to economic independence and manhood.[60]

Immigration, denser settlement near ironworks, and more subcontracting all translated into lower labor costs for ironmasters. Arthur Cecil Bining noted that ironworkers' real wages in Pennsylvania remained relatively stable for most of the colonial era, and Paul Paskoff has found that wages for colliers and woodcutters (whose labor accounted for most of an ironworks' fuel costs) fell slightly as the Revolution approached. Subcontracting probably helped to suppress wages. It certainly saved ironmasters from supervising many employees by outsourcing that responsibility to subcontractors. Given their refrain about "extravigant wages," ironmasters would have taken great satisfaction in controlling labor costs, especially when local iron prices fell and they felt compelled to organize a cartel in the 1770s to prop them up.[61]

Adventurers conspired to fix iron prices; did they also join forces to suppress ironworkers' wages? No evidence indicates that they did. Could they have held such an alliance together? Probably not. By the 1760s, many workers lived in established households in the immediate vicinity of most ironworks. They were unlikely to seek employment far from home. Unless several ironworks were concentrated nearby, few ironmasters would bother to compete for their services. Free tradesmen, especially those who worked with metal, were more mobile and more committed to remaining within the industry. But as Allen and Turner lamented, adventurers often poached artisans and they even suspected one another of sheltering each other's runaway servants. They did not trust one another enough to cooperate effectively on wages or hiring.

Then again, they may not have had to collude to tailor the labor market to suit their needs. By the Revolution, Pennsylvania and New Jersey adventurers had woven a thick web of ties that bound them to one another. Many entered the iron business by marrying into it, as did Henry William Stiegel, John Patton, and Robert Coleman. Adventurers' families intermarried so frequently and so strategically that a few kin-based alliances dominated the region's iron industry by 1775.[62]

No one used marriage more skillfully than did Thomas Potts and his heirs, who assembled colonial North America's largest ironmaking empire in southeastern Pennsylvania. Potts, a butcher and English immigrant, became an ironmaster in the 1720s. He expanded his influence by helping his three sons, John, Thomas, and David, to marry women who were daughters or granddaughters of ironmasters. Most of his grandchildren also sought marriage partners with blood ties to adventurers. By the

Revolution, the Potts family owned or held shares in two furnaces and seven forges.[63]

For adventurers, alliances made good business sense. They concentrated ownership, capital, and expertise within their hands. Owning shares in furnaces ensured forge owners a supply of pig iron. Kin ties provided a hedge against business failure by providing sources of credit and buyers for castings and pig iron. Such bonds certainly helped the region's ironmasters cooperate in buoying the domestic iron market.[64]

What did the ties that bound adventurers mean for ironworkers? To some the situation may have seemed advantageous. It promised, especially if they were tradesmen who had demonstrated their skill and reliability, security—steady work and steady income without having to travel far. Such workers could move easily between ironworks, their path smoothed by having the same employer or by a recommendation passed between in-laws.

It is more likely, though, that adventurers' kinship networks put ironworkers at a disadvantage. Control of so many ironworks in so few hands probably helped to hold down wages, particularly between 1750 and 1776, when the number of ironworks in Pennsylvania and New Jersey more than doubled. Wages within the region's iron industry varied little, another indication that concentration of ownership depressed what mobile ironworkers might have earned if there had been more competition for their services. The combined power of adventurers may have constricted workers' options in another way. The same networks that might open doors for them could also bar them from finding work elsewhere if they displeased their employers.

Debt gave adventurers another way to limit ironworkers' mobility. As the Potts family built its empire, many of its employees piled up debts which constrained them from moving on. Throughout the 1740s most of the hired hands at Mount Pleasant Furnace and Pine Forge owed money to John Potts or to his father, Thomas. Few carried negative balances of more than £20, perhaps by their employers' design. A debt that seemed too large might provoke a worker to flee and force the Potts family to write off his unpaid bills. In 1750, Robert Templeton abandoned his coaling pits, Colebrookdale Furnace, and Thomas Potts, to whom he owed nearly £47. Others did not get away so easily. In 1747, Henry Arringberger died. He had farmed near Colebrookdale, sold some of his produce there, and worked there after he had finished harvesting his fields. To settle his debts and get a badly needed cash advance, his widow bound herself to Thomas Potts for one year, during which she whitened linen, raised fowl, and churned butter.[65]

The range of goods available for purchase at ironworks' stores helped to lure some ironworkers into debt and it gave others a powerful incentive to be industrious. Well-stocked shelves attracted neighbors to ironworks, where they could exchange their produce and their labor for items that they might not get otherwise. Adventurers saw opportunity in the Anglo American world's growing rage to consume.[66] Peter Hasenclever claimed that "the American Iron Masters have an advantage which compensates, in some measure, for the exorbitant wages, which is, the selling of goods and provisions to the people." William Kirby, who worked at Allen and Turner's Andover Iron Works in the 1760s, echoed Hasenclever. Kirby recalled that "the wood chopper piled his wood so as to cheat the collier. The collier put his charcoal into baskets in such a manner as to deceive the iron master; and the iron master, not to be outdone, sold his provisions to the men at an extortionate price." For Kirby, adventurers, by preying upon workers when they shopped in the company store, closed the cycle of greed and deception that characterized the iron business.[67]

William Allen and Joseph Turner characterized the stores at their ironworks as burdens as well as profitable ventures. In their view, the quantity and variety of what they stocked owed more to worker demand than to what they wished to supply. In 1760, they ordered "a few Linnens for our Iron Works," largely because ironworkers had obliged them "to Keep an Assortment of Goods for" their "Conveniency." Allen and Turner feared that they would refuse to "Stay with us, for they cannot Spare time to run about the Country to buy Flour & Provisions & the Necessary Cloathing for their Familys" if the stores did not provide a wide selection.[68]

As Allen and Turner reckoned things, the loudest cries for convenience and variety came not from ironworkers but from their wives. They had to "have a great deal of finery, they Cannot do without Tea, Coffee Chocolate Loaf Sugar & C without which they Could not expect the high wages there husbands have." Henpecked husbands who would not or could not control their wives' consumption, the adventurers claimed, forced them to pay exorbitant wages and to search far and wide to stock their stores. Wives with excessive and frivolous tastes also stood between their husbands and independence, Allen and Turner insisted, since "Such of them as are frugall Lay up large Sums of Mony." The proprietors did not note that men might have demanded such goods. Nor did they mention that the frugal wives whom they held up as exemplars might ultimately cause them to lose the services of some of their best hands.[69]

Indeed, Allen and Turner placed the responsibility for promoting industry and thrift upon women's shoulders. Those in positions of visibility and authority should set a good example for those beneath them. In 1765, Allen sought to replace the manager of the Union Iron Works largely because his wife was "quite unfit for Iron Works, as she appears to be a fine Lady, & expects to live with a Delicacy not common in these parts of the World, especially at Iron-Works, either in England or here. The Mistress of such a Family as ours ought not to wear Silks," he insisted, ". . . nor spend much of her time in decking her person, or dressing her head, but rather by her Case indeavour all She can to promote Oeconomy & Frugality." Because few women, Allen believed, could pull off such a feat, he preferred to hire "a young unmarried man" to oversee Union, but would accept a married candidate, as long as his wife was "not a fine Lady."[70]

To adventurers, one form of consumption mattered most—how much ironworkers drank. Initially, Pennsylvania ironmasters clamored for government help to stem workers' drinking. In 1726, they won "An Act for the Better Regulating the Retailers of Liquors Near the Iron Works and Elsewhere," which began by noting that "the selling of rum and other strong liquors near the furnaces lately erected . . . have already proved prejudicial and injurious to the undertakers." The law empowered ironmasters to control workers' access to alcohol by barring anyone from selling liquor within two miles of any standing or future furnace unless specifically licensed by a majority of the furnace's owners. Violators were to pay a fine of forty shillings. Ten years later, the act expired and several tavernkeepers immediately obtained permits to operate near ironworks. Concerned adventurers accused the new taverns of "giving Shelter and Entertainment to their Servants, and detaining them from their Business." They petitioned the Assembly to revive and revise the defunct regulations so that no tavern could open within six miles of an ironworks without their permission. Under the law that did pass, the ironmasters had to settle for a three-mile limit.[71]

New Jersey adventurers also demanded laws to regulate where and how much ironworkers drank. In 1765, Charles Read petitioned the province's governor, council, and general assembly to deny licenses to taverns located within three miles of an ironworks without the ironmasters' approval, and to limit adventurers' liability for employees' tavern debts to five shillings. Four years later, the New Jersey Assembly authorized owners of the Hibernia Iron Works and of ironworks in Evesham and Northampton townships, Burlington County, to "deliver out to the Persons in" their "actual Employ and Service . . . Rum, or other strong Liquor, in such Quantity as they shall from Experience find necessary." It banned anyone else within four miles from entertaining "in or about their House, in Idling or Drinking" or from selling "any strong Drink, to any Wood-Cutter, Collier or Workman employed at said Works."[72]

Ironmasters never expected nor aspired to stop ironworkers from drinking. They adhered to the tradition of treating workers while they undertook particularly arduous tasks. In 1734, Thomas Potts gave a crew a "Customary Allowance off Rum . . . when Getting Inwall Stones over Schuylkill." Five years later, Thomas Yorke distributed rum "among the Fforge Men had from Birds" and Thomas Potts paid for one pint of rum "among the Mine diggers." Other workers negotiated contracts which included payments in alcohol. In 1738, Colebrookdale Furnace credited James Cannon for "breaking mine" for three months and gave him "1 Bottle Rum p agreement."[73]

If ironworkers had to have rum or whiskey, ironmasters could perhaps turn the situation to their advantage by influencing when, where, and how much they drank. That was the purpose of the licensing regulations that they wanted and obtained; they effectively made ironworks the principal legal distributors of alcohol to their employees. This brought more business to their stores and it afforded workers fewer ways to dodge their duties. Fewer taverns also meant fewer places for ironworkers to socialize free of their employers' supervision. Through political action ironmasters kept their workers' drinking habits on the books. They could tally who bought what, earn some money, and limit the damage that drinking might cause them.

By 1750, Pennsylvania adventurers had determined that they no longer needed government to help them manage workers' alcohol consumption. They did not try to renew the second "Act for Regulating Retailers of Liquors near the Iron Works." Instead, they implemented new, private means to measure ironworkers' drinking. Warwick Furnace tracked alcohol purchases by recording them separately in a "Rum Book." William Bird did the same in Roxborough Furnace's "Whiskey Book."[74] Other ironworks, such as the Andover Iron Works, kept quarterly tallies of how many quarts of rum employees had bought. Sometimes ironmasters tried incentives to temper hired hands' drinking. In July 1762, John Shaw agreed to stock the upper forge at New Pine Forge for one year. Shaw was to receive £18 and a pair of shoes for "faithfull performance" of his duties. His employers worried that he would not honor the bargain. According to their "Rum Book," Shaw had recently bought more than ten quarts of rum in about two months. To encourage him to curb his drinking, New Pine Forge promised Shaw "a pair of Stocking[s]" and board "if he does not get drunk above once in three months."[75]

By the 1770s, ironmasters began to penalize workers when they determined that drinking had impaired their job performance. For June 7, 1773, Cornwall Furnace's clerk wrote in the daybook: "Philips Whig & Armstrong all Absent from the Works last Saturday & Sunday Came Home

325

Andover Furnace 1st February 1774

No.	Entry		Amount	Total
196	Sundry Accounts Dr. to Store			
166	John Emmott sundries to Sarah Christy 4/2 . 1 quart rum 1/6		.. 5..8	
65	Abijah Chambers ½ pound Tea		.. 3..6	
219	Samuel Price Senr ½ pound do 1/9 . 2 pound Sugar 1/4 . 5 quarts rum 7/6		.. 10..7	
187	Samuel Patrick 6 pound Sugar 4/ and 3 quarts Rum 4/6		.. 8..6	
230	Arch Stewart 1 yard Lincey 2/ 1½ yard Shaling 3/ ½ yard peleng 11		.. 5..11	
148	James Willgas 2½ yards Lincey		.. 7..6	
182	Edward Dunlap 1 pair Gloves 4/3 . and 3 yards Shalloon 11/3		.. 15..6	
212	John Young ½ pound Tea 4/9 . and 13½ quarts rum at Sundry Times 20/3 . 1		.. 2–	
89	Edward McQuank 1½ yard drilling 7/6 . ½ lb Tea 3/6 . 1 handk 2/9 . thr 4½		.. 14..1½	
229	Moses Tharp 1 blankett		.. 14–	
221	Frederick McCafferty 4 quarts Salt 9 . 1 blankett 10/		.. 10..9	
120	John Seabott 1 pair Shoes 9/ . and ½ yard Check 7		.. 9..7	
208	Thomas Beaty ½ pound Tea		.. 3..6	
213	Abraham House ⅛ pounds ditto		 11	
123	Azar Reed remainder linen 4/9 . ½ yard Cambrick 3/	.. 7..9		
	1 Check handkf 3/ . ½ pound Tea 3/6 . and Tape 4	.. 6..10	.. 14..7	
154	Pigmetal Sent to Philada for Sundries gave Wm Murry for carting 2 Tonns pigmetal to Kirkendalls Landing		2 .. –	
165	David Willis 2 quarts rum		.. 3–	
220	John Shippard . . . 2 quarts ditto		.. 3–	
203	John Vennett . . . 1 . ditto		.. 1..6	
193	Samuel Thatcher . . . 6 . ditto		.. 9–	
118	Edward Goucher . . . 15½ ditto		1..3..3	
204	Thomas Price . . . 18 . ditto		1..7–	
200	Joseph Augustus . . . 1 . ditto		.. 1..6	
134	Benjamin Bedell . . . 8 ditto		.. 12–	
210	Garrett Farrell . . . 3½ ditto		.. 5..3	
227	Joseph Thurston . . . 17½ ditto		1..6..3	
153	Thomas Moore . . . 4 . ditto		.. 6–	
209	John Robinson . . . 25 . ditto		1..17..6	
218	James Knox . . . 3½ ditto		.. 5..3	
129	Higgins Coppinger Junr . . . 9 . ditto		.. 13..6	
201	John James . . . 13½ ditto		1..1..3	
146	Benjamin Moore . . . 2 . ditto		.. 3–	
214	John Sheals . . . 14 . ditto		1..1–	
225	Mathew Whealin . . . 2½ ditto		.. 3..9	
214	John Kelly . . . 2½ ditto		.. 3..9	
177	Bryan Horgan . . . 2 . ditto		.. 3–	
216	Anthony Roberts . . . 8½ ditto		.. 12..9	
197	Andrew Willson . . . 2 ditto		.. 3–	
211	Joseph Handrock . . . ½ ditto		 9	
223	William Coppinger . . . 2 ditto		.. 3–	
124	John Hull . . . 3 . ditto		.. 4..6	
178	James Feagan . . . 5½ ditto		.. 8..3	
98	William Mahony . . . 9 . ditto		.. 13..6	
186	John Lowe . . . 2 . ditto		.. 3–	
206	Juno Kelly . . . 11 . ditto		.. 16..6	
221	Frederick McCafferty . . . 29 . ditto		2..3..6	
207	Higgins Coppinger Senr . . . 4 . ditto		.. 6–	
217	Thomas Christy . . . 31½ ditto		2..17..3	
144	George Oynsall . . . 31 . ditto		2..16..6	32..3..10½

Figure 8. Andover Furnace Day Book/Journal, May 19, 1773–November 24, 1777. Taylor Family (of Hunterdon County, New Jersey) Papers, 1769–1882, MC 885. Courtesy of Special Collections and University Archives, Rutgers University Libraries, New Brunswick, New Jersey.

Drunk and Beat Patt McGuire very ill." His observations may have prompted Curtis Grubb to make an example of others. Five months later, Cornwall fined teamster John Bryans a day's wages "for neglecting of 1 days Work by being in Liquor." Cornwall's management paid more attention to Daniel Bryans. He lost a day's pay in April 1774 for "1 days loss of his team through his neglect by gettg drunk & oversettg Waggon." The following September Bryans was fined for "oversetting Waggon when Drunk" and for turning his horses loose and "going to Mulberry's & not bring.g Coals." His superiors recognized that Bryans had a problem; he could not control himself. To resolve it, and send others a message, they fined Lawrence Doyle "for making Bryans drunk."[76]

Fines for getting drunk or getting someone else drunk were new to ironworkers. Fines for failing to work to an ironmaster's satisfaction were not. Around 1760, adventurers throughout southeastern Pennsylvania began to demand diligence of ironworkers by noting their infractions and debiting their accounts accordingly. Colliers and teamsters commanded most of their attention. In 1766, Warwick Furnace fined collier Samuel Laverly fifteen shillings "for disapointing 2 Teams"—in other words, for not having charcoal ready when they arrived to haul it away. No one resorted to fines to discipline workers as often as Curtis Grubb did. Between 1768 and 1774 Grubb meted out seventy-seven fines to Cornwall employees: forty-two to colliers and twenty-five to teamsters. Generally, colliers had failed to have a full load of charcoal ready when a teamster arrived for it, although sometimes they had allegedly neglected their duties in other ways. In 1768, John Rice paid 15s "for a disapointmt of a load Coals it being afire & burned the Coal Box" and another 5s "for a Waggon being burnt by Coals being too hot." Grubb penalized colliers heavily when they did not make good charcoal. In 1771, Cornwall debited John Donahue's account more than £5 "for the Deficiency of 4 Loads of Coals in Coleing a 39 Cord Pitt" and James Cochran's £7 "For Coles Sent in by him from 26th August To 31st 15 Load not Good Coles." The teamsters whom Grubb fined usually had not arrived in time to retrieve charcoal or had caused their teams to remain idle.[77]

Financial penalties, ironmasters hoped, would make ironworkers more industrious and push them to develop a sense of time discipline. Teamsters and colliers were natural targets for such an initiative. Colliers had to have charcoal ready to go when teamsters came to claim it, both to keep fuel stocks from running low and to avoid wasting teamsters' time. Teamsters knit together the often distant (some furnaces, after all, claimed over ten thousand acres) but interdependent corners of an ironworks. Berkshire Furnace needed men like Michael Bertlin to be attentive to their duties. In 1768, Bertlin agreed to work as a carter for one year. He pledged "to be carefull of all the Gears, and all other matters entrusted to his Charge" and to "be Industrious in feeding his Horses and not lose any time except by sickness or accident." Workers' time was money to ironmasters, especially just before the Revolution, an era of stagnant or falling iron prices. Through contracts and fines they sought to make their employees understand it.[78]

Some ironworkers developed a sense of time discipline. In 1768, Cornwall Furnace's keeper David Ramsay bought a watch. Six years later, Cornwall moulder George Foulk did as well. The watches would have represented status symbols to Ramsay and Foulk—relatively few colonists, and even fewer artisans, could afford them. More important, watches served both men well. Furnaces followed twelve-hour cycles. For Ramsey, charged with monitoring the blast's progress, clock time helped him determine when and how to charge the furnace, when to tap it, and when to summon the founder. A watch helped Foulk pace his efforts. He would have a better idea when his superiors would tap the furnace, so he could gauge more accurately when he would need his moulds ready. By owning watches, Ramsay and Foulk gained access to knowledge that their employer, Curtis Grubb, mostly reserved to himself. By knowing what time it was, they may have also adopted the values that Grubb hoped they would.[79]

As the Revolution approached, adventurers were succeeding in forging an industrious revolution on their terms. The fines, their rather abrupt emergence, and the frequency with which ironmasters imposed them surely reflected a desire to curb labor costs within an increasingly competitive industry. They also reflected the actions of entrepreneurs who had come to believe that they could dictate terms to most of their employees. When Berkshire Furnace fined its keeper Jacob Showers for "3 pair of Flatt Irons which he made before the Furnace blow'd out, and that without asking leave," Robert Patton signaled that he did not fear offending someone of Showers's importance. Nor did the owners of the Carlisle Iron Works when they ordered their manager to fire John Goucher or any other forgeman who refused to work as they wished. Four years later, Carlisle's adventurers directed that those "who are now loitering about the Works not Employ'd, and those whose services are not wanted, must be Immediately removed," and they instructed their agent Samuel Hay that "for the purpose it would be necessary to write out a List of the whole number, & to Employ those only who may be deemed absolutely useful to the prosecution of the business." To be sure, Carlisle's owners sought to trim payroll, partly because they were thinking of selling the ironworks. They also indicated that they held the power to determine whose services were needed—and that they intended to use it.[80]

A few of the ironworkers whose lives we can glimpse through ironmasters' papers acknowledged that power. In 1767, Francis Brezina of Elizabeth Furnace alerted Peter Grubb that he could not accept an offer of employment. In a slanted but steady hand he wrote

> Having spoke to Mr Stiegel about my Affair, I cant omit, Sir, to acquaint you of his answer. I must first of all tell you I am Something yet in debt to Mr Stiegel, and Secondly the Time I agree with Him is not yet expir'd. Consequently I observ'd by a Little Cloudy Look of his which he has been Casting at me, that he was not quite willing to part with me, and tho' I am Convinced with his Great Goodness Conferred on me, I am afrade to offend Him with the Least ungratitude; therefor, Sir, I am infinitly Oblidge to you for your Kind offer you was willing to bestow on me. I shall ever endeaver to acknowledge it, and when the said Time I agree with Mr Stiegel is expir'd, and my Debt is paid, I shall gladly offer my Service to you, if it should be acceptable to you.

Brezina feared Stiegel: he owed him money, he had a contract to honor, and the ironmaster probably watched over his shoulder as he wrote Grubb. He had every reason to defer to Stiegel, or at least to humor him. What is revealing is that Brezina showed Peter Grubb nearly the same respect he accorded Stiegel. He signed off as Grubb's "Most Humble Servant" and in a postscript asked him "to give my Humble Respect to your Brother." Brezina had concluded that he had few options. Stiegel controlled his immediate future, men like Grubb would shape his life after that. Brezina believed that he could not burn any bridges if he wanted to remain employable in the iron industry.[81]

Francis Brezina was not the only ironworker who was forced to bend to adventurers. In 1774, forgeman Samuel Jones renegotiated his contract to make bar iron at Hopewell Forge. Curtis Grubb was in no mood to bargain. After speaking with the forgeman, he advised his brother that if Jones

> does not Comply with your Terms or is too high, I am Sure I will not Perswade him to Try any more. Neither will I have any more to do with it, so I would desire you May Act According to your Own Judgement in Every thing that Yeald you Most Satisfaction. If one workman should be too Stubborn, there is Plenty to be had, it is time a day Not to be Impos'd upon.

Peter agreed. He and Jones quarreled; the next day the forgeman reconsidered and tried to patch things up. He regretted that he could not take

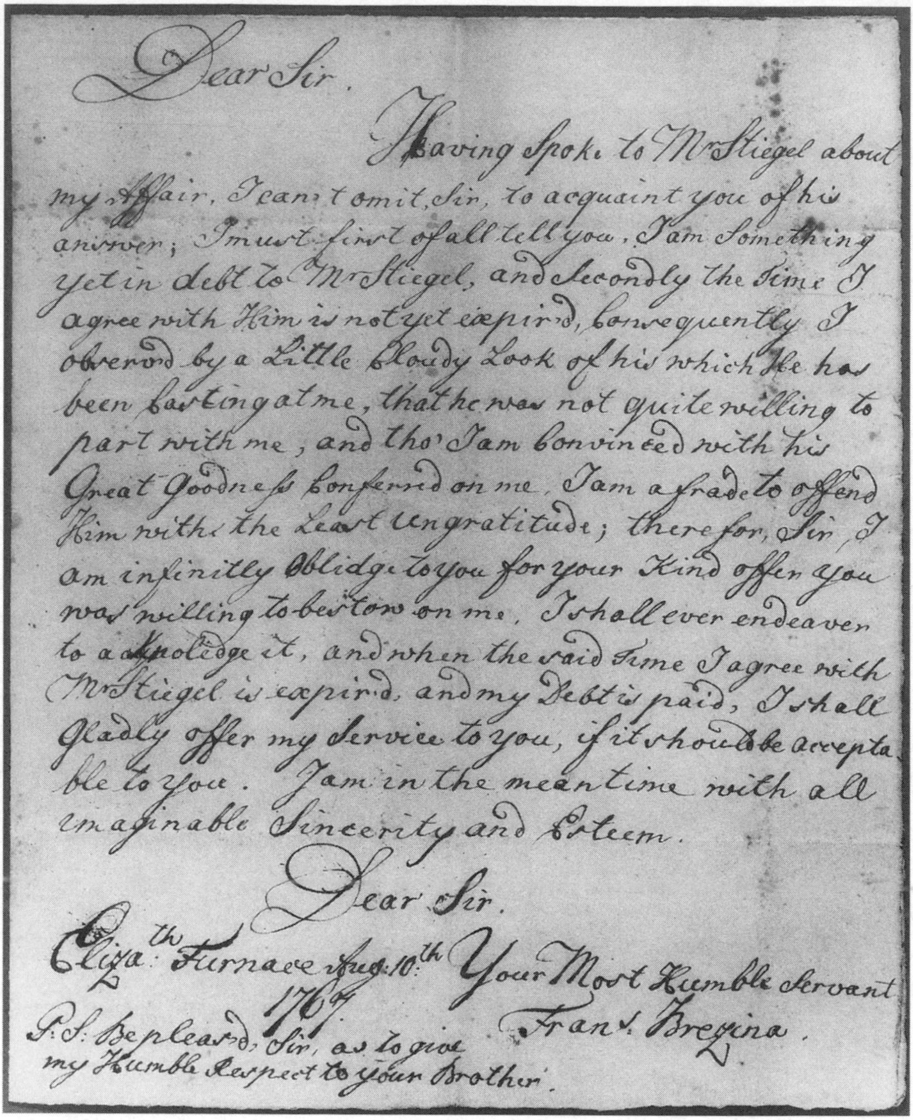

Dear Sir.

Having Spoke to Mr Stiegel about my Affair, I can't omit, Sir, to acquaint you of his answer; I must first of all tell you, I am Something yet in debt to Mr Stiegel, and secondly the Time I agree with Him is not yet expir'd, consequently I observ'd by a Little Cloudy Look of his which He has been casting at me, that he was not quite willing to part with me, and tho' I am convinced with his Great Goodness Conferr'd on me, I am a frade to offend Him with the Least Ungratitude; therefor, Sir, I am infinitly Oblidge to you for your Kind offer you was willing to bestow on me, I shall ever endeaven to acknolidge it, and when the said Time I agree with Mr Stiegel is expir'd, and my Debt is paid, I shall gladly offer my Service to you, if it shoulde be accepta ble to you. I am in the mean time with all imaginable Sincerity and Esteem.

Dear Sir.

Elizath Furnace Aug: 10th 1767

Your Most Humble Servant
Fran. Brezina.

P.S. Be pleas'd, Sir, as to give my Humble Respect to your Brother.

Figure 9. Francis Brezina to Peter Grubb, August 10, 1767. The Historical Society of Pennsylvania, Grubb Family Papers, Acc. #1967.

"yesterday back," noted that it was "the only time we ever fell out," and inserted "hope youl look over it" before declaring himself Grubb's "Most obedient Humble Servant." Within seventy-two hours Jones had complied with Grubb's terms. He remained at Hopewell for at least fifteen more years.[82]

On the eve of the Revolution, mid-Atlantic adventurers were largely defining the industrious revolution for ironworkers as they saw fit. They would not "be Impos'd upon" when there were "Plenty" of workers available. Heavy immigration and thickening settlement had helped to create such a situation. But ironmasters were hardly passive beneficiaries of an invisible hand operating in labor markets. Their hands were quite evident in the measures that they had created or adopted to discipline labor, among them the enslavement of hundreds of ironworkers. All played a key role in bringing to heel even skilled tradesmen like Samuel Jones. William Allen and Joseph Turner would have applauded. One of "that Sort of Gentry," a forgeman, had been put in his place.

But Samuel Jones was no victim of the industrious revolution. Its racialization of labor had helped to make him a master. During the 1780s Jones bought Mark, who was eighteen when Peter Grubb registered him in 1780 (see Figure 7), and he agreed to teach another slave to draw bar iron. He might have disputed that the mid-Atlantic was the best Poor Man's Country, but Jones could not deny that he had fared well there.[83]

IRON AND NATION

The Early Republic

5

INDUSTRIAL SLAVERY DOMESTICATED

As 1813 began, David Ross resolved to teach the slave watermen of Virginia's Oxford Iron Works a lesson they would never forget. "I must confess," he fumed upon hearing that Peter, Aaron, and Lewis had lost a load at Oxford's landing, "that I never before experienced such infidelity even in the worst of our Black servants." Peter fled to escape the punishment that awaited all three; Ross doubted that he would return. He directed his agents to recruit "some of the most respectable black people" to grab Aaron and Lewis and "carry them with ropes round their necks to the boat landing where the load was lost & there have them stript naked" so that each could suffer "39 stripes inflicted well-placed" on his bare back. Some "trusty servants" were to whip the watermen, for which Ross promised them "half a dollar . . . provided they do their duty." Every slave but Aaron and Lewis was to get new clothes because he considered them "the most unfaithful people attached to the Estate." Moreover, they had just boated only half a load to the ironworks and had taken nearly two weeks too long to do it. "Let them know I am perfectly acquainted with their rascally behaviour—I shall be well pleased if you can sell them both to the Carolina Hogg drivers—Such scoundrels," Ross concluded, "are not fit companions for honest servants."[1]

At first, Ross's tirade sounds like the next act of the colonial Chesapeake's industrious revolution. Disobedient slaves who shirked their duties served as examples to others by facing the lash, exposure to winter with no new clothes, and the threat of sale. Ross's order that other slaves whip Aaron and Lewis and his pledge to pay them for doing it properly seems a sadistic expansion of overwork—wages of slavery earned literally on other slaves' backs.

Another look at Ross's response and at slavery at Oxford reveals that quite a bit had changed since the Revolution. To judge from Ross's letters to his agents, whippings were rare events there—Aaron and Lewis were the only slaves to confront the lash. Perhaps that helps to explain why Ross tailored their punishment so carefully and so theatrically by marching them to the scene of their "crime" and having them flogged there. He wanted witnesses to remember what they saw, including who inflicted the punishment. By having other slaves lash the watermen and by paying them for it, Ross invited divisions within Oxford's slave community which might make it harder for it to unite against him. There was theater too in the distribution of clothing from which Ross pointedly excluded Aaron and Lewis. But they had more to fear than lashes or old tattered garments; Ross had threatened to sell them. To Aaron, married and with three children, that seemed particularly terrifying. Fewer whips, more sundered families—both characterized the South's industrious revolution in the new nation.[2]

So did the some of the sources of Ross's anger. The watermen had neglected their responsibilities, defied him, and worst of all, betrayed him. Peter had won his confidence and then deceived him, though Ross could accept that. Aaron and Lewis were another matter. "Surely better might have been expected from our own native born people," he scolded. Why did they fail him? Perhaps, Ross surmised reluctantly, it was in their lineage: "It would be as ungenerous, as painful to think or to say that the inborn depravity of the African cannot be eradicated in the third or fourth generation." Colonial slaveholders would have agreed, but they seldom worried whether they could or should eradicate it. Masters like Ross thought it their duty to try. That too made for a different industrious revolution.[3]

Industrial slavery became a domesticated institution after the Revolution. The Revolution entrenched slavery as the Chesapeake's and the South's preferred labor system. It also helped adventurers to view slaves, born in North America and raised in bondage, as more than just human assets. Slaves were "their people," who deserved and needed what they regarded as their benevolent guidance. Slaves could change for the better, ironmasters believed, and they considered it their obligation to help their slaves do so. It fell to masters to attend to slaves' hearts and souls, as well as to their bodies and to the ledger balances of their ironworks. Concerned stewards of what many masters had begun to reimagine as a humane institution would do no less, especially to direct slaves who aspired to more comfortable and secure lives.

The industrious revolution assumed new meanings for slave ironworkers as well. For the first time, slave men participated in it explicitly on terms that more closely resembled those of free white men—as providers to families. The development of a regional market in hired slave labor—partly the product of entrepreneurs' rediscovery of the problems that white labor posed for them—complicated industrial slavery by forcing adventurers to negotiate the value and terms of work with masters and often the slaves whom they wanted to lease. That empowered slave employees, but at a high cost which included absent husbands and fathers and white workers' growing animosity. Overwork, greater access to trades, and industrial paternalism all bolstered slave men's efforts to provide for families, to structure their work, and to redefine what it meant to be a man under slavery. What success they enjoyed remained precarious. Masters reserved the right to sell slaves to the highest bidder, especially when they failed to act the part of grateful dependents. Adventurers, like other slaveholders, may have recast slavery and the industrious revolution by reimagining their relations to slaves in paternalistic terms. But there was no civic component to masters' paternalism, no sense that through it they were contributing to the construction of a nation to which slaves might belong. Industrial slaves might have become fully human to their masters, but they remained commodities.[4]

Revolutionary Consequences

In 1782, with the Revolution winding down, Ben Johnson fled Baltimore Furnace. Clement Brooke alerted newspaper readers that Johnson "was seen, some little time before he went off, at the French Camp in Baltimore, and very likely will try to pass for a free Negro." Brooke needed their attention; he suspected Johnson would "endeavour to cross Susquehannah with the troops now leaving Baltimore." He offered a quick way to identify the fugitive: "When he is spoke to sharp, it will be near a minute before he makes any motion with his lips, and after opening his mouth, stammers much." Ben Johnson was not what masters derisively called an "artful" slave; he had not yet learned to mask his discomfort among whites. Still, he seized what might have been his last chance at freedom. So did many of his peers. At least thirty-three slave ironworkers ran away during the war. Virginia governor Lord Dunmore's 1775 promise of freedom to slaves who enlisted with British forces may have lured some. "It is said since Lord Dunmore's Proclamation . . . ," Baltimore Company partner Robert Carter wrote in 1777, "that about 1500 black people have availed themselves thereof—about 200 negroes were recovered lately, on board some British Ships, lying in Chessapeak Bay. . . ." Others, including Ben Johnson, headed for nearby British or French troops or used their proximity to distract those who would pursue them.[5]

Ben Johnson had another reason to leave the Baltimore Iron Works; the enterprise was in dire straits. The State of Maryland had confiscated Daniel Dulany's share because he was a Loyalist. Most of the partners wanted out of the iron business; Robert Carter sold his interest in 1787. The Baltimore Company had brought declining returns for years. In 1785, it scaled back production and sold over forty slaves, "consisting of women, girls, and boys." The auction shredded families and a community that had just formed. Slave women and children had begun to play a prominent role at the Baltimore Iron Works just before the Revolution, thanks to the firm's decision to buy young slaves so that it would have to hire as few white hands as possible.[6]

The fate of the Baltimore Company's slaves reminds us that the Revolution brought more slave ironworkers an auction block rather than a chance to claim liberty. In 1781, Maryland's government confiscated British subjects' property, including the Principio and Nottingham Companies and the more than 250 slaves who belonged to them. It displayed them at auctions, one of which invited bids on "160 slaves of different ages and sexes, amongst whom are several valuable tradesmen, such as forgemen, colliers, blacksmiths, carpenters, &c." The sales fractured the ironworkers' communities. Thomas Russell bought eight slaves at one of the last Principio auctions, seven age sixty-one or older. Auctioneers and successful bidders seem to have tried to keep nuclear families, or at least mothers and children, together. Charles Ridgely bought ten Principio slaves and twenty-five men, six women, and twenty children seized from the Nottingham Company. Many remained at Nottingham's forge, which Ridgely also purchased.[7]

Why did Charles Ridgely buy so many slaves at once? The Nottingham Company owned sixteen slave forgemen. Such skilled men were hard to find; they would permit Ridgely to diversify the range of products that he sold. By purchasing so many slave women and children and by keeping them close to home, Ridgely minimized his risk of confronting many runaways. He also guaranteed his ironworks a slave labor force for the foreseeable future.[8]

Perhaps that mattered to Ridgely because the Revolution had made it far harder to bring slaves into the Chesapeake. The war stopped the Atlantic slave trade. New Maryland and Virginia laws effectively excluded it, and they barred or strongly discouraged interstate slave imports. The legislation owed less to moral scruples or egalitarian idealism than to economics and slave demographics. Planters were abandoning tobacco for less labor-intensive crops; the region's slave population was growing as black men and women started families at record rates. Thousands of masters had already manumitted slaves or soon would. In short, the Chesapeake already had all the slaves it needed. To bring in more courted danger. State authorities believed that slaves from Africa, the British Caribbean, or worst of all (after the Haitian Revolution began), Saint Domingue, might spark rebellion and bloodshed. Besides, they would drive down the price that masters might get for the slaves they considered "surplus." Each year thousands of them were sold, chained together in gangs, and marched away to power the plantation South's expansion.[9]

A bigger concern loomed for Ridgely as he bid on slaves; he could no longer buy transported felons. Ridgely had been one of the best customers for convict servants. They were ornery, but they were cheap. Many ran away during the war—some from Ridgely's furnace. Others enlisted to escape servitude early. Continental Army recruiters in Maryland conscripted convicts, partly to prevent the British from getting them first. Ridgely's forges employed several British prisoners as the war ended. They were the last group of prisoners to work for him. Peace and national independence terminated the convict trade to North America. British and Irish felons whom Chesapeake masters might have bought went instead to Australia, the British Empire's new penal colony.[10]

The end of convict servitude helped to revive white indentured servitude at Ridgely's ironworks. In 1786, Northampton Furnace bought thirty-four servants, eleven of them from Ireland. Twenty-one were to serve four years each. Not all stayed long. The furnace immediately sold three servants; Ridgely bought four for himself. Two years later, Northampton sold six who had from one to two and one-half years left to serve to North East Forge.[11]

Ridgely may have wished he had sold them all. Few imitated William Wirle, whose debts Ridgely and his partner forgave because Wirle "has been a good Old Servt & Hireling."[12] Instead, most proved almost as unruly as the convicts they replaced. They fled Northampton, often in large groups. In 1786, Ridgely bought "200 advertisements for yr 7 white servants." A band of twelve fugitive servants was seized deep in Pennsylvania, nearly 200 miles from Northampton, in 1787. Northampton Furnace charged six servants the following year for the eight days that each had missed plus the expenses of retrieving them, while Peter Coyl paid for "5 days runaway time & Come himself." Two weeks later, twenty-seven Northampton servants spent a total of eighty-five days at Baltimore County Court while authorities determined their punishment for having run away. The court found them liable to the ironworks for their time in court plus part of the expense for transporting and boarding them. They included seven of the eleven Irish servants whom the furnace bought in 1786.[13]

Northampton servants soon learned to challenge their bondage with the legal system as well as with their legs. Thomas Herrick, bought in 1792 for £15, went to court in 1796. He won nearly £20 "for 5 months Service for his Runaway Expences" and £5 for freedom dues. Herrick likely inspired Peter McDonough "to complain" at the Baltimore County Court. Northampton dispatched Nicholas Jessop to look for him and it charged McDonough for the day that he missed plus Jessop's time and expenses in finding him. The furnace had to write off what it spent on the eight days that Jessop stayed in Baltimore while the court heard McDonough's case. Charles Carnan Ridgely, Charles Ridgely's heir and son-in-law, had had enough. White indentured servants became an endangered species at Northampton as Ridgely turned to a mixed labor force of slaves he owned, slaves he leased, and free waged workers.[14]

By the time Herrick and McDonough went to court, Northampton servants had resisted their bondage by almost any means for at least thirty years. Changes prompted by the Revolution made them more likely to test its limits. Convicts may have blurred the line between slavery and servitude, but they allowed those who chose to sign indentures to distinguish themselves from those who could not. Their sudden absence compressed the social and cultural distance between forms of bound labor in the Chesapeake; it became easier to liken them all to enslavement. Growing numbers of "term slaves," whose masters promised to free them after they had served several years (which made them akin to indentured servants), gave whites in northern Maryland even more reason to distance themselves from indentured servitude. Under such circumstances, resisting and escaping servitude provided a way for bound white workers to distinguish and separate themselves from the black men and women who were trying to struggle up from slavery.[15]

Northampton's slaves watched servants flee. They knew of the growing community of free blacks in nearby Baltimore. If they harbored hopes of deliverance from bondage, they were frustrated. Charles Ridgely died in 1790; his will manumitted few. Neither did his wife Rebecca, a devout Methodist, to whom he willed some slaves and who hired out Toby to the furnace. Many Northampton slaves ran away. In 1791, "Seaser" fled six days after the ironworks bought him and Nicholas Jessop caught and returned five fugitive slaves—Bateman, "Company's Charles," and three other of the "Companys Negroes." Charles Carnan Ridgely footed half the bill for "150 Advertisements for the Co.s Seaser and your Neg.o Ben" the following year. Some probably sought refuge among Baltimore's black residents. In 1796, Jessop claimed "a Runaway Negro" from the city's jail. Three months later, the ironworks paid "for taking up Negro Tom in Baltimore."[16]

Northampton slaves were not the only slave ironworkers who awaited freedom in vain. No Chesapeake adventurer manumitted his slaves while in the iron business. In 1791, Robert Carter began to manumit hundreds of slaves, four years after he sold his stake in the Baltimore Company. Isaac Zane openly supported antislavery campaigns while he owned northern Virginia's Marlboro Iron Works, but he never freed anyone. In 1794, he died master of twenty slaves: eleven adults, six girls, and three boys. Zane's will sentenced the adults to more enslavement. The girls were to be manumitted when they were twenty-eight years old, the boys when they turned thirty-one. The same bondage awaited any children who might be born to slaves Beth or Hester. Fortunately, Zane's sister Sarah, who lived in Philadelphia, bought all of his slaves and black servants, shielded them from his creditors, and liberated them all within two years.[17]

Harry tried to capitalize on his connections to his former owner Sarah Zane and to local Quakers. In 1798, he successfully lobbied Hugh Holmes, the executor of a nearby estate, to purchase his wife Molly and their five-year-old son before their new master could take "her much agt. her will to the Lower parts of Virga." Holmes promised Harry that if Sarah Zane would buy his wife and child, Molly would serve seven years and his son would be freed when he turned twenty-one. "Harry says," Holmes added, "that some of the friends would hire her and child & he wd. work out the ballance in a year or two." Harry wanted to reunite his family on lands that Zane had set aside for him and for the others whom she had manumitted.[18]

Those lands led to conflict between Zane and Marlboro's freedpeople. In 1802, Robert Mackey informed Sarah Zane that "the Negroes on the Barrack Land have lately been much engaged in selling logs for sawing off it." Months later, he reported that the Barrack Lands were "so compleatly occupied by the Blacks that no person will bid for them. I am informed (for I never go there) that they have settled on the line between lots No 10: & 15, . . . so as to embrace both Lots, & use no other industry but selling Timber. You had better give them half the Land & circumscribe them," Mackey advised, "than let them wander over the Whole & destroy the Timber." Mackey had found takers for four lots "but on wishing to take possession the Negroes claimed them, & informed them that you had given them seven lots, & before you went away had said they should all have lots that they were now twenty one in Number & must each have one." One buyer expressed interest in purchasing "the improvement on Lots No. 10 & 16, if he thought he might do it safely. I told him I would not concern in moving or disturbing them. He said the Blacks told him I had nothing to do with the Business." Mackey agreed. When Zane authorized him "to remove or restrain the Negroes," he declined.[19]

The standoff between Sarah Zane and Marlboro's freedpeople exposed the tension between her benevolence and their desire for independence. Zane wished to sell the land for the highest price she could get. She had provided for the people she had manumitted and they had "stolen" twice from her: they cut and sold timber that she considered hers and they lowered her land's value by deforesting it and scaring off buyers. It also probably bothered her, as it did Mackey, that her former slaves had not chosen the path of improvement that many Quakers prescribed for African Americans and American Indians whom they sought to help.[20] Rather than establish productive farms, Marlboro's freedmen marketed timber, which demanded considerably less time and energy. Their conduct raised other concerns for Mackey and Quakers. After Gabriel's Conspiracy of 1800, most white Virginians eyed emancipation, free blacks, and the Society of Friends more suspiciously. To them, the standoff at Marlboro would have underscored the danger posed by all three.[21]

Harry and his neighbors clung to the land because they believed that they had earned it. It was compensation for their enslavement. Why should they improve their plots? Zane held the land in trust; they did not own it and would never recoup the value of their labor. Cutting timber allowed them to be as independent as possible with minimal effort. This was a sure sign of freedom to people accustomed to toiling for others for little reward; as was defying Sarah Zane. Besides, what did they care if she had trouble finding buyers? Fewer white landowners nearby meant more autonomy and more resources for them. Marlboro's freedpeople defended their lands and their right to timber because this was the best way to stay together and preserve what freedom they enjoyed. It was becoming harder to be free and black in Virginia. If they left Marlboro, they would sacrifice their autonomy and their community.

Marlboro's freedpeople placed a premium on staying together. In 1810,

Simon and Bristol, two of the oldest black men, were each promised twenty dollars "if they would rent places where they chose distant from" the Barrack Lands "provided there was only themselves with their wives—and they would not comply." The people whom Sarah Zane had manumitted also had to look after their dead. They insisted that the boundaries of a lot set aside for them be drawn to include "where Harrys brother Charles was buried." In 1816, Fleet Smith, a buyer of several lots, promised "that the grave yard shall never be disturbed by my self heirs or assigns forever." The living could not abandon their dead.[22]

How committed Zane was to their community was another matter. Her 1812 will reserved 100 acres in trust "for the use of Simon, Bristol, and Harry, . . . it being land whereon they reside, during their natural lives." It also gave 500 adjoining acres to three Quakers to improve and devote "the annual income, or interest to relieve the real Necessities" of Marlboro's freedpeople. The "3 old ones among them" had first claim to help. The others were "if in real Necessity . . . to be assisted, upon producing a copy of their bills of manumission." Zane made no provision for the children of those she named. Their hold on Marlboro was to slip after everyone she had freed had died.[23]

In 1816, Marlboro's freedpeople were still felling trees, even though Zane had "not given any one liberty to cut one stick." Dick and Venus were the most defiant. Dick dealt in lumber and charcoal and had acquired a reputation as a hog thief. Worse, he and Venus kept "a *house of bad fame*." Fleet Smith reported, "it being as I am credibly informed the resort of the lude, & Rogues of all colours ages & Sexs from Winchester and elsewhere & *shame to tell* some amongst the numbers call themselves *Gentlemen*." Sarah Zane sold most of the Barrack Lands to Smith. He and Venus agreed to preserve her lifetime interest in one lot. Smith hoped that removing Venus and Dick would "eventually tend to their reclamation . . . especially if they should live with Jacob Baker or any other respectable white man." By then, Zane viewed most of Marlboro's black residents as ingrates. In 1819, she revoked her previous wills and cut out all but Venus and Daniel. They might receive support, "shou'd it be necessary," from $500 that Zane left to the disposition of three trustees.[24]

Zane's final will reminded Marlboro's freedpeople of the limits of her benevolence and of the constraints on their freedom. They had done what they could to preserve their community for as long as possible. They had gained their liberty, strived for independence on their terms, and for a time forced local whites and their former mistress to accommodate them. Marlboro's freedpeople owed their freedom and the constraints on it to revolutionary consequences. But they were the lucky ones. For most African Americans, the Revolution and industrial employment reforged their enslavement.

Hirelings

In January 1829, forty-four slave men crossed the Blue Ridge Mountains and headed for the Bath Iron Works, where they joined ten others whom William Weaver had leased for the year. They were among hundreds of slaves who marked the new year by leaving home and family and heading to work for the iron industry and among thousands who took on temporary employers that year. Most went home for Christmas, spent at most two weeks there, and then went back to work. Many who Weaver hired did not return to Bath, mostly because they and their masters opted for employment elsewhere.[25]

The system that brought so many slaves to Bath represented important changes for the iron industry, for the industrious revolution, and for slavery. Hired slave labor was not new—colonial Chesapeake ironworks were among its most significant patrons. After the Revolution, a regional system arose to match masters with entrepreneurs who wanted to lease slaves. By the early nineteenth century, a sophisticated market annually circulated thousands of slaves between masters and employers. They constituted a flexible, mobile, relatively affordable unfree labor force that gave employers respite from the demands of free white labor.[26]

Hired slaves were especially attractive to Chesapeake ironmasters after the Revolution. They allowed adventurers beset by competition from northern and foreign-made iron to stabilize their labor force each January and adjust its size the following December. Most who entered the region's iron business after the Revolution were transplanted northerners. Few owned slaves when they became ironmasters; most needed years to accumulate the capital to buy many. Leasing permitted them access to slave labor and it gave them experience in managing large numbers of slaves. The hiring system had its drawbacks for adventurers. It bound them to a labor market over which they had little control and so shaped how they organized iron production. It also required flexibility, diplomacy, and patience of ironmasters, largely because demand for leased slave labor often empowered slaves as well as their masters.[27]

This echoes what many historians who have recently studied the practice of hiring slaves have argued. Some scholars, particularly those of urban slavery, have contended that hired slaves gained enough autonomy

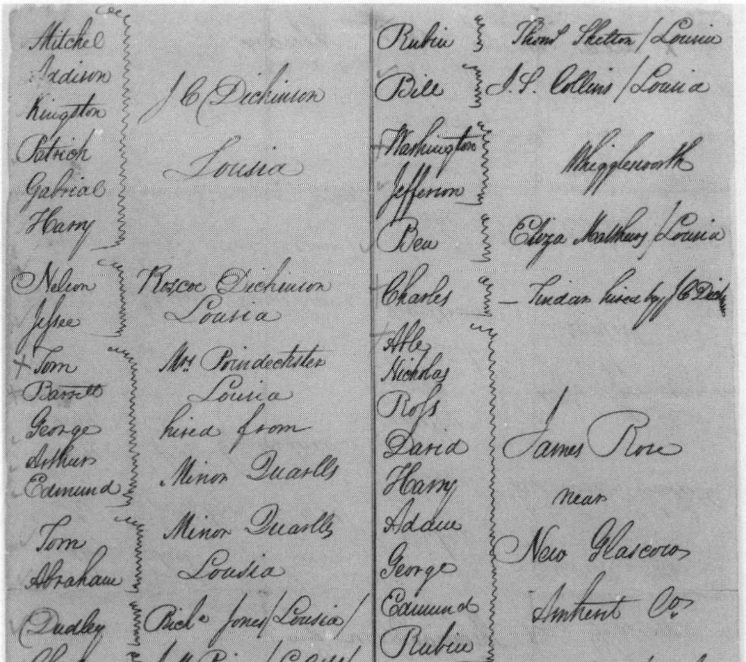

Mitchel, Addison, Kingston, Patrick, Gabriel, Harry — J C Dickinson Louisa

Nelson, Jesse — Roscoe Dickinson Louisa

+Tom, Barrett, George, Arthur, Edmund — Mrs Poindexter Louisa hired from Minor Quarles

Tom, Abraham — Minor Quarles Louisa

Dudley — Richd Jones/Louisa/

Charles — J M Price (L C H)/

+Henry, Anthony, Thornton, Davy — Coln Crutchfield

Charles — Miss Susan Harris/Louisa/

Jerry — George Tidal /Louisa/

Davy — Henry Dickinson/Caroline/

Prince, Jacob — Tindar Louisa

Harry, Davy — John S May Louisa

Rubin — Thoms Shelton/Louisa

Bill — J.S. Collins/Louisa

Washington, Jefferson — Wigglesworth

Ben — Eliza Matthews/Louisa

Charles — Tindar hired by J C Dick

Abe, Nicholas, Ross, David, Harry, Adam, George, Edmund, Rubin — James Rose near New Glasgow Amherst Co

Emanuel — Sam Turner/Amherst

Timothy, Dick, Washington, Anderson, Sam — Mrs Spencer Amherst

Blake — S. Pendleton/N. Glasgow

Edmund — Mrs Crawford/Amherst

Frederick + — Henry Knight/N Glasgow

Brutus, Dan — Wm A Caruthers Lexington

Figure 10. "Memorandum of hired Negroes at Bath Iron Works for 1829." in the William Weaver Papers, located in the Duke University Rare Book, Manuscript, and Special Collections Library, Durham, North Carolina. This list includes hired slaves' masters and their home counties in Virginia. Men with a cross by their names died in a dysentery outbreak at the Bath Iron Works.

to challenge and weaken their bondage, if not the institution of slavery itself.[28] Hiring often enabled slave men to redefine their enslavement by asserting themselves as participants in negotiating the terms of their hire. Employment under temporary masters offered slave men opportunities to earn income and to act as providers to those who awaited their return each December.

But the rewards of getting hired out exacted a high price. The ability to lease slaves' services discouraged masters from pursuing private manumission. The hiring system restricted the role of slave men and it reinforced the authority of women within the families that stayed behind. It also imposed a cruel irony: the Chesapeake's market in hired slave labor became one of the principal means by which slave men could keep their families together. Were it not for the income that they earned for masters, they, their wives, or their children would likely have been sold off. If the industry of hired slaves preserved slave families, particularly after Virginia tightened its private manumission laws in 1806, it also preserved slavery as a vibrant institution that could be molded and reshaped to fit the needs of the region's farmers, artisans, and manufacturers. Slave hire strengthened slavery, and it encouraged white workers to think of themselves first as white and second as workers whose welfare depended on exclusion of blacks.[29]

The annual rhythms of slave hiring structured the routines of ironmasters and of ironworks. It took careful organization and networking to lease slaves. Each autumn ironmasters started recruiting for the following year. William Weaver devoted about one month to finding and contracting for the nearly sixty slaves he leased for the Bath Iron Works in 1828 and 1830. Furnaces particularly had to synchronize with the hiring calendar. Etna Furnace assistant manager James Brawley estimated that "there is from two to three weeks of time lost every winter" when hired slaves returned home. In 1830, managers of Bath considered running the furnace "through the Christmass holy days and going on as long as possible" but with few remaining hands they decided to "stop up for a short time during Christmass and blow as long as possible."[30]

Ironmasters enjoyed some advantages in competing to lease slaves, largely because they could address masters' concerns and accommodate slaves' aspirations. The iron industry promised masters more security than did urban employers, and it had long offered hired slaves opportunities to gain more autonomy. Urban jobs attracted many slaves because they could hire out their services far more easily in cities. The iron industry's overwork system was a form of self-hire, but it was one that was easy to monitor. Employers in cities often permitted leased slaves to find their own housing, often among free blacks. Ironmasters housed slave employees and kept them under closer supervision. Hired slaves employed as colliers, woodcutters, or teamsters likely experienced less oversight while on the job than did slaves who worked in urban factories and workshops or on small plantations.[31]

Still, adventurers found the process of hiring slaves frustrating. "This system of hireing negroes is a very bad and uncertain one," Thomas Mayburry complained in 1817. He had discovered what all employers of temporary slave labor soon learned: they could not reliably foresee who they could hire or how much they might have to pay for slaves' services. Availability and cost of leased slave labor fluctuated with tobacco and wheat prices. Adventurers competed with urban artisans, farmers, canal builders, and other manufacturers for slave employees. In January 1828, William Weaver noted that "several persons have already sent more than I expected. This will be at least 60 Negroes—and the probability is that there will be several more, so that it will not be necessary for you to hire any in the Neighborhood unless it is Clarkes man." The ironmaster's luck ran out. Within a year rising produce prices had pushed the cost of leasing slaves "higher than you are in the habit of giving," James C. Dickinson reported, "& I have made no positive engagement." A month later John Wigglesworth echoed Dickinson—"I could not hire any Negroes for you. Those I expected you to get went over $50 which I considerd your limit." In December 1830, John Chew expected that "the price of negro men will advance" around Fredericksburg because hundreds were "required for the Gold mines in this section of the country. Should that be the case," he asked Weaver, "will you be willing to give as much as others?"[32]

Thomas Mayburry was right: hired slave labor was often a headache for employers. That was a consequence of taking on temporary workers who were someone else's property. Leased slaves lacked powers that courts generally attributed to free workers, including the ability to make legally binding contracts, to leave a job, or to look after their own interests or those of fellow employees. All made hired slave labor far more appealing than free labor to most southern industrial entrepreneurs. But engaging employees who had so few legal powers brought risks. Courts throughout the antebellum South consistently safeguarded the property rights of masters in hired slaves by ruling that employers bore most of the responsibility for what slave employees did or for what happened to them. By 1836, Virginia courts had concluded that employers assumed liability for leased slaves and for what damages or injuries they caused to others or to themselves. In short, hired slaves posed dilemmas that employers never faced with free workers.[33]

One dilemma was that leased slaves recognized that they were property and used that knowledge to shape and even direct the hiring process to serve their interests whenever they could. To be sure, they often had no say over whether they would be leased or over who would lease them. In 1794, slave master Rezin Hammond decided that Ben and Will should spend several months at the Northampton Iron Works because he wanted them "accustomed to more Obedience and Discipline than they have hitherto been used to and no Place is so proper to effect a Reformation as one at a Distance from their Connections and where they can be constantly confined and their Labour steady and uniform." To Hammond, a stint as ironworkers would make Ben and Will better slaves. They would learn to work hard, to value the lives that they had, and to respect their master's power.[34]

Most of the slaves William Weaver aimed to hire had far more room to negotiate than did Ben and Will. Demand for their services often enabled slaves to bargain with master and employer. In 1828, Brandus refused to work for Weaver because he wished to remain with his current employer. His master refunded the bond for his hire and he reminded Weaver that the deal had always hinged on Brandus's approval; he could not "think of compelling" him "to go any where it is not his wish, as that has always been my rule, as I told you he could have his choice."[35]

Leased slaves sometimes determined which jobs they would perform. In 1825, John Jordan hired Davy, Charles, and John. Their master noted that "Davy does not wish to blow Rock and I promised him that he should not" and he underscored the point by dropping that "I was offered several Dollars more for Davy & John." Jordan honored Davy's request. Four years later, Davy returned and again tried to limit where Jordan could employ him: "Davy Says that Working in the furnace is ruinous to his Eyes, therefore I do not Wish him to work there against his will." Nancy Matthews was "very unwilling" for Phill "to work in the ore or blowing Rock as he has been so much injured by it and he is very dissattesfied at it—but he is willing to work at anything that thear is not so much danger." Davy and Phill knew Brooks and Matthews had their investment to protect. They capitalized on their status as property to assert some control over their bodies, their time, and their industry.[36]

Adventurer and master had reason to heed the wishes of hired slaves; they might flee and deprive both of their services. In 1829, Abram quit the Bath Iron Works and went home early. His master, W. E. Dickinson, sent Abram back with apologies for not returning him sooner and for his "almost unpardonable" conduct. Since Abram's "principle complaint seemed to be lodged against your overseer," John Doyle, Dickinson asked that Weaver "put him to cutting wood"—a task that would permit Abram to work with little supervision.[37]

Abram was the first of many hired slaves to speak poorly of Bath. A dysentery outbreak that killed six hired slaves in 1829 may have been the final straw for many. Davy Carter refused to let Weaver employ him again. In March 1830, Ben returned home to Elizabeth Mathews nine months early. He left "in consequence of not being fed well & other harsh treatment & If he dont suit you," she promised, "I will take him back again if you will pay for the time he has been with you." Two months earlier, William Staples informed Weaver that Sam "was unwilling to return" to Bath, "but says he would have no objection, provided, he could live at your own establishment." Like Davy Carter, Isaac wanted no more of Weaver. He "expressed such an unwillingness to return to you" that Staples "feard should I send him over he would run away, and perhaps be of little or no service to you during the year." Isaac persuaded his master "to hire him in Amhirst where he is willing to stay, for the same you were to give."[38]

By then, word of life and death at the Bath Iron Works had made it over the Blue Ridge Mountains and spread through slave networks. Robert Crutchfield reported from Spotsylvania County that "I have not been able to get you any hands in consequence of their being unwilling generally to go over the mountains (as they call it)," largely because "some of the hands (I know not whom) have made somewhat an unfavorable impression on the negroes in the neighborhood as to the treatment at the place, for which they are wanting." Weaver had to address conditions at Bath and make sure that prospective employees learned about those changes before he could enjoy better luck at engaging slaves to work there.[39]

Leasing skilled slaves was even more difficult for Weaver. By 1827, he owned six forgemen, having realized that "it is all important particularly in this country" since he could seldom find such slaves for sale or for lease. In 1829, he tried to renew the contracts of some young slave forgemen early. Their master, Moses McCue, warned Weaver that he should not count on their services in future because "I shall be at liberty to take the highest offer made me. I think *you* would do the same." McCue asked Weaver to submit his best offer "so that I may act consistently with other applicants." Four years later, Weaver bonded for three slave forgemen. His former partner Thomas Mayburry then "bribed both negro & master, hired the main hand that you wanted, and left the other two for you." The experiences only hardened Weaver's resolve to staff his forges with slaves whom he owned.[40]

Weaver's decision to minimize his forge's dependence on hired slaves illustrates one way in which the hiring system shaped slavery throughout the region's iron industry. Because a disproportionate number of hired slaves were men, they exacerbated the gendered division of labor that already existed within industrial slavery, as slave employees filled jobs that slave women once worked or might have worked. Fragments of circumstantial evidence suggest as much: the Oxford Iron Works employed few hired slaves and it seems to have involved far more slave women directly in the process of making iron than other ironworks did. This was not completely negative for women whom ironmasters owned. It spared them some of the worst jobs that the iron business had to offer. But it also barred them from most avenues to earn income, confined them to a narrower range of jobs which brought them under closer supervision than slave men, and prevented them from working alongside husbands, fathers, brothers, and male children. In short, slave hiring, a product of the Chesapeake's economic diversification during the revolutionary and early national eras, reinforced on ironworks what it was doing throughout the region—expanding job opportunities and prospects for autonomy for slave men while constricting both for slave women.[41]

Slave men who worked in the Chesapeake's iron industry gained tangible benefits from slave hiring. Slaves who stayed behind when hired slaves headed home each winter sometimes could demand a break. John Jordan stopped Bath's furnace when the hired slaves left, largely because the slaves who remained were "not willing to be closely confined." The system of hiring slaves improved all slave ironworkers' access to networks that slaves used to communicate. Those networks probably also enabled slave ironworkers to influence their relations with ironmasters. Adventurers knew that their reputations mattered when they wanted to recruit slave employees; those reputations depended on what hired slaves said, heard, and observed. Who knew them better than the people whom they owned? Permanent slave ironworkers could exercise some influence over the market in hired slave labor by indirectly shaping how adventurers related to them. Commodification of temporary slave labor, in other words, might have enabled slave ironworkers to compel ironmasters to do more to honor their paternalistic view of themselves in deed as well as in word.[42]

Hired slaves helped to transform industrial slavery in a far more significant way: they wanted and received cash for overwork. Most leased slaves who had positive balances carried away paper money when they settled accounts before journeying home. By 1800, the slaves whom ironmasters owned also received cash for their labor. In 1791, Northampton Furnace's slaves began to earn cash regularly for overwork or for extraordinary service, as did Caesar for "Molding Open sand Castings" and for moulding in December. Nearly every slave who performed overwork at Ridwell Furnace drew his pay in cash. By the early nineteenth century, it had become customary for slaves to choose between cash, goods, or some combination of the two when it came time to settle accounts. Nearly all took away some cash, if not exclusively cash. In 1837, fifty-three slaves collected cash from William Weaver.[43]

Slaves liked cash because it allowed them greater autonomy. Cash permitted them more flexibility to consume what they could afford and to buy from whoever would sell to them. In an economic culture characterized by transactions on credit between people who knew one another intimately, cash permitted more anonymity. For slaves this meant that they did not need to buy from their master or use personal notes, the value of which hinged on his reputation, if they purchased from someone else. Currency offered slave men more opportunity to establish an identity independent of their masters and more ways to reap the fruits of their labor. It potentially limited what they owed white men because it enabled them to pay off debts quickly rather than continue an extended relationship with a creditor. Together, these factors allowed slave men to participate in a market economy on terms that they found more favorable. It was no coincidence that masters who wished to direct slaves' economic activities tried to keep cash out of their hands.[44]

Hiring enabled many slave men to achieve a bit more control over their lives. As they asserted themselves through participation in negotiating where and for whom they would work, they gained access to a wider variety of jobs and a sense of their own value. Their mobility, dramatized by annual journeys to and from work, distinguished them from other slave men and from the slave women who had to stay home. Their journeys made them conduits through which information passed between slave communities and through which slave men at home learned about prospective employers. Most significantly, the cash they earned from overwork allowed hired slave men to provide for their families, a role that they could assume when they returned home each year to celebrate Christmas. The Christmas holidays allowed men to reunite with their families, to give their families money and gifts, and to compete with masters who distributed presents to slaves at Christmas.[45]

However, the Christmas holidays also gave slave men more time to reflect on what the hiring system cost them. They spent all but two weeks a year away from home. Although getting hired out and hiring themselves out by undertaking overwork allowed them a chance to realize the value of their time and labor, it also reminded them that masters and employers confiscated most of that value. Getting leased was certainly better than getting sold off. A week or two with family each year was better than a lifetime apart. But such options underscore just how little power slave men had to protect their families or to determine their futures. Hiring reminded them that they were property, commodities whose lives could be broken down and rented out in days, months, or years to the bidder whom their masters considered most suitable. Leased slaves also inadvertently made slavery more politically viable by increasing the number of people who benefited directly from it.[46]

They also allowed employers to get by with fewer free white hands. That was largely why adventurers and so many other southern entrepreneurs rented so many slaves. In 1826, Ludwell Diggs, manager of Cloverdale Furnace, echoed most Chesapeake ironmasters and most southern entrepreneurs when he answered the question "Are not works carryed on with much more facility with Slaves than with white hands?" with "I would rather have Slaves; they are less trouble."[47]

Diggs's declaration would have resonated with colonial adventurers like John England or Charles Carroll of Carrollton. But for them, "white" labor incorporated bound labor and free labor. After the Revolution white labor increasingly meant free labor. In 1795, Virginia ironmaster Francis Preston noted that "as to hiring negroes I am more than ever engaged in it as Servants here cannot be procured by any means." Leased slaves plugged a niche in the early national Chesapeake's bound labor market that transported felons and white indentured servants had once occupied. They became the region's pool of unfree temporary labor from which entrepreneurs could draw when free white workers proved too scarce, too expensive, or too assertive for their liking.[48]

Chesapeake adventurers, especially those with personal and business connections to the North, hoped to attract white workers from Pennsylvania and New Jersey. Many were sorely disappointed. William Weaver and his agents spent years trying to recruit Pennsylvania ironworkers, especially forgemen, to little avail. In 1829, John Doyle reported that a hammerman would soon arrive at Bath Iron Works from Pennsylvania. "When this man gets here," Doyle exulted, "the ice will be broken and there will be no dificulty in geting what Forgeman we want." The ice never broke. A long move south was burdensome and it stretched the networks through which free white tradesmen found new jobs.[49]

To most Chesapeake ironmasters, free white workers from the North seemed more reliable than local whites. David Ross recruited four white tradesmen from Connecticut rather than look around Virginia. In 1825, William Weaver stated that "no reliance could be placed in the free White laborers who are employed about Iron Works in this country." They were "generally very poor and in moments of greatest pressure & necessity, the proprietor must either make them advances which they will never repay, or they leave his service to the ruin of his business." Subsequent reports from the Bath Iron Works did little to change Weaver's views. In January 1828, founder Dixon Hall agreed to "Keep sober and Steady and on no account absent himself from the Furnace day or night" except at meals or after alerting manager John Doyle that he would be gone for one or more days. Nine months later, Doyle fined Hall twenty-five dollars for getting drunk and neglecting the furnace. In 1829, Doyle charged teamster John Mellon ten dollars for "neglecting [to] hall his coal by which the Forge was Stoped one day" and collier Isaac Peterson sixty-five dollars: ten dollars for sending off teams without charcoal and fifty-five dollars for wood destroyed because he had neglected his pits. Doyle derided the white forgemen who worked under him as "consummate bunglers." They included a hammerman who "last week drew but one Journey each day and this week he draws none, pretending to be sick nor is there the least prospect of his doing better." Weaver responded by gradually replacing white forgemen with slaves at his forges. No white forgeman drew bar iron at Buffalo Forge after November 1840.[50]

For Weaver, the issue was control—slave labor allowed him more of it than free white labor did. So it was for David Ross, who pointed to smith shop supervisor Thomas Gray as an example of what might happen if Oxford employed more white tradesmen. Had the ironworks been "entirely dependent on Gray or at all dependent upon him," Ross asserted,

> you may be assured he would have lain down his hammer & would not have taken it up again untill every thing he ask'd for was complied with, but he wisely considered & well knew that he was of no consequence beyond his own labour, and that his absence would not have been felt for it would not have stopt a wheelbarrow. This is the true reason he did not come down as he threatened. . . . 'Twas his own insignificance in the scale of the iron works that prevented it and not any favour to the estate.

Ross was even more explicit when he castigated Reuben Smith, one of his plantation overseers.

> Take I say again and again one of my most faith[ful] servants, give him some encouragement or a fourth part of what you must give a white lad. You'll find him ten times better than any you can hire. He will labour day by day. . . . He will receive your instructions with patience & humility & if a reprimand becomes necessary he will receive it without pulling out your eyes.

To Ross, whiteness was antithetical to the discipline that his enterprise required, so he did his best to exclude white workers from it.[51]

Ross's views beg key questions: What impact did industrial slavery have on white workers after the Revolution? How did white workers view industrial slavery and industrial slaves? It seems certain that white ironworkers benefited from slavery less than they had during the colonial era. Rising prices, especially for skilled slaves, put mastery beyond the grasp of most white ironworkers of the early republic. Men like John Cauldwell, a forgeman who owned Jerry and leased his services to Pine Forge, were rarer than they were before the Revolution. To judge from surviving account books, white employees seldom hired slaves from ironmasters by the early nineteenth century.[52]

By the 1830s, more white workers had begun to conclude that slavery—or to be more exact, slaves for hire—threatened them. The low cost and flexibility of hired slave labor encouraged many employers to replace white hands with leased slaves. The engagement of slaves to construct the U.S. Navy's dry dock at Norfolk, Virginia, led the white stonecutters whom they displaced to petition the President and the Congress to exclude hired slaves from federal projects. In 1847, white puddlers struck the Tredegar Iron Works after its owners trained hired slaves to staff a new rolling mill. In neither case did the white workers oppose slavery. Instead, they explicitly cast their grievances in racial terms—that employers were leasing black slaves to deny white men a living. White racism, it seems, figured among the wages of slavery for hire.[53]

By 1830, some white ironworkers were expressing doubts that slaves could master their trade. William Norcross, a hammerman from New Jersey who drew iron at Buffalo Forge, contended in 1826 that Bill Hunt was not "a first rate work man," though he conceded that Hunt was "a first rate underhand for a Slave." Norcross's view of slaves did not improve with time. In 1840, he asserted that they "were not as good workmen and do not take as good care or as much pains as white workmen." To Norcross, it seems, the issue eventually boiled down to race—slaves could not hold their own with white men.[54]

By the 1830s, some of Norcross's peers had resorted to violence against slave ironworkers. In 1836, a riot erupted at Maryland's Antietam Iron Works. According to Antietam's owner, John McPherson Brien, who related the story to Thomas Henry but did not witness the incident, the conflict began after some slave ironworkers exchanged words with his agent and some white employees. Enraged, they threatened to seize the slaves and whip them. The slaves stood their ground; a fight broke out;

and the manager summoned the militia. The slaves fled and hid until Brien returned.[55]

Brien believed that his appointment of James Reeder, "a very faithful colored man," to supply the puddlers triggered the incident. Reeder's promotion honed the "spirit of animosity" that "the white help" bore for Brien's slaves "because they were so well treated." They, Brien claimed, enjoyed "more privileges than any white man had on his place." He offered them "all the refuse from his mills, which was an immense quantity of fuel, which no white man on the premises could disturb." Brien insisted that whenever "his white employees wanted any work done, it should be done by his men's wives, that they might make all the extra money that could possibly be made." Each Saturday night slave ironworkers lined up with white employees to receive their overwork pay.

According to Brien, his enlightened mastery inverted the order that Antietam's white workers considered natural. In seeking the greatest possible control over his enterprise, he distinguished between workers on the basis of race, and white employees unexpectedly found themselves disempowered and deeply resentful. Brien's demand that they hire slaves' wives, many of whom were free, deprived their households of income. James Reeder was the last straw: a black slave who wielded authority over them dramatized their dependence on Brien too vividly. They turned on Antietam's slave men when he was gone. When the slaves fought back, they tried to invoke whites' right to bear arms to punish them. To Brien's delight, which his slaves surely shared, they failed. They failed because the slaves empowered Brien to turn aside their challenge, just as white workers throughout the antebellum South often lost when they opposed employers' right to own slaves, hire them, and employ them as they wished. They also failed because of the new relationship that had developed between slave ironworkers and their masters, one that bound men like John Brien far more tightly to his slaves than to his white employees.

All in the Family

When John Brien told Thomas Henry about the battle at Antietam, he was the story's center—everything revolved around his policies, his decisions, his absence. He was the indulgent master who granted his slaves authority and autonomy. He shielded them from the militia after they streamed down from the hills "coming to me like wild cattle." Brien reminded his agent that "no man had authority to strike any of his hands, and if they have done anything that conflicts with the law, I will settle that myself" and then convened his slaves "and settled with them as he thought best." The ironmaster almost certainly tailored his account for Henry's ears. Henry, a freed slave and an African Methodist Episcopal minister, had preached throughout western Maryland for years. Brien's father gave him "a present of a church" at Catoctin Furnace and Brien welcomed Henry to Antietam by telling him that "I am very glad that you have come among [my slaves] to teach them the way to live." He invited Henry to come and go at will and he instructed James Reeder to "hoop-polp" any man who "misbehaved" at Henry's meetings.[56]

Brien emphasized another point to Henry: he was master of accomplished and assertive men. He entrusted Antietam to "a very fine set of young men." They determined where and how they would build their houses. Brien encouraged them to marry free women, which meant that they might father free children. His insistence that white hands hire his slaves' wives generated income for slave households. Antietam's slave ironworkers were strong—Brien delighted that his white hands had learned the hard way that "his men could not be taken." They seized the opportunities available to them to stake their claim as men to roles that enslavement often denied them.

They did so within clear limits. Brien encouraged and praised their success because he believed that it reflected well on him—it proved what an enlightened and benevolent master he must be to have such dependents. Brien lauded the "privileges" that Antietam's slaves enjoyed and its white employees did not, but the whites were free to leave. The slave ironworkers were his "boys" when they returned from hiding; they needed his protection from the militia and the law. Brien would neither protect them from his creditors nor grant them freedom. In 1841, he mortgaged fifty-four slaves. Seven years later, he went bankrupt and sold Antietam and the forty-nine slaves who lived there. After learning of the sale, many approached Brien, told him of their "unwillingness to remain with me," and threatened to flee to Pennsylvania. Brien thought he was the aggrieved party. His slaves were guilty of "gross ingratitude" because he had "*always* treated them most kindly."[57]

John Brien's relationship with Antietam's slaves reveals how Chesapeake adventurers after the Revolution promoted and exploited the domestication of industrial slavery. Colonial ironmasters rarely expressed concern for slaves' welfare or souls. To Brien and his peers, slaves were more than assets or economic actors; they were their "people," their "men." They believed that they owed them something and felt obliged to accommodate their unquantifiable aspirations: for stable families, for more autonomy, for salvation. The paternalism that emerged at Chesapeake ironworks was a tacit bargain founded upon the needs of adventurers and slave ironworkers to assert and define themselves as men. For masters, it reflected their desire to care for and transform their slaves as part of what they considered an enlightened enterprise. Slave men saw in industrial paternalism room to assert more autonomy at work and at home. The limits of paternalism, however, were clear. What masters gave they could take away. What slaves painstakingly built masters could destroy. Slaves remained commodities and outsiders. Their masters aspired to make them better workers, better slaves, and perhaps even better people with fuller lives, but never neighbors or citizens of the new republic.[58]

The new world that adventurers and ironworkers made together stemmed partly from slave demography. Industrial slavery became a domesticated institution in two senses after the Revolution. The end of the Atlantic slave trade ensured that nearly all slave ironworkers would be born and raised in North America. Second, families became an integral feature of industrial slavery. Both redefined how slaves viewed their work and their relations to their masters by recasting and grounding slave manhood in the roles of husband and father. Both encouraged ironmasters to reconceptualize their relations to their slaves and to accommodate new demands from them, a process which gave new meanings to the industrious revolution.[59]

As before the Revolution, overwork was the foundation of adventurers' efforts to make slaves more industrious. Ironmasters still considered it the best way to adapt slavery to industry; slaves, if anything, embraced overwork more eagerly, though for new reasons and with new expectations. Those expectations, rooted in slaves' desire for autonomy within family and community, prompted slaves to renegotiate and sometimes win new terms for overwork, such as the right to demand payment in cash.

The flexible use of time that overwork encouraged allowed adventurers and slaves to dodge confrontations. In 1832, Bath Iron Works manager William Davis claimed that Davy shirked his duties because he drank. Davis had few options for disciplining the teamster, having learned that "since I reprimanded Davy for drinking and carrying whiskey on the place here that he is very apt to make for the woods when taken to task severely for misconduct." Though Davy missed loads during the week, he made up the work on Sundays (probably without pay), so Davis decided to let the matter slide. He promised Weaver that "there will be an eye kept" on Davy "in the score of his drinking &c.," but did not pledge to take any action against him. Davy's Sunday work allowed both men to wriggle out of a potentially worse predicament. Davis avoided losing Davy's services indefinitely; the teamster knew the woods and the surrounding area far better than he did. Davy evaded a dressing down or worse from the manager.[60]

Ironmasters held the upper hand in such negotiations, and they imposed new conditions on overwork. Before the Revolution adventurers often paid slaves for extra or unusual service with alcohol. They still did so sporadically into the early nineteenth century. In 1783, Northampton Furnace paid six slaves with whiskey, including Lux's Dan "for bringing home Goaller Tom." A few years later Harry earned three pints of rum for "Cutg wood Over his task." Cumberland Forge gave Nat whiskey "for Taking Care of the Forge in the Carps. absence." Thereafter the practice nearly ceased. Drinking threatened productivity and ironmasters had little reason to promote it.[61]

Adventurers sometimes converted overwork into forced labor. David Ross calculated that Oxford Iron Works teamsters cost him nearly $7,000 a year because, he believed, they wasted up to one-third of their time. He mandated that every driver report by 4:00 A.M. and work Sundays, for which each would earn the going rate for a day's labor. Ross left room for compromise: his overseer was "to consult with the most sensible of the Waggoners" and "if there be any thing unreasonable in the hour I have mentioned, I am willing on being so convinced that it may be altered." Work on Sundays, the issue that mattered most to the teamsters, was not negotiable. Ross acknowledged that he owed them for it. But in no other sense was Sunday "their time" to him. He compelled them to work on a day that slaves by custom spent as they wished. For the teamsters, getting paid for surrendering their day off perhaps only underscored their enslavement.[62]

Ross and Oxford's teamsters aside, the vast majority of slaves who participated in overwork did so willingly. If anything, changes within the iron industry after the Revolution may have enhanced overwork's voluntary character. In theory, ironmasters could have extracted overwork by skimping on slaves' rations or clothing, though most would have considered that counterproductive. Many also must have realized that deliberately worsening the material conditions of their slaves' lives would have undermined the paternalist rhetoric by which they justified enslavement and to which slaves tried to hold them. For slaves, overwork met their masters' demands and enabled them to pursue their own goals. By the nineteenth century, in the iron industry and on plantations, this meant creating and sustaining an internal slave economy which centered on families and focused on achieving the most liberty possible within slavery. For slave men, it often meant embracing the role of provider, which recast their relationship to the industrious revolution.[63]

Slave artisans stood in a particularly good position to define the role of provider on their terms. In July 1829, Lynchburg merchant John Schoolfield ordered two sets each of tire iron drawn three inches, two and one-half inches, and two inches wide and one-half inch thick from William Weaver's Buffalo Forge. Schoolfield instructed Weaver to "make Sol gage them or else people will not have them. . . . You may promise Sol that if he will draw Iron nicely to suit my orders that I will give him a beautiful Calico dress for his wife" for Christmas. To Schoolfield's dismay, Sol Fleming did not deliver, but he still expected the dress. In early December, Schoolfield demanded that Weaver tell the forgeman that he "had not forgotten the Dress I promised him but he has not done any thing for me to earn it." Sol Fleming "must not expect me to give him a Dress promised on a condition with which he did not comply."[64]

Why did Sol Fleming refuse to fill Schoolfield's order? Why did he request the dress anyway? Weaver stayed out of the matter. He could have ordered his slave to comply, but that might have offended him and caused his forgeman to draw shoddy iron. Sol Fleming wanted to give his wife the dress, but decided that Schoolfield's price was too steep. Schoolfield should have known better than to offer him so little for so much work. White forgemen balked at drawing iron to three-quarters of an inch thick and demanded extra money to do it because it took so long. Schoolfield wanted his order drawn to half an inch thick. Did Schoolfield think he would take the bait? Did he think that Sol Fleming did not know the value of his time and expertise? If the forgeman wanted a dress for his wife, he could have bought the material himself and paid someone to make it. He did not need Schoolfield's custom. But Sol Fleming had a point to make, so he asked for the dress without first filling the order. To the merchant, the request was exasperating and audacious. Did the slave not understand the deal? No iron, no dress!

Sol Fleming understood. He thought little of the deal and he wanted to bargain. Sol Fleming may have been a slave, but he was also an artisan. Schoolfield had placed a special order that only he could fill. He demanded respect. When he failed to receive it, he balked and left Schoolfield empty-handed. He may have known something of the merchant's predicament. By late September, Schoolfield had impatient customers who wanted tire iron in sizes that he did not have in stock. If the merchant needed the iron that badly, maybe he would send the dress as an advance? Sol Fleming lost nothing by asking, so he stood his ground and then advanced. Perhaps it was coincidence, but a few months later Schoolfield complained of the iron that Buffalo Forge had recently sent him. Over the next few years, Sol Fleming regularly bought cloth with his overwork earnings, some of which likely found its way into dresses for his wife.[65]

Schoolfield and Weaver were undoubtedly not Sol Fleming's only audience. Other slaves surely knew of Schoolfield's offer and of the forgeman's response to it. Sol Fleming's conduct was likely as much for them as for the white men directly concerned. Slaves seldom held the upper hand when dealing with white men. John Schoolfield was no stranger to many slaves at Buffalo Forge. Sol Fleming's defiance offered Buffalo Forge's slaves vicarious enjoyment and only heightened his reputation among them—a reputation that he had earned largely because he knew how to work iron.

The respect which men like Sol Fleming enjoyed likely had African as well as American roots. While the cosmology that surrounded ironmaking in Africa may not have survived in North America as a belief system, elements of it probably did. Blacksmiths formed a disproportionate number of slave conspiracy leaders and of itinerant preachers such as Thomas Henry who evangelized the region. Slavery may have perpetuated and even bolstered such beliefs. Forgemen and smiths commanded a rare skill which enhanced their value to masters and afforded them opportunities and status beyond the reach of most slaves. At the Oxford Iron Works, tradesmen who handled metal stood out by what they wore. Every smith, forgeman, and furnace hand received "a scarlet waistcoat with sleeves" and items made of blue cloth in addition to the garments allotted to other slaves. To David Ross, their distinctive clothing signified who he considered most valuable. To Oxford's metalworkers, their wardrobes displayed their skill and status to other slaves.[66]

Anglo American craft consciousness and African American traditions reinforced one another when slave ironworkers taught their skills to their sons. By the early nineteenth century, many slave ironworkers were at least second generation artisans. In 1811, "Little Natt" worked with his father Ben Gilmore in Oxford's forge. "Shop Phill" and Sharper instructed their sons Saunders and Edmund in the smith shop. William Weaver acquired skilled slaves who had boys in the correct belief that he would gain at least two generations of forgemen. Weaver's shrewd purchases illustrate the benefits that masters reaped from having slave tradesmen train their sons, not least of which was that it pleased the fathers. They supervised their sons and passed down their knowledge to them, which they could use to improve their lives and those of their families.[67] Training sons gave slave artisans a legacy which bolstered their claims to manliness. Slaves inherited bondage from their mothers. A career as a forgeman, smith, or master collier was what male artisans could impart to their sons which would distinguish them from other slaves. If they could not undo slavery for their sons, they might at least ameliorate it and stake a stronger claim for themselves as providers for their heirs.

Slave ironworkers usually trained their sons within workplaces over which they asserted considerable influence. Forgemen and smiths at Oxford, David Ross lamented, had too much control. He accused the forgemen of getting away with "much waste, much plundering, and great idleness" because the forge was "principally left to the management of unprincipled Slaves." Nor could Ross use drastic measures to address the situation because the "idleness" there was "entrenched & must be corrected by degrees." The Oxford smiths were, if anything, more zealous about determining how their shop operated. Thomas Gray, their white supervisor, resigned because "differences" between him and "the People" had made his job "disagreeable." Gray's troubles surfaced during an initiative to pinpoint problems within the smith shop. Accounts of work done over the previous ten months revealed that "one Shop did little by having a cholic, another did as little on account of the rheumatism, a third did less on account of sore eyes." Were the smiths digging in their heels by feigning sickness and nursing injuries? Ross thought so. He recalled that the last campaign to "reduce this part of the Estate into some order" had provoked "much trouble and vexation" because "the ignorant Servants opposed it."[68]

What made slaves "ignorant"? Why did they oppose his campaign? Ross faulted slavery as he practiced it. The isolation that he imposed upon slaves prevented them from altering their techniques. Oxford's forgemen "were as good workmen 20 years ago as they are now. They have had no chance to improve—they have not an opportunity of traveling to see other works and the annual improvements" and Ross did not allow any "travelling Iron Works people" to visit Oxford. The result was that "my people in every branch have remained as it were stationary."[69]

Ross surely had a point. For centuries white forgemen had circulated freely throughout the Anglo Atlantic world, exchanged insights, and diffused knowledge. Indeed, ironmasters traditionally relied on them and their willingness to make minute improvements to boost productivity and efficiency. By excluding his slaves from the industry's networks, Ross knew that he denied them and Oxford the benefits of those networks.[70]

Some in the iron business thought that slaves were inherently ignorant. John Doyle claimed that it cost more to forge iron with slave labor than with free labor because slaves could not understand the refining process and so wasted "more metal and coal to greatly overbalance the difference in other respects." When he managed the Bath Iron Works, Doyle recalled that a white forgeman, "a free man from the North," used half as much charcoal as Hal Hunt, whose father taught him at Oxford, to make the same amount of iron. Ross did not share Doyle's conviction that slaves could not comprehend the mysteries of forging iron, but he did believe that Oxford forgemen wasted fuel and iron and that his tradesmen were less productive than white artisans.[71]

Of course slaves were less productive, many economists and historians have argued. Higher productivity meant little to them because it would change nothing for them. Besides, masters invested so much capital in slaves that they had less to devote to innovation. Still, southern manufacturers adopted new technologies when they could afford them and thought they would be worth the expense.[72] So did ironmasters. William Weaver made improvements to Buffalo Forge's hearths while he was increasing the number of slave forgemen who worked at them. In 1831, the owners of Cloverdale Iron Works dispatched their manager and their most experienced slave finer to Buffalo to observe its refinery fires. They also asked Weaver to send someone to demonstrate the technique at Cloverdale or to allow their slave to train at Buffalo. Indeed, Cloverdale's adventurers may have sought improvements in part because they believed that their slave forgemen were shirking their duties. In 1832, George P. Tayloe, dissatisfied "with the work our black hands are performing," requested that Weaver help him find a white hammerman "(*who is first rate*)" to man his chaffery forge for least two weeks. Perhaps Tayloe thought the temporary hand would make his slave forgemen more industrious by demonstrating new techniques to them. Perhaps too he wanted to scare them by installing a white employee and by showing them that they were replaceable.[73]

Did slave ironworkers oppose innovation? Not necessarily, so long as they concluded that it did not come at their expense. Slaves viewed new technology warily. It upset routines that they had struggled to establish and it usually meant more work for them.[74] Overwork, like task systems more generally, may have reinforced conservatism by enabling slave ironworkers (especially tradesmen) to cap how much work adventurers could demand from them. It would have been difficult—though not impossible—for ironmasters to alter overwork's terms, for example by devising a system to reward smiths and forgemen who conserved charcoal and metal. In a tight iron market, ironmasters could ill afford the backlash that such changes would likely spark. Slave tradesmen desired and expected control over their work. The skills that they had mastered and the relative autonomy that those skills provided were central to their identities as men.

Their ability to conserve knowledge of their craft, pass it on to their sons, and withhold it from masters—all mattered far more than modest gains that they might realize from improvements. All were their property, which they wished to secure. The industrious revolution that industrial slaves made probably helped stymie the South's industrial revolution.[75]

So did ironmasters' paternalism, which slave ironworkers often deftly manipulated to set limits on their bondage. Reconsider the struggle between David Ross and the smiths over how their shop should operate. Ross cast them as ingrates who acted in a manner "injurious to themselves, ruinous to their Master and disgracefull to that Estate of which they were members & from whence they drew protection & Support."[76] He had spent over a year trying to buy the wife and children of smith James Noble and had threatened a lawsuit when his efforts were frustrated.[77] What perhaps bothered Ross most was that the smiths' conduct reflected badly on him as ironmaster. His reputation in the iron market and among other white men rested on their workmanship. Ross vociferously defended the smiths in a dispute over a set of bank doors: "I never will refer the workmanship of my people to be judged of by ignorant men who can merely form a piece of iron into the shape of a Grubbing hoe . . . no—The Oxford Smiths have served a long & severe apprenticeship to their Trade and I will never disgrace them." Their disgrace was his; aspersions on their handiwork struck at their masculinity and at his.[78]

So did internal disputes. Forgeman Ben Gilmore, of whom the ironworks' inventory said "full of Complaints," told Ross's daughter that the smith shops were "a ruinous business because their workmanship does not sell, the iron they used would sell & they might be usefully employed at the hoe & the ax." Ross largely agreed; he cited Gilmore's comments approvingly in two subsequent letters. But he could not let the slave's indirect criticism of him go unchallenged, especially because he was a leader of what Ross derisively called "our Swaggering Forge men." He derided Ben Gilmore for raising the issue with his daughter, who "he knew was no judge & he was safe" and then counselled him to mind his own business: "I wish Ben would turn his attention to the proper object the Forge fire under his care," because accounts showed that "every smiths fire yields to me twice as much weekly as his Forge fire. This," Ross added, "is done quietly and without insolence or turbulence." Decisions about what the smiths should do were his to make, he reminded Ben Gilmore. Ross learned from the forgeman's comments; he thought little of the smiths, so little that he wished to dispatch them to the woods and to the mines. Such work would be drudgery and a humiliating demotion. Oxford's slaves knew of Susan, a weaver "who took retrograde course." Ross sent her to the mines, where "the digging of iron ore & raking it has greatly enlighten'd her weaving talents. She has returned & [is] doing pretty well."[79]

Oxford's ironmaster did not implement Ben Gilmore's suggestion for the smiths. He had another way to put them in what he considered their place. Ross saw in Edmund, Sharper's teenaged son, "great ambition which is a very laudable virtue when properly guided." He sought to stoke and hone Edmund's ambition by having him cast small ware exclusively, a job "best fitted for his strength and years and best calculated to make him a good workman." Ross suggested that Edmund get "a stamp to distinguish his workmanship so that I may know it from the others." When it came time to distribute clothing, Edmund was "to be particularly attended to—give him pantaloons also of blue cloth." No other Oxford slave received similar pants; they would literally make Edmund stand out.[80]

In singling out Edmund, Ross challenged the world his artisans had made. He subverted the hierarchy of the shop and the slave quarters by awarding Edmund privileges that the young slave's elders did not enjoy—distinctive clothes and the exclusive right to have his handiwork marked. Ross removed Edmund from the "idleness" and "plundering" of the smith shop before it dulled his ambition or he learned to redirect it against his master. He also took Edmund away from his father, Sharper, who found his role as teacher and provider to his son diminished, though he could take comfort that Edmund seemed to have a promising future. But he had to worry for Edmund, lest Ross's favor prove a mixed blessing. Other slaves, particularly older artisans, might resent a teenager who leapfrogged the ranks.

Ross sometimes put himself between fathers and sons while they were at work, but he and other ironmasters usually preferred to let them be when they went home. Off the job, slave men and adventurers seem to have inhabited worlds that seldom intersected. Evidence of the distance between them lies in their bones. The remains of men buried between 1790 and 1840 in Catoctin Furnace's slave cemetery contain far less lead than those of the women there, who likely accumulated it as domestic workers by handling lead-glazed ceramics or pewters or by consuming food and drink in which they were served. The gap in lead levels between Catoctin's slave women and men are among the highest recorded in early African American populations, slave or free, and they suggest social separation between masters and slave men even greater than that found on plantations.[81]

Religious beliefs likely narrowed the cultural divide between some ironmasters and slaves. Colonial Chesapeake adventurers cared little for the souls of those who worked for them. But itinerant evangelists frequently stopped at Maryland ironworks during the late eighteenth century. Charles Ridgely's wife Rebecca hosted Methodist preachers and services in her home that slaves and free hands alike attended. Rebecca Ridgely was one of Methodism's staunchest supporters; Bishop Francis Asbury considered her one of his greatest patrons. Preachers also exhorted slaves at Catoctin, where John Brien erected a small stone chapel on furnace grounds in 1827. He later invited Thomas Henry to live and serve at the nearby Antietam Iron Works.[82]

Rebecca Ridgely and John Brien believed deeply in spreading their faith to all who would hear and accept it. David Ross harbored reservations about evangelical Christianity and the enthusiastic response that it evoked from white and black converts, but he "never laid any injunctions on" Oxford slaves' "worshipping the deity agreeably to their own minds (so far as rational) in any manner they please." William Weaver said even less about his slaves' religion. In 1834, the chaffery at Buffalo Forge shut down so that workers could attend a camp meeting. Joe, one of Weaver's slaves, studied the Bible and conducted wedding services at the forge during the 1850s. Stories told by descendants recall prayer meetings in Buffalo Forge's slave quarters. Whatever slaves said at such meetings, they wished to keep it secret from their masters.[83]

Nor do many records survive of what itinerants preached to slave ironworkers. Moravian Brother John Frederick Schlegel stopped at Catoctin Furnace in 1799. He visited with part-owner James Johnson, his family, "and particularly with the poor Negroes whose inward and outward conditions are troubled." Some of them, Schlegel recalled,

> gathered around me at the top of the furnace opening (cavity). I depicted the Saviour as He redeemed them from sins upon the cross through His suffering and death. (I told them) how many of their countrymen in the West Indies, through belief in the Saviour, have achieved bliss (happiness) through His death. They wept very much because they were bound to work so hard during the week as well as on Sunday in the iron smelter and thus were seldom able to hear the Word of God.

Schlegel's "conversation came to an end, the signal was given for the pouring and each of them had to go back to work."[84]

Johnson allowed Schlegel to preach because slaves' souls concerned him. So did their fidelity to him. By the 1790s, Moravians had fully accommodated their faith and their consciences to slavery. Their missionaries in the United States and the Caribbean looked to convert slaves and secure their acquiescence to bondage. If Moravian evangelizing made slaves more dutiful and industrious, as it often did in West Indian colonies wracked by resistance and rebellion, so much the better. Schlegel's message at Catoctin spared Johnson any direct responsibility for his slaves' well-being. He did not master them, the furnace did. Their duty to minister to it plagued "their inward and outward conditions" by keeping them from hearing the gospel and taking it to heart. Schlegel's talk gave them emotional release and a break from the consuming demands of Catoctin Furnace.[85]

But his audience could easily have interpreted his presentation differently. What he said and where he said it invited slaves to assess their bondage critically. Schlegel perched them over the furnace's torrid mouth as he spoke of what Jesus had endured to save them and so invited some to equate their suffering to that of Jesus. He invoked the joy that knowledge of Christ's death and resurrection had brought to slaves in the West Indies, but he also reminded them that bondage had brought them to central Maryland and stretched their ties to the rest of the African diaspora. All may have provoked some of their weeping, which struck Schlegel forcibly. His preaching had clearly touched them and that gratified him. But the slaves he wished to convert also shaped his presentation and how he recorded it. Their "conversation" with him, in words and in tears, encouraged Schlegel to conclude that their toil was what troubled them, mostly because it separated them from God. What he did not see or chose not to confide to his diary was what Catoctin's slaves knew but could not voice—their masters, who forced them to tend the furnace, were ultimately responsible for their torment in this world, if not the next.[86]

Catoctin's ironmasters eventually took some of that message to heart by demanding that slaves conform to their cultural expectations. Catoctin's slave cemetery suggests a community that had adopted the values of its masters, at least when under close supervision. Nearly all put to rest there were buried as whites were—in coffins, their bodies extended with their heads pointed to the west. None of the graves was decorated with bottles, shells, or broken crockery—any or all of which commonly adorned African American graves of the era. None of the dead wore necklaces, beads, bracelets, or rings. Neither knives nor tobacco pipes accompanied them. If Catoctin's masters permitted slaves leeway in many aspects of life outside work, they seem to have allowed them little when they interred their dead.[87]

Slave ironworkers enjoyed considerable autonomy within their households. David Ross largely left Oxford slaves "free only under such controul as was most congenial to their happiness." He urged that "the young people might connect themselves in marriage, with the consent of their parents, who were the best judges." Unlike John Brien, who encouraged Antietam slaves to marry free women, Ross frowned upon "connections out of the estate and particularly with free people of colour because I was certain twould be injurious to my people but I have used no violence to prevent those connections for tis well known that now they exist upon my estate and have not been expelled. . . ." Above all, his slaves "should connect themselves in a decent manner and behave as a religious people ought to do."[88]

The policies that Ross and many adventurers adopted recognized and perhaps even reinforced the authority of slave men. Oxford's inventory grouped slaves by family, with each headed by a father or a husband. Unmarried slave women like Fanny, who abandoned her newborn baby in the woods, threatened the moral order of Ross's estate. Fortunately, the "innocent child" survived and spared Fanny from being "prosecuted . . . unto the ignominious death of the Gallows." Word of her deed triggered his exasperated recitation of the spiritual and romantic latitude that he permitted slaves. Fanny, he fumed,

> was under no restraint as to a husband whom soever she might choose. What then could induce her to such an inhuman act? She had little to fear from her earthly master and still less to fear from heaven. . . . Her going astray was most certainly wrong, but an attempt to destroy the most innocent infant was horrid *damnable*. As to Fanny's tale of being delivered in the woods, it is not to be credited. My conduct to pregnant women requires no comment. You know that no pregnant woman is put to labour within some weeks of her delivery and as long afterward. 'Tis not my desire to make this poor woman unhappy, quite otherwise, but 'tis my earnest wish that nothing may be wanting to the preservation of this exposed child that my estate can furnish.

Ross ended his tirade by telling Richardson that "you will please me and I hope to be pleased yourself in exposing such an abominable conduct—the disgrace of all religion and most abominable to me as the master."[89]

Why did Fanny abandon her child? The question bothered Ross, who never quite answered it. He was certain that she was to blame and he was certain what permitted her to do it. She could not control her sexual urges. She refused to choose a husband. Fanny needed a man to supervise and contain her. The Oxford inventory identifies every adult slave by occupation or physical condition, except Fanny, who was only described as having had "no husband." If Ross ever asked Fanny why she did it, he

Oxford Iron Works — in Families &c

No.	Name	Age	Employment
34	Big Fanny ..	36	infirm, Briton's wife
35	Joe Briton	19	Collier Sickly
36	Jim Briton	15	Woodhauler
37	Little Briton	14	do
38	Rachail —	11	Nurse
39	Claiburne —	10	
40	merideth —	7	
41	Davy —	5	
42	Judy —	2	
43	Ben Gilmore	48	Refinery, full of Complaints
44	Sibby —	34	his wife, infirm
45	Little Rose	22	Furnace, Alex. wife
46	Little Matt	18	Refinery
47	anderson —	13	with Carpenter George
48	Obey —	11	nurse with L. Rose
49	Delphia —	8	
50	Forge Charles	45	Refinery
51	Sally	28	Furnace his wife
52	Pleasants	18	Stable
53	Sukey —	16	House
54	milly —	12	Nurse
55	Frank	8	
56	Nancy	6	
57	Wyatt	2	
58	Charley —		3 months Old

Figure 11. A page from "List of Slaves at the Oxford Iron Works in Families and Their Employment Taken 15 January 1811." in the William Bolling Papers, located in the Duke University Rare Book, Manuscript, and Special Collections Library, Durham, North Carolina. Note the comment of the inventory's composer about Ben Gilmore.

probably did not write down her reply. We can at best speculate on why she abandoned her baby. Fanny was thirty-three and already had five children, ages three to fifteen. Maybe she did not wish to raise another child, though she would not have had to do it alone. She had her parents and five adult siblings to help her. Perhaps she could not bear the thought of bringing another child into the world as a slave.[90]

Therein, it seems, lay Ross's inability to answer why Fanny abandoned her child and his silence regarding the matter thereafter. His religious beliefs and his sense of propriety led him to condemn her actions. Fanny's decision angered him for other reasons. She had defied his authority and nearly deprived him of property. Ultimately, her action implicated him because it called into question Ross's self-proclaimed status as the protector of his "people," especially of the women and children. By leaving her child, Fanny had challenged the fatherly image that Ross tried to cultivate for himself among slaves, especially when she seemed to conclude that death for her child was preferable to enslavement by him.

Defense of his conduct as master obsessed Ross. He cast himself as an unappreciated father burdened by debt who nonetheless struggled to provide for his children. "If I was to regulate my conduct by the rule of profit & Loss as to their services," Ross responded to news that many needed clothing, "they have no claim on me." Yet they did, because he had put them "under the authority of temporary masters who have command over them, altho' less understanding & sometimes less integrity than those poor blacks." For Ross, his agents bore responsibility for any suffering his slaves endured; he remained blameless. They underfed Oxford's ironworkers; they clothed them poorly; they overworked them while he tended to his affairs in Richmond.[91]

Oxford's slaves capitalized on Ross's invocations of paternalism. Patsey intervened in a brawl between clerk William Dunn and John Pearce, overseer at an Oxford plantation. Pearce sent her to Lynchburg to be whipped. Ross had Patsey rescued from jail and gave her a shawl "as compensation for trouble in being carried to Lynchburg." She had forced him to walk a tightrope. In no way did Ross wish to convey that he was rewarding slaves for defying "their Superiors," but he muddied the point by contending that sometimes slaves could exercise discretion "to do good & prevent evil in a judicious manner" so long as they used violence only for "self preservation." Ross defended Patsey even more forcefully to Pearce. He told his overseer that he had sent the shawl "with an admonition." Ross admonished Pearce in what were likely far stronger terms: "If such conduct would have been meritorious in a white man or a white woman, was it not still more so in a Slave?" Pearce, Ross implied, had no honor or shame. It ate at him that his overseer would have had Patsey "degraded by a public punishment," especially since Pearce "by his birth rights" as a white man "had superior advantages in law to herself." Pearce had publicly shamed Patsey and so had shamed Ross, who concluded that Pearce was to blame. Patsey was "unoffending" and had acted within his guidelines. "When ever any of my Slaves shall come bold forward, and prevent murder, robbery, or any outrage against the laws and Peace of the Society," Ross saw "such Slave as acting a meritorious part in the Community." Patsey and other Oxford slaves could easily have heard another message—that they reserved the right to judge their superiors and take action against them, with their master's blessing.[92]

The leeway that Ross allowed Oxford's slaves encouraged them to play the ironmaster and his agents against each other. They fanned his suspicion of the agents, sometimes by appealing to him directly. "Carpenter George" wrote Ross to complain. "We wish you were here," he began, "to see the present situation of the dam we are about. . . . We do little or nothing under our present arrangements and management." George claimed that he and others wanted to dodge blame "for what we cannot help" and to decry management's poor planning. "It seems intended," he continued, "that the whole blame of delay & perhaps of miscarriage & ruin must fall upon innocent men. We can do the Work which belongs to us to do," but "we have no authority to command waggons, to force provisions, or to provide the materials absolutely necessary. This is the duty of our Superiors." The news infuriated Ross. He summoned George to Richmond, sent a copy of the slave's letter to his clerk, and warned Richardson that he might hire someone to oversee Oxford and report directly to him. George's words resonated with Ross because they echoed his frequent and often withering criticism of his agents and confirmed his suspicions that they, like their predecessors, believed that "the main Strength, blood & Sweat of the Servants would do the Work & draw a veil over [his agents'] misconduct."[93]

Carpenter George and all Oxford's slaves had every reason to reinforce that impression. They could evade blame and turn Ross's mistrust of his agents to their advantage. If Ross recognized that his slaves exploited his troubled relationship with Oxford's supervisors, he did not say so. Slaves, he expected, would pursue their own agendas; it was his agents' job to keep them focused on the needs of his enterprise. Mismanagement had allowed his ironworkers to fall into bad habits. Besides, as Ross often complained, he could not pinpoint where Oxford lost money unless his agents ran experiments more carefully and kept their accounts more faithfully.[94]

Results from tests of the furnace brought disturbing news; the amount of iron that it smelted fluctuated wildly. Why? Power was not the problem; there was plenty of water. Ross could see only two causes and both made him worry that Oxford might "have rascals in the Furnace as well as the Forges." Either the filler and the keepers who worked the night shift had conspired to ignore the furnace so that they could nap or the fillers had followed their father's bad example (what Ross derisively named "Hunt's trick") by exaggerating how often they had charged it. Either scenario implicated his founder, "the faithful Abram," the slave in whom he placed the most trust. Ross suggested that Abram be informed that he suspected "great neglect in some of the keepers & fillers." How could Abram have been ignorant? If he knew nothing, then he was not the supervisor Ross thought he was. If Abram was aware, then he concealed and condoned their actions. Either prospect unsettled Ross, an old man, hounded by creditors, who badly needed to trust someone.[95]

The furnace hands, and indeed all Oxford's slaves, nursed grievances against Ross. Their supervisors' bungling and their master's precarious finances had worsened their lives. Ross admitted that his debts had helped to leave Oxford slaves "undercloathed for two years" and with less food than what they considered "their legal right." His promises to "the faithfull" Abram and other slaves that he would address when he was able "whatever wants they may have suffer'd in our present situation" fell mostly on deaf ears. Ross had violated their bargain and undermined further the myth that he sought only to protect them and promote their happiness. It was his duty to fulfill their basic "wants." He had failed.[96]

Ross had already failed what to slaves was the key test of paternalism. Lydia and Charity, fourteen and eight years old respectively in 1811, could commiserate that Ross had sold their mothers. Perhaps they knew why. We cannot because the records do not say. Ross did exercise his power to sell slaves and break apart their families when they angered him enough. The watermen, he thought, deserved to be dealt to "Carolina Hogg Drivers." Ross ordered that Solomon be sold for running away, for perceived ingratitude, and for having shamed his master. Solomon fled Oxford to be near his wife after Ross closed a plantation and moved its hands to the ironworks. He acceded to Solomon's wishes by leasing him to Edmund Sherman, a local planter. Solomon ran away from Sherman at least twice, which to Ross made him a "disgrace" and a "hypocritical canting rascal." The $300, preferably in cash, "young Cows, Beef Cattle, or in young work horses or mares," that Ross thought the sale of Solomon should bring would address some of Oxford's greatest needs and dramatize his authority, especially if "the Kentucky horse merchants" bought Solomon and marched him west along paths that thousands of other Virginia slaves had trod.[97]

To be sure, Ross tried hard to preserve or even reunite families. But he died in 1817 and was powerless to protect the families that Oxford's slaves had so painstakingly constructed. They paid for his debts. To satisfy Ross's creditors, his executors broke apart Oxford and its slave families. In 1816, Oxford paid taxes on 145 slaves over age 16. Three years later, there were 25. Neither paternalism nor control over their work could save Oxford's ironworkers from what they dreaded most.[98]

Oxford's slave artisans circulated in the iron industry, which prized their skill and helped them to maintain contact with each other. William Weaver pursued as many Oxford forgemen and smiths as he could afford. He acquired five: Billy and Hal Hunt, Billy Goochland, Phill Easton, and Ben Gilmore. To get them, Weaver and his agents worked through communication networks that Oxford's slaves had developed to stay in touch. Billy Goochland, who "was appraised the highest of any Negroe in Ross's Estate," came to Buffalo Forge after he decided "to visit his friends ownd by you and to see you also, in order that he may make up his mind how fair he woud be pleased with an Exchange of residence." Goochland's friends helped persuade him to consent to the sale. While engaged in Phill Easton's purchase, Weaver learned that Aaron, Oxford's master forgeman, had "gone to the western country," another victim of the domestic slave trade who thereafter probably lost touch with his crew. Phill Easton also had information about Ben Gilmore, who was "in the neighbourhood and is a stout old man sufficiently strong to draw Iron pretty well. Consult Phill about his qualities and let me hear from you as I shall probably hear from Bens master shortly whether he can be had." Weaver bought Ben Gilmore. In 1830, he gave him permission to return to the vicinity of Oxford, where the forgeman was to try to hire out his services so that he could buy his own freedom. Oxford's forgemen remained proud; they valued and cultivated the bonds that they had forged to each other, and they nurtured dreams of freedom which they hoped their industry would eventually bring to life.[99]

Ben Gilmore, sixty-seven years old when he left Buffalo Forge, probably never won his freedom. Had he earned it, old age, law, and racism would have rendered him virtually a "slave without a master." To David Ross, Ben Gilmore, like Patsey and their fellow Oxford slaves, were members of the "community," inferior to whites, but with particular rights and responsibilities nonetheless. They were members of society, in slavery or in freedom. But they did not belong to the republican nation as whites defined it. They were civically dead.

This created confusion when southern ironmasters participated in national campaigns for higher tariffs. In 1831, George B. Pennybacker told William Weaver how much iron Pine Forge had produced from 1828 to 1830 for a report that Weaver was compiling on Virginia's iron industry for the Society of Iron Manufacturers of Pennsylvania. That was easy. Another crucial issue was far more complex: exactly who, Pennybacker wondered, did the Society consider a "dependent" of the iron industry? "Not knowing whether Negroes will be taken into the estimate," he alerted Weaver, "I have not put down the number of persons dependent on the Works—therefore must submit that you, & request that you will insert whatever you may conceive to be right, there is at this time 68 whites & 50 Slaves."[100]

Pennybacker raised a fundamental question: did slaves "count" to the Society of Iron Manufacturers and, by extension, to Congress? Clearly, they were dependents before the law. They were politically useful dependents when it came time to apportion congressional representation and electoral votes—for such purposes the U.S. Constitution declared that each slave was three-fifths of a white person. The issue was far more ambiguous when it came time to lobby elected officials in Washington. It carried no political weight there to argue that slaves would lose their jobs if Congress did not shield the iron industry from overseas competitors—they were property, not people who deserved the federal government's protection. Indeed, the few Chesapeake ironworks that responded to the federal 1820 manufacturing census—among them William Weaver's and Thomas Mayburry's Union Forge, the Antietam Iron Works, and the Northampton Iron Works—never acknowledged that many of their ironworkers were slaves. Richard Green, Charles Carnan Ridgely's agent, came closest by stating that Northampton "would do a losing business under the pressure of the existing times" if Ridgely did not furnish "nearly all its materials, laborers, stock & c. & c." Slaves were dependents, but they were not of the political nation and never could be, unlike the men whom members of the Society of Iron Manufacturers of Pennsylvania employed. They counted to Congress, and that had made them potentially invaluable allies to adventurers.[101]

6

MANUFACTURING FREE LABOR

In 1785, thirty-two forgeowners petitioned Pennsylvania's General Assembly to impose duties on foreign bar iron. They recalled that manufacturing bar iron "was always considered a Source of public Wealth and Benefit" because it kept money within the state and furnished exports that stimulated commerce. To sharpen the argument that their forges sustained the commonwealth, adventurers invoked their employees. Their iron could not compete with imports, even though it was "equal in Quality if not superior to any that is made in Europe" because "the higher Price of Labour in this Country" made it more expensive than "in those Countries where the Labourers are but little removed above the Condition of Slaves." The ironmasters reminded the Assembly that they had employed "some thousands of Hands, . . . who were thereby enabled to maintain themselves and their Families and to contribute to the Support of Government by the Payment of Taxes."[1]

What is good for us is good for our employees and good for Pennsylvania, the forgeowners argued. Protect our enterprises so that free ironworkers may continue to support themselves, their dependents, and our commonwealth. Few in the Assembly could have missed the irony. Curtis and Peter Grubb, George Ege, Robert Coleman, and William Bird—some of the state's biggest slaveholders—were begging protection from competitors who unfairly undersold them because they oppressed their workers. The adventurers were not exactly "disowning slavery," but they were not owning up to it either. After all, slaves had no public identity in the new commonwealth. The ironworkers whom they hired did; they were free men who merited the state's protection. Despite another 773 signatures to back their petition, the Pennsylvania forgeowners did not persuade their representatives. Still, they had learned something valuable. "Free" labor, so often an imposition to them before the Revolution, had become a political asset.[2]

Adventurers and ironworkers, those whom ironmasters hired and those whom they owned but refused to acknowledge, created a new order in the mid-Atlantic countryside between the American Revolution and 1830. By the 1790s, all confronted slavery's slow demise. A few ironmasters applauded gradual abolition and promoted it; most clung to slaves and black indentured servants. With a whiter and freer labor force, adventurers in Pennsylvania, New Jersey, and Delaware pursued new ways to recast the industrious revolution by seeking to transform ironworkers' behavior on and off the job. They promoted evangelical Christianity, education, and temperance and hoped that such initiatives would create employees who would be more industrious and more deferential. Ironmasters needed to pay closer attention to the men whose work determined what price their product could command, especially when they weathered stormy iron markets. That was not all. Most adventurers believed that it was their duty to lift up those workers who wished to rise.

They had another reason to campaign for ironworkers' hearts and minds: the health of their enterprises depended on government action. The iron industry had to be politically viable if it was to survive and prosper. Employees were natural allies in that mission—if ironmasters could get workers to think and act civically, if they could appeal to their employees as fellow free men. Ironworkers could help adventurers forge the symbolic bond between iron and nation for elected officials and for voters.

Some never had the chance. Gradual abolition delayed freedom for most black ironworkers and helped give rise to the racism that circumscribed it. Every black child born into indentured servitude reinforced the link between bondage and blackness. Blackness also came to signify denial of opportunity; by 1830 the disproportionate share of African American men in skilled positions had become a memory. Black ironworkers joined growing ranks of unskilled and semi-skilled hands who found themselves in an increasingly precarious position within the region's labor market. Perched uneasily above them were tradesmen who worked with metal and who insisted on job security and on more autonomy at work. Seldom did ironworkers' concerns translate into collective action against adventurers. Nor did they join together to challenge the initiatives that ironmasters promoted to transform them. Some participated wholeheartedly; few resisted openly; while others chose to interpret those initiatives in ways that served their needs rather than those of employers. The industrious revolution generally continued on terms that favored adventurers, largely because they had mustered enough support from workers to keep their enterprises viable in a world of swifter political and economic change.

Rusting Chains

The Revolution created a crisis for mid-Atlantic adventurers. Their region and their enterprises were battlegrounds; their land and their products had enormous strategic value. War and national independence challenged ironmasters' ability to determine who would work for them and on what terms they would work. Both eroded slavery yet perpetuated bondage for most black men, women, and children. By 1820, indentured servitude had become closely identified as the legacy of slavery, a state of unfreedom born of black women, from which whites chose, and many free black men tried, to distance themselves.[3]

The war strained relations between adventurers and hired hands. Disrupted supply lines left New Jersey's Hibernia Iron Works unable to stock enough goods to keep employees happy. In February 1776, manager Joseph Hoff tried to mollify them by sending a collier to New York so that he could "see the Scarsity & extraordinary prices of goods that he may give his fellows here the need full information, that they may not complain of my charges." By May, Hibernia's workers had so tired of promises that they told Hoff "almost in plain English I am a Lyar." Hoff warned that "it will be impossible for me to keep the Men at work" if he did not soon receive what he had ordered. A week later he reported that Hibernia had lost good workers—"because they were naked and could work no longer"—to John Jacob Faesch, who had "plenty of Oznabrigs and many other suitable Goods for his People."[4]

Ironmasters also worried that military service might claim their employees, so they lobbied for measures to get them excused from it. Pennsylvania's Council of Safety ruled that furnace hands involved in casting munitions could neither join militia marches nor leave their posts without its permission. Hibernia agent Charles Hoff, Jr., asked New Jersey's governor to exempt workmen from a militia draft, pleading that he could not fill munitions orders "for want of Stock of Coals & Oare." Citing the custom of state authorities to excuse skilled ironworkers from duty, Hoff urged Lord Stirling to seek the release of keeper Robert Minnis for at least the current blast.[5]

Hoff discovered that the power to shield ironworkers from military duty was a mixed blessing. Upon hearing that the state legislature might no longer call out the militia, he observed that

> our haveing it in our power at this time, to give exemption to 25 Men, is the only thing . . . that induces the greater part of the Men to work here that we now have; as they are Farmers, & have left their Farms & come here solely to be clear of the Militia & from no other motive. This I have experienced as I find they are determin'd to Shuffle away the time they are exempt, and do as little business as they possibly can. Shou'd that exemption be revok'd, I don't see how wee shou'd be supply'd with Workmen.

A partial solution, Hoff thought, lay in engaging men who had already served and had to go where they were told. He requested that Sterling recruit thirty to forty "of the Regular & Hessian Deserters" who refused to enlist with the Continental Army. That summer, he dispatched agents to Philadelphia to hire some.[6]

Captive labor temporarily revived white servitude in the mid-Atlantic states. Ironmasters had strong claims to British and Hessian prisoners of war. In 1776, Pennsylvania's Council of Safety authorized furnaces that cast "Canon or shot for the publick service" in Chester, Lancaster, and Berks Counties to employ soldiers imprisoned at Lancaster or Reading. Elizabeth Furnace, Mary Ann Furnace, Durham Iron Works, Cornwall Furnace, Charming Forge, and Hopewell Forge all leased Hessians from the Continental Army. They often paid for them with iron, as did George Ege, who in 1777 bought the services of thirty-four Hessians who dug a channel to power Charming Forge's slitting mill. In 1777, Cornwall Furnace built barracks for Hessian prisoners; at least twenty-six worked there that year. Five years later, Hessians were mining ore at Cornwall. By December 1784, most had left; ten men received their "Liberation money" that month.[7]

Some slave ironworkers in Pennsylvania, as in the Chesapeake region, saw the Revolution as a chance to escape bondage. Six of James Morgan's seven slaves fled the Durham Iron Works and were "supposed to be with the enemy in New York." Abel, one of Peter Grubb's slaves, made a bid for freedom too. In 1780, Grubb reported to Lancaster County that Abel, then twenty-two years old, had "run away" (see Figure 7).[8]

Perhaps Abel left Grubb because he was following Pennsylvania politics and knew that state legislators had imposed a commuted death sentence on slavery. Under the Gradual Abolition Act of 1780, all children born to slave mothers were to be bound until they were twenty-eight years old. The act manumitted no one save those slaves whose masters failed to register them in time with county authorities. Indeed, by reporting runaways, Morgan and Grubb established their legal right to re-enslave them if captured. Not surprisingly, most Pennsylvania ironmasters, having invested heavily in slaves before the Revolution, seem to have opposed gradual emancipation. But as long as they complied with its terms, the law safeguarded their property and guaranteed them the services of as yet unborn children of slave mothers well into their adulthood.[9]

Only a few adventurers, nearly all of them Quakers, stood publicly against slavery. In 1786, Philadelphia merchant Henry Drinker informed North Carolinian Richard Blackledge that, when he bought a share of New Jersey's Atsion Iron Works, he had resolved "to have nothing to do with Slaves," though he acknowledged that "this perhaps will not be imitated in your Country, tho' it is devoutly to be wish'd that this unchristian practice was universally rejected." In part, Drinker opposed slavery because he believed that it bred white sloth. The proof, in his eyes, lay in southern society. Blackledge's proposed ironworks would likely fail because "to the Southward few Men will go thro' an equal Quantity of Labour with those living more northerly." Drinker recalled that the owners of two Maryland ironworks had told him that "their principal difficulty & final failure of Success was owing to this Cause. Where it is not the fashion and turn of the People in any Country to be diligent & industrious," he warned, "it is hard work, if it be effected at all, to produce a change in any considerable degree for the better." Drinker also attacked slavery on humanitarian and moral grounds. In 1790, he lamented that "the hardness of Heart & insensibility which remains among many of the Southern People & the grevious oppressive conduct on various instances is cause of sorrow & mourning." Anyone complicit in slavery's survival, Drinker worried, might face God's wrath.[10]

Henry Drinker tried to save himself and the nation from divine punishment. He thought globally, acted locally, and hoped to turn a tidy profit. Drinker owned tens of thousands of acres in northeastern Pennsylvania, many heavily timbered with maples. A maple sugar industry would boost the value of those lands, provide a market for kettles cast at Atsion, and stem demand for slave-grown Caribbean cane sugar. Tapping "American Sugar Cane," Drinker stressed, would reap humanitarian and social benefits. "I could have employed our Furnace to better advantage," he observed while trying to dispose of some three hundred kettles, "but a strong desire to promote the making of American Sugar by the labour of Freemen, & to guard against the Want of so material an article as Boilers, led me to provide more than I should otherwise have done."[11]

Drinker's blow at the Atlantic slave trade and at West Indian slavery glanced off its targets. Maple sugar proved difficult to sell in markets already glutted with Caribbean cane sugar. For that, Drinker bore some blame. A maple sugar industry was but one of several new markets for iron that he pursued. Another was Cuba, where sugar cultivation and plantation slavery had taken off after the Seven Years' War. Drinker had received "several applications" from the island for mill cases (to squeeze juice from cane) and for sugar boilers because "perhaps I am the only person who has undertaken this new business." An untimely shipwreck, technical difficulties in casting the items, and confusion over Spanish trade regulations all prevented him from realizing a profit from selling iron to Cuba and from further sacrificing his principles and his vision to his need to sell what Atsion made.[12]

Abolitionist ironmasters were the exception, not the rule. Pennsylvania adventurers complied with the Gradual Abolition Act by rushing to register their slaves with county authorities lest they be freed by fiat. The struggles of Pennsylvania slaves to build and maintain families before the Revolution paid off for their owners. Gradual emancipation converted children who had recently been born to slave mothers into the greatest potential human assets at masters' disposal. Slave women remained potential generators of bound labor.

Birth records testify to the new value that masters placed upon slave women and the children born to them. In 1788, a new Pennsylvania law which required that masters report children born to slave women (or have them emancipated) generated a flurry of activity at county courthouses to document slave women's reproductive activities over the previous eight years. Jasper Yeates registered three girls on behalf of Peter Grubb's heirs: Nancy's daughter Hannah and Amy's daughters Fanny and Amy. James Old named seven children who ranged in age from nine months to nearly six years old. In 1807, Joseph Miles registered Charles, "born in our Family," and two years later he and John Miles entered that Jerry had been "born in the Family at Milesborough Forge."[13]

The identification of bound black children as "family" suggests the lengths to which adventurers might go to secure their claim to them. In 1813, John Anderson wrote attorney William Orbison concerning Phoebe's son. He explained that Phoebe had accompanied him to Cincinnati. Ohio state law entitled her to freedom, but Phoebe either did not know her rights or chose not to invoke them. She returned to Pennsylvania with Anderson, who then manumitted her. While they were in Cincinnati, Phoebe became pregnant "by the west wind, or something else" and gave birth to a boy. Anderson came to the point: "how am I to have the Boy bound to me until he arrives at the age of 21?"[14]

Few ironmasters sought advice of counsel to determine their property rights over black children, but most took pains to validate their legal claim to bound black labor. Indeed, the paper trails that followed unfree black people literally underwrote commerce in them. Before Robert Coleman would buy Juba, a twenty-two year old slave, he demanded that Joseph Krebs prove that he had properly registered the slave with county authorities. In 1801, Matthew Brooke of Birdsboro Forge bought Sarah, fifteen years old, from ironmaster Benjamin Morris, who had purchased her from Elizabeth Old six years earlier. In 1810, Buckley & Brooke bought Davy, twelve years old, from George Ege, and Brooke purchased Sampson, then fifteen, who was to serve him "to 28 years of age to have no freedoms nor schooling."[15]

Court challenges to slavery necessitated such documentation. Members of the Pennsylvania Abolition Society combed through county records looking to file freedom suits against masters who had failed to comply with gradual emancipation laws. Nat might have been a good candidate for their attention. Peter Grubb bought Nat in 1781. Within three years Grubb was leasing his services to John Hare and Samuel Jones for thirteen shillings per ton of bar iron. Nat fled Hopewell Forge sometime before Grubb died in 1786. Although he was returned, Jasper Yeates, chief executor of Grubb's estate and a guardian of Grubb's children, worried that a legal loophole might eventually free Nat. Yeates verified that Joseph Wharton, Nat's master in 1780, had properly registered him. He also asked Lancaster County's Clerk of the Peace to attest in writing that "Negro Nat was entered by Peter Grubb in his Life Time in your office as a run away Slave." His efforts succeeded; Nat's name appears on a 1789 inventory taken just before Grubb's estate leased Mount Hope Furnace and Hopewell Forge.[16]

Nat embodied Peter Grubb's commitment to slavery. In 1780, he bought Mark, eighteen years old, and the following year put him to work with Samuel Jones, who later purchased the slave forgeman.[17] After Grubb died, the guardians of his estate and his principal heir, Henry Bates Grubb, trained more slaves to forge iron. The guardians signed over Bill, then seventeen, to John Hare for four years in 1787, and in 1788 they contracted George out to Samuel Jones for five years so that he would learn to draw bar iron. In 1795, Henry Bates Grubb signed over Jasper's services to John Jones for nine months. Jones pledged to "do all that in his Power Lays to give" Jasper "a competant Knowledge of the business." Grubb promised that Jones could continue to lease Jasper for "as long as Mr. Coleman let Mayberry have his Slave or Slaves" for ten shillings a ton after the nine months had expired. In 1801, Grubb acquired Tom, a slave forgeman at Charming Forge then in his late forties, and a woman named Giff for ten tons of pig iron.[18]

Hopewell was hardly the only enterprise which relied heavily on slave forgemen. In 1795, Union and New Market Forges employed at least two each. Slave forgemen outnumbered white forgemen at Charming Forge between the Revolution and 1800. All earned income, either as underhands to whites or by hammering out small amounts of iron themselves. In 1795, "Mulattoe Ben" credited James and Jem a total of nearly £118 for helping to make almost 118 tons of anchonies. The wages of slavery did not buy them freedom; they subsidized their bondage. In 1790, Jack, Tom, Mulatto Ben, and James saw their earnings siphoned into Charming Forge's "Profit & Loss" account. All footed the bill for their board.[19]

The Revolution made slave labor even more indispensable to Cornwall Furnace. In 1780, Curtis Grubb owned more slaves than anyone else in Lancaster County: fourteen men, four women, and seven children. The number of slaves who worked at Cornwall, as measured by furnace timebooks, rose from five to sixteen between 1776 and 1786, even though Grubb sold two slaves and three others died during that decade.[20]

What explains the growing reliance on slave labor at Cornwall? Some slaves who were children in 1780 had grown old enough to join the workforce. Besides, white indentured servants had become a terrible alternative to slaves. By 1785, most Hessian prisoners had left. Curtis Grubb bought white indentured servants to replace them. In Fall 1784, twelve began work at Cornwall. Few stayed long. John Dignan, Richard Prendergast, and James Roark ran away the following February; only Dignan returned. Daniel Dillon, Alexander Stewart, John Garvens, and John Murray fled four months later. All but Stewart escaped.[21]

The runaways probably hated their work. In 1783, Johann David Schoepf toured Cornwall Ore Bank with a miner who had recently immigrated from Ireland. He "seemed very dissatisfied," Schoepf recalled, "deceived in his expectations of America; 50 shillings a month and keep hardly seemed to him worth the trouble of exchanging dear Ireland for America." Breaking rock in Cornwall's massive quarry must have given Schoepf's guide a poor impression of his new home. It certainly would give servants reason to flee.[22]

The pariah status of their coworkers also gave indentured servants reason to flee. The servants worked with slaves and they feared that they might become identified with the slaves. In the Chesapeake region, the end of the convict trade and the rise of term slavery invited whites to view servitude as unfree labor akin to slavery. Indentured servitude in Pennsylvania became more deeply associated with childishness, dependency, and blackness with every baby born to a slave mother. Servants who left Cornwall's mines also tried to avoid becoming the social and cultural equals of the black men who toiled beside them. Their employer sometimes conflated them; Cornwall accounts in 1787 lumped together wood cut by "Negroes & Servants."[23]

Curtis Grubb did not care why the servants ran away, just that their flight made them an unacceptable risk. Neither he nor his successor, Robert Coleman, employed white indentured servants in large numbers again. The transatlantic trade in servants abruptly collapsed in the early nineteenth century. By then, most indentured whites were debtors or children in apprenticeships, which only strengthened the link between servitude and servility. Indentured servitude had become the antithesis of white male freedom, thanks largely to gradual emancipation.[24]

This generally made unfree black labor even more valuable to ironmasters. The value of slave labor, though, gradually shrank at Cornwall. On March 5, 1790, Cornwall's timebook recorded the work of fifteen slaves. That was the last day that clerks compiled regular monthly tallies of time worked by slaves. April 1794 was the last month that Cornwall clerks used timebooks to record how many days slaves worked. Brief notes jotted down in the margins of timebooks suggest that owner Robert Coleman let slavery expire gradually. Harry died in December 1791, Sampson a few months later, Cato in 1808. Cornwall still added to its staff of blacks in bondage. In 1799 the furnace "paid for Entering a female Negroe Child" and in 1804 it purchased a "Molatto Boy named Jack." Others stayed on as slaves. Dick and Beck, both in their twenties when registered in 1780, remained at Cornwall for more than forty years.[25]

Although black unfree labor did not disappear at Cornwall, it slipped off timebooks and into account books, where it became divorced from production. "Negroe Accounts" received credit for over fifteen months' work of "kitchen work p. Dinah & Beck" and for 150 cords of wood that Cato, Dick, and Toney cut in 1797. On March 3, 1798, Cornwall Furnace noted "work" done by slaves or black servants for the last time. Thereafter black unfree labor became mostly anonymous and it became associated primarily with consumption. "Negroe Accounts" persisted into the 1830s in furnace bookkeeping, usually invoked when unfree black workers, most of whom were female, received shoes. The feminization of black bondage rendered it far less visible, both to Cornwall Furnace and to us.[26]

The growing share of black women and children among the unfree influenced how black men behaved. Sampson fled Birdsboro Forge in January 1816, forcing his owners to send "to Phila. after him & other places at considerable expence, advertising him & c." In the fall, Sampson "at last came home of his own accord." For running away, he earned nearly seven months in Reading's jail. Sampson bided his time at Birdsboro until he found a better way out. In 1821, he arose one day to make "an ankony & a bloom" and "waisted 6 or 8 baskets Coal" in doing it. That was the last straw for Matthew Brooke. He freed Sampson three months later, more than a year before his servitude was to have expired, and awarded him "customary wages in service of the estate for one year."[27]

In freedom, Sampson enjoyed some relative advantages: he was a young man who knew how to handle a forge hammer. The skill he had learned in bondage offered him entry to the lucrative employment that many free black men had enjoyed for more than a generation. In the 1790s, Sam Benn worked as a forgeman at Coventry Forge, where he earned nearly as much per ton of iron as his white peers did. At Birdsboro Sampson may have heard stories about black forgemen who preceded him; nine of the forty-eight forgemen there between 1800 and 1808 were African American. They comprised one-third of the black ironworkers at Birdsboro; white forgemen represented 13 percent of white employees. But black men were nearly absent from other trades at Birdsboro; there was but one among sixty-nine whites.[28]

His skill made Sampson more employable and it afforded him more freedom to move. The previous generation of free black Birdsboro employees did not stay long, and they passed through more quickly than white hands did. Forge ledgers show more than one payment to only eight of twenty-seven African American workers between 1800 and 1808 (30 percent); 123 of 293 white employees (42 percent) were paid more than once.[29]

Most Birdsboro employees, black or white, passed through quickly. Lower mobility rates for whites reflect that proportionally far more of them owned land nearby or belonged to households that did. But most whites had long taken freedom to leave for granted. Sampson and his peers could not—for most of them bondage was either a personal or familial memory. They could see reminders all around them in those who were slaves or servants. Between absconding and jail, Sampson had spent more than a year away from Birdsboro. Why would he stay once he could go? Besides, another forge might hire him, and on better terms than he could get from Matthew Brooke. Freedom meant the right to move—to strike out in new directions and hopefully to outrun the popular perception that anyone black who had to work for someone else was a servant. As the Pennsylvania Supreme Court noted in 1817, the term "servant," as it was commonly understood, applied exclusively to "indented servants or coloured hirelings."[30]

Indeed, for some black men freedom of movement trumped inducements to set down roots. In September 1819, Frank Paul engaged to drive a team at Birdsboro for eight dollars for one month "but if [he] continued 10$." The following month, "Negro David" agreed to the same terms. Both left after one month.[31] Perhaps they earned what they needed and quit. Perhaps they disliked Birdsboro. Perhaps they had come to associate mobility with black manhood. Gradual emancipation had increasingly feminized and infantilized black bondage. For some, especially those in rural areas without ties to large communities of free blacks, the best way to assert themselves as free black men was to keep moving and avoid becoming too dependent on one employer.

Liberty and mobility were almost mutually exclusive for some married black men and for most single black women with children. To Harry Wilson, a teamster at Delaware Furnace in Delaware, freedom brought the opportunity and the obligation to earn money to purchase his enslaved wife and keep his family together. A job at Delaware Furnace was probably as good as life could get, at least until his wife was free. For Flora, freedom meant domestic service at Coventry Forge and a constant struggle to support her two children. In 1792, Flora earned nearly four shillings a week cleaning house. This allowed her to supplement income that she earned for: three days whitewashing and one of washing, ten chickens, two beef tongues, over three gallons of soap, a live hog, and a half-bushel each of potatoes and of unpeeled dried apples. Flora was relatively fortunate; she at least had access to enough land to feed her family and have something left over to sell. But life was a constant struggle, especially after Coventry had her repay nearly £1 "over Credited . . . this to reduce her wages to 5/ p week on ye first 7 weeks without her Children & all her time afterwards to 3/ p week with her two Children." Domestic service, poverty, and gender conventions similarly circumscribed the freedom of black women on ironworks and in cities throughout the North during the early republic.[32]

The dependency that free black women endured increasingly plagued free black men of Sampson's generation. To be sure, ironworks still offered them jobs. Most emerging industries excluded them. Within the iron business they encountered a labor market much like that which thousands of free black mariners experienced. Their skin color mostly limited them to work that was dirtier, more arduous, and paid less. By 1830, black forgemen could find work, but there were far fewer of them than there had been twenty-five years earlier. Among them were forgeman Robert Yost and forge carpenter Jonah Harmon—the only black tradesmen of sixty-one men whose names and occupations appear in Cumberland Forge's worker provisions accounts for 1828 to 1835. Seven black men only cut wood and one was a laborer. Still, ironmasters were far more likely to employ black men than were other manufacturers. Black ironworkers probably encountered more animosity outside the iron industry than within it.[33]

The declining fortunes of black ironworkers owed partly to employers' attitudes. In 1831, Gardiner H. Wright hired a young black man to help him run the store at Delaware Furnace. Wright's decision horrified his partner, William Waples, who opined that "if he was to be a Negro, to put a Collar on him and be done with it." Waples, one of the area's largest slaveholders, embarrassed Wright, who resolved to discharge his new assistant soon. He felt pressured to agree with Waples—minding the store was no job for a black man.[34]

By then, many, if not most, white ironworkers would likely have seconded Waples's declaration. At Martha Furnace, some white employees targeted black colleagues and their families for abuse or worse. In 1808, James McGilligan "made a violent attempt on the chastity of Miss Druky Trusty ye African." He remained on staff for the next two years. Attempted rape, or at least the attempted rape of a black female, was no grounds for dismissal.[35] But the actions of Jack Johnson, a black man, were. In 1808, he bent a handbar to "spite" Richard Phillips, who had missed his turn at the furnace. It was a Sunday; the furnace was short-handed; and Johnson did not notify founder Michael Mick that he was leaving it unattended. Mick fired him and Johnson was charged for the damaged handbar. Martha Furnace rehired Johnson, who became a marked man. The next year he fought twice with white employees. Some whites who associated with him ran into trouble. James Nash got into an altercation with another white worker on the same day that he brought in a load of ore with Johnson and got drunk with him. Jack Johnson was an outsider, especially because few blacks worked as furnace hands. Unlike other black workers at Martha, he lacked religious ties to white colleagues or to his employers. Most significantly, he openly challenged white men. That isolated him in life, illness, and death. In 1810, John Howard brought a doctor to Johnson's deathbed. Only Howard and one team from Martha attended his burial two days later.[36]

White ironworkers might resent and target blacks for many reasons. For Irish immigrants such as James McGilligan, such violence might stake a claim to employment and to the privileges of whiteness, whether by forcing sex from black women, or by attacking men who challenged them or were potential competitors for jobs. All white men stood to benefit from racial preferences in hiring. Slavery's lingering death ultimately left white workers with little to gain from black labor. Fewer could aspire to become masters by hiring or owning slaves or black servants. They could claim and defend their property in skill and in whiteness, both of which they could leave as legacies to their children.[37]

Gradual abolition bought adventurers time to adjust to the realities of making iron in a new political economy and a labor market that had become whiter, freer, and more mobile. The tortured legacies of gradual abolition enabled them to invoke implicitly a common bond that might unite them to white ironworkers; together they were freemen who could exercise political and social clout to preserve their industry against forces that might destroy it.

Plotting Free Labor

In 1786, Henry Drinker believed that he had seen the future and he wanted to share his vision with Richard Blackledge. By avoiding slavery on principle, he had become a pioneer on the frontier of a whiter and freer industrial revolution. A sober hand was an industrious hand, so Drinker hired "no intemperate drunken Person" and fired those "found to be so." He paid employees promptly and did not require them to patronize Atsion's store. Those who did bought provisions "on reasonable terms and as low as any other would sell," except for rum. "Much pains has been taken to break them" from drinking it, a campaign which "has succeeded in part but not wholly and it is the only article I make them pay a high price for, & knowing my motives they submit to it without grumbling." Drinker reaped "the Fruits of these regulations" in the men he had recruited and retained: "I have divers Workmen that have continued with me from ten to twelve Years, . . . while other Iron Works, within a few miles have frequently suffered largely for want of hands we have turn'd many away & scarcely ever knew the want of them."[38]

Drinker hoped to convince Blackledge of the merits of free labor, which led him to idealize free labor and project his fantasies onto it. Colonial adventurers' struggle to regulate where and how much ironworkers drank tempers Drinker's claim that Atsion employees knowingly paid more for rum "without grumbling," especially when he could not overhear them. The local labor market favored Drinker when he wrote Blackledge. He apologized to customers in 1793 for delaying a shipment of pig iron; he could not say when Atsion's furnace would fire up again because "the various public works such as Canals, Turn pike Roads & c which are now under way, seem as if they would employ most of the Labouring Men—so that we know not whether a number sufficient for our various Services can be engaged."[39] Atsion's forgemen and moulders reminded Drinker that their interests and his did not always coincide. They forced him to tell customers that they determined what he could sell and when he could sell it. Forgemen "very unwillingly draw Bar Iron so small as ¾ Inch square" and customarily demanded extra money for making cart tire or plough share moulds. Drinker had to consult the moulders before he could commit to an order for sugar boilers because they found them "a difficult troublesome article to mould & cast" and might not "agree to undertake that business again."[40] He depended on his tradesmen, so he grudgingly had to accommodate them. Drinker found it "very difficult to keep our Forge-men to a due observance of their duty," and his manager's patience was "frequently tried with their negligent conduct. But this," he resigned, "I believe is almost universally the case at Iron Works & perhaps we are as well off as to workmen as almost any other Iron Masters."[41]

Drinker's adventure with free labor reflects the concerns of mid-Atlantic ironmasters after the Revolution. He tried to regulate drinking; he carefully managed Atsion's store; and he honored ironworkers' demands for prompt payment and for fair store prices. What made those policies successful was that Drinker had attracted and retained industrious and dutiful men, which freed him from the labor shortages and high turnover rates that plagued competitors. Such policies came at a price. He had to acknowledge that he depended on tradesmen and show them respect. The correspondence of other ironworks tells a similar story of an industrious revolution in a world with slavery and servitude in decline—a world in which plotting free labor became a key issue for ironmasters and ironworkers.

Adventurers plotted labor most clearly in timebooks, volumes composed mostly of monthly grids of names, dates, and numbers. Names of workers line the tops of most pages, with dates written along their left margins. Most of the boxes formed where names and dates intersect contain a "1" to denote a full day's work. By adding columns a clerk or manager could easily total how many days each hand had worked that month. Flipping through the grids indicated who an ironworks had employed and for how long it had employed them. Men were graphed against the calendar, their time quantified in fractions of days. Women remained mostly off the grid, making only cameos of a few days to harvest hay and grain.

Timebooks reflected adventurers' aspirations nearly as much as they measured what employees contributed to ironworks. They were fantasies of ironmasters' industrious revolution as well as tools to help them achieve it. Adventurers hoped to see fields of boxes filled with "1" as they perused timebooks. They did not when clerks replaced numbers with words to indicate that they knew a worker to be "sick" or "drunk" or "frolicking" for one day or several. Numbers could also be fantasies. Sometimes "1" concealed that someone had not actually put in a full day. Timebooks, moreover, were incomplete by design; they did not quantify all the time invested in making iron. Ironmasters paid most tradesmen by what they produced. What they did and when they did it seldom figured into the monthly grids that clerks plotted; their time was usually off the books.

June 1808

	Days of the Month	Samuel Stewart	John Hedger	Phineas Wolbert	James McEntire	Owen Hedger	Michl. Mich Junr.	Thomas Estell	John Howard	Peter Cox	John Cumming	Asa Lanning	William Fitzgerald	King & Co. Team	Luther & Co. Team	Loads of Coals	Loads of Ore	Henry Shimer's Team	Jacob Trusty
Wednesday	1	1	1	1	1	1	David Carson	1	1	1	1	1	1	1	1	6	4		
Thursday	2	1	1	1	1	1		1	1	1	1	1	1	1	1	6	4		
Friday	3	1	1	1	1	1		1	1	1	1	1	1	1	1	6	4		
Saturday	4	1	1	1	1	.		1	1	1	1	1	1	1	1	6	4		
Sunday	5	..	..	..	..	..	..	..	..	..	..	..	..	..	..	..	..		
Monday	6	1	1	1	1	.	1	1	1	1	1	1	1	1	1	6	4		
Tuesday	7	1	1	1	1	.	1	1	1	1	1	1	1	1	1	6	4		
Wednesday	8	1	1	1	3/4	.	1	1	1	1	1	1	1	1	1	6	4		
Thursday	9	1	1	1	1	.	1	1	1	1	1	1	1	1	1	6	2	.	
Friday	10	1	1	1/2	1/4		1	1	1/2	1	1	1	..	1	1/2	4	..		
Saturday	11	1	1	1	frolicking		1	1	1	1	1	1	1	1	1	5	.		
Sunday	12	..	..	..	..	..	..	..	..	..	..	..	..	..	..	..	..		
Monday	13	1	1	1	1		1	1	1	1	1	1	1	1	1	6	2		
Tuesday	14	1	1	1	1		1	1	at Home	1	1	1	1	1	1	5	2		
Wednesday	15	1	1	1	1		1	1	Do	1	1	1	1	1	1	6	6	1	
Thursday	16	1	1	1	1		1	1	1	1	1	1	1	1	1	6	6		
Friday	17	1	1	1	1		1	1	1	1	1	lame	1	1	1	5 3/4	6		
Saturday	18	1	1	1	1		1	1	1	1	1	Do	1	1	1	6	6		1
Sunday	19	..	..	..	..	..	..	..	..	..	..	..	..	..	..	..	..	..	..
Monday	20	1	1	1	1	.	1	1	1	1	1	Do	1	1	1	6	3		1
Tuesday	21	1	1	1	1		1	1	1	1	1	Do	1	1	1	6	4		1
Wednesday	22	1	1	1	1		1	1	1	1	1	Do	1	1	1	6	6	1	1
Thursday	23	1	1	1	1		1	1	1	1	1	Do	1	1	1	5 1/2	6	1	1
Friday	24	1	1	1	1		1	1	1	1	1	Do	1	1	1	6	3		1
Saturday	25	1	1	1	1		1	1	1	1	1	Do	1	1	1	5 3/4	2		1
Sunday	26	..	..	..	..	..	..	..	..	..	..	..	..	..	..	..	..	..	..
Monday	27	1	1	1	1		1	1	1	1	1	Do	1/2	1	1	6	6		1
Tuesday	28	1	1	1	1		1	1	1	1	1	Do	1	1	1	6	6		1
Wednesday	29	1	1	1	Sick		1	1	1	1	1	Do	1	1	1	6	6		1
Thursday	30	1	1	1	1		1	1	1	1	1	Do	1	1	1	6	6		1
Total		26	26	25 1/2	23	3	23	26	23 1/2	26	26	14	24 1/2	26	25 1/2	151	106	3	11

Figure 12. Timesheet for June 1808, Martha Furnace Daybook and Timebook, 1808–1815, Acc. 339. Courtesy of the Hagley Museum and Library, Greenville, Delaware. James McEntire skipped work Saturday, June 11, because he was out "frolicking"—that is, on a drinking spree. Asa Lanning injured his thumb and missed work from Friday, June 17, through the end of the month.

Ore from Pappore

1 George Taylor sent the Constable after James Mc. Entire, Thos. Estell & Owen Hedger hauling
2 Owen Hedger & Thos. Estell brought 1 Load Ore each from Pappore in forenoon & 1 Do. from Speedwell Creek in Aftn.
3 a very Sultry Day, put the new Stampers into ye Mill excellent Coal Comeing in.
4 Owen quit driving Team David Van Commenced with it very warm
5 Sultry weather every thing going on well
6 Robert Donnegan & Co. an order balance his a/c Mr. Evans Survey'd a tract Land joining Old French }
7 nothing remarkable this Day every thing going on well
8 James Mc. Entire worked 1/4 Day in his Garden | Shells at the Landing
9 Thos. Estell & David Vann hauled Iron in Aftn. Isaac Cramer del.d a boat & Scow Load }
10 Training at Bodines was very fully attended Journey works most of the Hands about the bank ret.d orderly
11 Elisha Fenemores Team Came in Empty
Teams hauled Holloware Sets ye Landing James Mc. Entire at Bodines again 45 Minutes past 2. P.M. }
12 by John Hedger that he was gone to Mr. Jenkins, got John King & Jacob Stilling to right it, got under way at 45 Minutes Past 11 A.M. one of the Stirrups gave way, sent for Mc. Entire to repair it who ret. for Ans.
13 Teams hauling Holloware to the Landing in aftn. Ore hauled in forenoon
14 finished sending the Holloware to the Landing, finished hauling Coals from further Job Stewart & Luker each 1/2 Load }
(Howards wife ill He staid at home to attend Her)
15 Commenced hauling from near Job Luker hauling Ore Henry Shinn hauling wood to the House with 2 H. Team
16 The Moulders ret.d from the Beach very moist weather = gone to Tuckertown
17 Asa Lanning Idle with a sore thumb John Hedger filled his Turns Mr. Evans = }
18 Ditto Jacob Trusty filling for him
19 remarkably Cold & raw for the time of Year the Furnace Made a small Puff
20 Thos. Estell hauled 1/2 day for M. Meck Junr. & Jno. Luker hauled 1/2 Day wood for Geo. Hunter Estell hauled Iron to the Landing 1/2 day }
21 Thomas Estell hauling Iron to the Landing John King went for Phila.a yesterday afternoon
22 Henry Shinn hauled 1/3 day for G. Hunter & 2/3 day for Jno. Howard. Mr. Evans ret.d from Tuck.on
23 2 Loads Coals Short of Complement (see mem. book) Henry Shinn hauled Iron to ye Landing
24 John King ret.d from Phila.a with a newly purchased Mare Teams hauled Iron to the Landing in aftn.
25 King & Co. Team hauld Iron all day John Kings new Mare in Team, the Other Teams =
26 hauled Iron in forenoon & Ore in afternoon (26th. Sultry & Hot)
27 Wm. Fitzgerald absent in forenoon a very hot Day.
28 Owen & Joseph Hedger quit Chopping, Dan.l Farrell quit Oreraising very warm
29 James Mc. Entire sick, pain in his Back (Elisha Fenemore quit in the Coaling sick)
30 the Furnace doing well very warm weather Luker left old grey out to rest

Many of their actions were recorded in timebooks in dated comments that clerks wrote along the margins of timesheets or on the pages that faced them. Birdsboro Forge's timebooks recite tales of ironworkers whose habits vexed their employers. In 1810, on a Wednesday evening in May, John Colgan returned from a drunken spree that had begun the previous Saturday. On July 3, forgeman George Horn "threw his fire down last Saturday went off drinking Johnson put it up Monday morning." A few weeks later, Horn got drunk in the afternoon and was ill the next morning, which forced another hand to take a double turn. George North "Cut the Gray mare on the back Eating on her with an ax in his hand a careless trick which he ought to pay well for (if he was able he should)." In the summer of 1816, teamster Jim Cook seemed to attend his duties when it suited him. On July 29, Cook "was away yesterday & did not come till 8 this morng." He "left his team stand all day" on August 21. Some three weeks later, Cook skipped an entire day "& came home drunk." He took off the next day and left Birdsboro for good. In January 1817, George Moyer "did nothing but grease 2 pair gears—was requested to grease them all he did not do it." In July 1821, Andrew Nixon obtained a half-gallon of whiskey and "kept drinking for 8 days he then began to work—made one bloom & said the helve would not do." The next September he left the forge grounds and "promised to be back Sunday but did not come till Tuesday & did not work till Friday."[42]

Brief notes on who was drunk or on why someone had put in only half a day were ways of distinguishing employees. They also served an emotional purpose for their authors and their intended audience. With them clerks and managers expressed longing for more compliant workers, stated their exasperation with those they had, floated plans for making workers more dutiful, and uttered fantasies of what they might do to recalcitrant or reprobate hands. In short, they were at times to adventurers and their agents what diaries have often been to individuals, means to vent, to prescribe, to proscribe, and to fantasize.

Timebooks tell a story about the industrious revolution and the meanings of work. They allowed proprietors (and they permit historians) to chart workers' movements. Timebooks reveal how adventurers valued some workers far more than others. Above all, they testify to struggles between adventurers and free labor over the organization of work, struggles in which timebooks provided ironmasters with a tool to combat behavior that they deemed harmful to them. Unlike account books, timebooks did not directly translate time into money, but they indicate greater concern with time, especially in an environment in which adventurers perceived their industry to be under siege.

Labor turnover concerned ironmasters. Timebooks of two Pennsylvania furnaces suggest that ironmasters often succeeded in maintaining a stable corps of workers. Well over half the employees listed in the timebooks of Cornwall Furnace and Pine Grove Furnace between 1810 and 1830 remained for at least one year. Both furnaces retained between 30 and 40 percent of their employees for at least four years and around 30 percent for at least five years. In addition, the average share of Cornwall employees who stayed for at least five years increased steadily over time.[43]

What explains such persistence rates? Concentrated ownership of ironworks likely played a role. The Potts family and the Grubb family owned several ironworks well into the nineteenth century. Cornwall Furnace and Pine Grove Furnace each anchored a chain of family-owned and operated ironworks which had begun to form during the revolutionary era. Between 1810 and the 1830s, the Coleman family owned two furnaces and two forges in Lancaster and Lebanon Counties in addition to Cornwall; the Ege family owned two ironworks and a forge in Cumberland County to complement Pine Grove. By 1840, Peter Schoenberger owned seven furnaces and three forges in Pennsylvania's Juniata Valley. The Richards family dominated southern New Jersey's iron industry for most of the early nineteenth century.[44]

Many ironworks owned by few families had advantages and drawbacks for ironworkers and adventurers. It offered many workers steadier employment, sometimes year-round. Examples of such men abound in the accounts and timebooks of the Richards family's enterprises. Such security came at a potential cost to employees—less mobility and more need to avoid antagonizing their employers, especially when jobs were scarce. Workers who left Cornwall or Pine Grove and wished to stay within the iron industry had to leave the area to escape the influence of their former ironmasters. For adventurers, the ability to offer work elsewhere sometimes seemed burdensome, especially when ironworkers expected a job at another of their furnaces or forges and had the leverage to demand it.

More often, the perks of ownership of several ironworks outweighed the disadvantages for adventurers. It partially shielded them from labor markets; they complained less of shortages of hands than did other ironmasters. In 1793, internal improvement projects drained Henry Drinker's labor pool, in part because he had no work to offer outside of the Atsion Iron Works. Ironworks that belonged to a family network probably suffered less from exposure to the agrarian calendar, a problem that plagued Delaware Furnace. William Waples observed in 1821 that "we have found hard work to get hands, it being in harvest time and such bad weather that every person seems to be engaged." Two months later, he griped that "those we hire hear has some farm to attend To and they are breaking of at times which confines our Work." Waples needed men from Philadelphia to assist in coaling "for we can not get Such hear at present that will do us. . . . It will be impossible to get four loads per day until we can procure some more hands."[45]

Waples was fortunate. He owned a furnace located near water transportation to Philadelphia and he had a partner in the city to recruit hands. Delaware Furnace and other nearby ironworks could find many laborers in Philadelphia. Urban unemployment and underemployment rates soared during winter, precisely when most ironworks laid in cordwood for spring. Urban migrants insured against local labor uncertainties, as Delaware manager John G. Smith noted when he requested that ten to twelve woodcutters be dispatched "immediately. There is no dependance on those I have here."[46]

Growing ranks of landless men also provided labor. Ironmasters whose enterprises claimed tens of thousands of acres frequently traded access to land for service. Daniel Conrad promised Curtis Grubb in 1784 that he would "Cut 150 Cords wood at 2/6 p Cord Per Year, for the Rent of Plantation he lives now on or haul wood to the amount thereof." In the 1790s, Henry B. Grubb began to strike sharecropping or tenancy agreements that enabled him to summon workers when he needed them most. In 1798, Peter Bright began a three-year lease on "Hagys Place," located near Hopewell Forge. Grubb reserved a new orchard for his own use, prohibited Bright from farming the plot except "for a little Indian Corn for his own Use & Flax, Potatoes & c.," and required him to maintain the fences. Bright was to pay the rent by hauling pig iron from mid-May "Yearly & every year untill after seeding." Bright's lease was one of many Grubb signed, most of which stipulated that the tenant cut no live timber and that he prevent others from poaching it. By leasing land, Grubb protected the fuel supply of his ironworks and sometimes provisioned ironworkers and livestock.[47]

Grubb and other ironmasters also lured labor with the promise of landownership. In 1796, George Petz bought twenty acres from Grubb, "to be paid by Mason Work at Mount Hope Furnace & Hopewell Forge as" Grubb directed "the Work to be done by him as reasonable as it can be had from others." Grubb bound Petz to work "at anytime he may want him at either place." Petz's path to land ownership came with fewer strings attached than did Jacob Greenwalt's. Greenwalt agreed to pay for two acres by working exclusively for Cumberland Forge owner Jacob M. Haldeman, except "a few days in harvest or hay making," and was "always to consider himself in the employ of" Haldeman. If Greenwalt were ever unwilling to work for Haldeman "as is right & just," the ironmaster was "to pay him back for the land, and for the improvements . . . what any honest, disinterested & upright person is willing to give." Haldeman's obligations were vague: who was an "honest, disinterested, & upright person," and what value would such a person assign to the labor Greenwalt had put into the land? Greenwalt's obligations were clear. His labor effectively belonged to Haldeman, who put it to good use. By 1810, Greenwalt had cut more than 2,500 cords of wood, helped to build a road, made hay, repaired the forge, made a hog pen and fences, boarded a barn, and put down a barn floor. Land ownership promised independence and the realization of full manhood. Men like Grubb and Haldeman guarded the path to them. The price that Petz and Greenwalt paid was greater dependence—at least in the short term.[48]

What Petz and Greenwalt owed for their land literally grounded them to the men who sold it to them. This echoes a familiar story; debt often enabled colonial adventurers to keep ironworkers in place. But it seems that ironmasters after the Revolution tried to minimize how many indebted hands they employed. Cornwall Furnace's "Workmen's Accounts" for 1806 to 1816 contain 153 names. The overwhelming majority, 133 (87 percent), had positive balances. Only eighteen owed Cornwall money—ten dollars each on average. Cornwall owed creditor workers an average of nearly thirty-one dollars, six of them more than one hundred dollars each.[49]

The Coleman family probably saw several advantages in seeing a ledger full of creditor employees. Tracking down debtors in flight was burdensome. In 1785, Samuel Laverly sent Peter Grubb a list of six men who owed him a total of nearly £16 and asked Grubb to send payment "if you intend to Continue them in your Imploy as I would not wish to disoblidge you by puting them to trouble." Henry Drinker asked an associate in 1792 to search southeastern Pennsylvania for Henry Parsons so that he could collect on a four-year-old bond worth over £30. Timebooks noted that some hands opted to outrun their debts; in 1810, William McCloskey left Birdsboro Forge "clandestinely (in debt)."[50]

Sometimes debtors who stayed vexed adventurers more than the ones who ran. Delaware Furnace manager Derick Barnard was sure that some employees piled up arrears to keep a job. He asserted that "the majority of our moulders had always been a curse to the furnace, but they have been in debt from the start, and I wished if possible to extricate them." They "are always in debt if they can get so (and they never want opportunity here) consequently there is no check upon them. You may take such as they give you or nothing." To Barnard, employee debt tied his hands. If the moulders set down roots, he could do little to change their ways. Whether he fired them or they left, the furnace picked up the tab.[51]

Therein likely lay the story behind Cornwall's workmen's accounts. The furnace stood a better chance of retaining employees to whom it owed money. Carrying a small amount of book credit for most employees gave the Colemans the option of weeding out those they no longer wanted; they could simply settle and let them go. Fewer hands in arrears meant fewer feet on the run and the expenditure of less time and money to track them down or write them off. George Petz and Jacob Greenwalt had to stay involved in the iron business; they could not take their land with them. Positive balances on their accounts gave Cornwall hands reason to stay.

Adventurers could also regulate turnover by making ironworkers pay for leaving. Mount Hope Furnace charged Godfrey Grindal nearly £4 in 1786 for "Neglecting his Business and Going Off without serving Notice." In 1798, Hopewell Forge fined teamster Charles Harding "for running away And disapointing the team and Not Complying with contract." The next year, Hopewell charged John McGinly "for not Complying with Contract having engaged for 1 Year & staid but Half." Codorus Forge deducted half the cost of nine months' boarding from George West's account in 1819. West would have received one dollar a week "Provided he workd a year but went away before his year was up on acct of which he forfeited 50 p week."[52]

Enforcing contracts with fines was not new. Adventurers began to penalize teamsters and colliers in the 1760s for failing to discharge their duties. Fines continued to fall most heavily on them after the Revolution. Teamsters came under particular scrutiny at Martha Furnace, where agents documented each team's daily activities.[53] Adventurers extended the use of fines to discipline furnace and forge staff. Mount Hope Furnace charged William Roberts "for Getting drunk on his Turn." In 1796, Moses Jackson paid £1 to Hopewell Forge "for Refusing to put up the Chafry Hammer." Christian Miller, assistant to Codorus Forge hammerman George West, doled out $2.50 in 1820—$1 to West for being "too drunk" to help him and $1.50 to the forge for charcoal that his hearth burned while he was stupefied. Miller's employers may have worried that West had become a bad influence on his underhand. In 1819, West forfeited the cost of nine months' boarding by leaving early and he paid one dollar for "not coming to work on monday morning," during which Miller let the charcoal burn. Some three years later, Codorus fined West five dollars: three dollars for getting drunk at work and two dollars for not manning his post when called.[54]

Many ironworkers' fines stemmed from adventurers' efforts to curb their drinking. Colonial ironmasters seldom used fines to regulate consumption of alcohol, preferring instead to restrict workers' ability to purchase it. Fines usually targeted problematic drinking. In 1811, following a series of incidents, Martha Furnace manager Jesse Evans "made Sollemn resolution" that anyone "bringing Liquor to the work enough to make drunk . . . shall be liable to a fine." His policy perhaps succeeded; the furnace's timebook mentions fewer incidents of intoxicated employees over the next three years. Martha Furnace also discharged some hands when their drinking resulted in chronic absenteeism or otherwise badly impaired their performance.[55]

Such campaigns often foundered against the culture of ironworkers. Evans narrowly defined employee liability to fines because a broader policy would have failed. Most ironworkers would have disregarded it; alcohol was an entrenched feature of their lives. At Martha, most hands celebrated when the furnace stopped in 1809 and 1811 by becoming drunk. Ironmasters treated ironworkers with alcohol well into the nineteenth century because their workers demanded it. In 1783, Cornwall manager John Campbell requested more whiskey for workers "as they Want Some at Christmas." New District Forge gave workers at the head race and the "new Coal house" twenty-one quarts of whiskey in 1799. Greenwood Forge woodcutters received three quarts of cider and one mug of beer as a reward for their labor. In 1821, Codorus Forge distributed whiskey to forgemen, whose drinking it tried to curb with fines. On occasion adventurers tried to deny ironworkers direct access to alcohol by refusing to stock it in their stores, as did Hopewell Furnace, which sold no whiskey from 1817 to 1821 and ceased carrying it in 1826.[56]

The labor market also influenced how adventurers addressed workers' drinking. Skill often determined which workers were disciplined for drinking that impaired their job performance. The most notorious and disruptive abusers of alcohol at Martha Furnace were carpenter James McEntire and smith Solomon Reeve, yet furnace managers never took either to task for it. Delaware Furnace manager Derick Barnard reported in late 1822 that he had "plenty of hands *of all kinds* and shall soon have some to spare." As a result, he planned to fire master collier James Anderson because he found him "very dissipated and consequently neglectful. I should have done it long since if I could have been assured of Mr. Westcoats staying." By discharging Anderson, Barnard began to implement a policy which would enforce owner Samuel Wright's views regarding "ardent spirits and profane language."[57]

Delaware Furnace's campaign to promote temperance soon unraveled. In 1824, Barnard complained that "there appears to be something peste-

Figure 13. "From the tavern beyond Ege's forge," Berks [today Schuylkill] County, Pennsylvania, 1802. Watercolor on paper by Benjamin Henry Latrobe. Courtesy of the Maryland Historical Society, Baltimore, Maryland. The tavern was located near George Ege's Schuylkill Forge. The men in front of the tavern are likely ironworkers.

ferous in the air of this place and if a *tolerable* sober man come here he generally goes away a sot." He attempted to contain the problem. In 1825, he discharged a moulder who had "been a fire brand ever since he arrived" as well as "a poor workman and a *complete sot.*" Two years later, Barnard refused to hire another moulder because he suspected that he drank too much. The Wrights persisted in trying to dissuade workers from drinking—to little avail. In 1833, Gardiner Wright denied rumors that the moulders at Millville Furnace in New Jersey drank excessively: "[O]ut of 23 men employ'd in the furnace but 3 drink to excess, and they but rarely. We have 13 moulders and but 2 of them drink at all." Moreover, "of the five men employ'd of his work, not one tasted spirits in any form, and have not the whole blast!" Wright's defense rings hollow. He was swimming against a strong tide and he knew it. In the early nineteenth century, Americans consumed more alcohol than at any time in their history. If ironworkers' drinking habits were cultural, then adventurers had to reform the culture of ironworkers.[58]

The campaign by adventurers to combat drinking reveals one facet of the industrious revolution in the new nation. Detailed reports from Delaware Furnace to Samuel Wright, mostly from Derick Barnard, vividly document how adventurers and ironworkers struggled to redefine their relationship. Adventurers resorted to discipline, even mass termination, to ensure worker compliance when they judged that the labor market favored them. Ironworkers often facilitated such efforts, sometimes deliberately, when they deemed it in their individual or trade's interest to do so. Ironworkers shaped adventurers' policies, though their influence rested largely on the degree to which they had mastered a trade. Skilled hands, especially those who cast or forged iron, could compel ironmasters to accommodate their desire for workplace autonomy and job security.

Derick Barnard arrived at Delaware Furnace in the spring of 1822. His initial assessment of the situation there echoed that of John England at Principio Iron Works a century earlier; he had to show unruly men unaccustomed to discipline who was boss. Barnard soon learned that his predecessor had put off paying the men. They surrounded him "like a pack of hungry wolves, determined to have something, not asking it as a *favor*, but demanding it *as a right.*" Barnard refused to concede that "to have something" was their right, but he tried to placate the workers by paying them as much cash as he could. That did not satisfy some. "Last friday," he alerted Wright, "I discharged one of the best men in the company *for an example* because he was calling for the bulk of his wages in cash." Nor was Barnard finished with making examples of Delaware employees. That summer he repeatedly found ore in places that the miners claimed they had exhausted. He instructed the supervisor "to raise every stone before he left." The miners, Barnard observed,

> did well for 10 or 12 days at first but I found they began to be careless, I then gave them positive directions to leave *nothing.* I told them that ore was too valuable to be wasted, and I was determined it should not be done. A few days after I was out and found my expostulations had not produced the effect I desired, and I discharged every man on the spot.

"Since that," he noted, "I have had many fair promises from the master ore raiser, but I have not yet formed a new company."[59]

Some resented the new order that Barnard brought to Delaware Furnace and tried to change it. They fared no better. In July 1822, the carpenters, in Barnard's words, "formed a combination for the purpose of extorting more than the worth of" a new coal house he planned to erect. He fired them and engaged a group of moulders to replace them. "This plan defeats them," he congratulated himself, "[it] gives our moulders employment and keeps them together and out of debt."[60]

Five months later, Barnard faced down a more significant challenge.

Five furnace hands deserted their posts and the furnace grounds. Had they stayed out long, they would have crippled the furnace. Some broke almost instantly; filler Thomas Charlesworth and keeper William Stitser returned once "they found their efforts unavailing." They insisted that the others had convinced them to walk and they promised that they would return to work and never cross Barnard again. Barnard threatened them with legal action "to shew them the enormity of their conduct" but relented and took them back since they seemed "truly penitent."[61]

The other three did not give in so easily. Hugh McMenomy and Thomas Adams "continued obstinate, and Robert Downs was scarcely prevented from beating Stitser for his desertion of his cause." Barnard decided "to make a signal example of them, not more to punish them than to convince others that they were endeavouring to coerce authority that must be respected" by issuing writs against them. "Thomas Adams has since repented in sackcloth & ashes," he gloated, "he has pleaded, begged, and cryed for admittance to his birth, but all in vain. I have not yet punished him sufficiently." Adams campaigned to get rehired. He asked William Waples to intercede and he presented Barnard with a petition signed by Joseph Harvey and all the moulders which requested his reinstatement. Barnard complied—after he hauled Adams before the sheriff and secured a bond of $500 from him. The others proved harder to cow. Robert Downs and Hugh McMenomy "have been constantly throwing out threats. Robert said yesterday that he would be 'd—d if he did not change the working of the furnace before he left the ground'—they shall suffer severely." Barnard fired McMenomy and sent him to jail. He had to agree to pay the costs of his imprisonment and acknowledge before he could get out that he had " 'freely and voluntarily left the service of Saml. G. Wright Esq. without cause, and without fulfilling my engagements, thereby subjecting him to damage.' " This was music to Barnard's ears; he happily reported that his handling of the affair "has had an excellent effect upon all the hands in the furnace."

Barnard's success rested largely on a local surplus of labor. The workers who demanded payment when he arrived knew it and chose to stay. Where else would they go and how else would they collect what was owed them? After Barnard dismissed one of his best workers for demanding cash, he noted that "his place was immediately filled by a good hand." In November 1822, Barnard had "plenty of hands *of all kinds* and shall soon have some to spare" so he could fire James Anderson and begin to enforce Wright's views on "ardent spirits and profane language."[62]

Barnard's victory also lay in his employees' inability, or their refusal, to stick together. This owed to the local labor market and to other circumstances, many virtually unique to the iron industry by the early nineteenth century. Consider the failed walkout that Delaware's furnace hands staged. Why did it collapse so quickly? Perhaps financial hardship forced Charlesworth and Stitser to return so soon, though it is hard to tell since detailed furnace accounts no longer exist. Barnard may have been right; they recognized immediately that they would lose and threw themselves upon the manager's mercy, though jail may have scared them even more. Many employers summoned the power of law against defiant workers during the nineteenth century, as did Martha Furnace's manager when he took out a warrant in 1811 against collier Richard Rose to address his "indolence & sauce." Fear of legal consequences certainly allowed Barnard to compel Hugh McMemony to state that he had been completely at fault.[63]

Three other features of ironmaking culture bolstered Barnard's hand. Ironworkers were more inclined to battle one another than to battle adventurers. Accounts of employees fighting one another, sometimes on the job, sometimes in or near local taverns, fill Martha Furnace's Diary. Ironworkers had a reputation for a "rough" masculine culture similar to that of canal workers, in which hard drinking, quick tempers, and physical prowess were badges of honor. The Martha Diary is replete with stories of domestic violence, most vividly in 1813 when Joseph Camp got drunk, "abused his wife," and the next day was "in high order and threat'ed to kill himself and family." Ethnic tensions only added to the violence that ironworkers so often directed at women or at one another.[64]

What thwarted collective action even more, at Delaware and in the mid-Atlantic iron industry, were the contracts that ironworkers hammered out with adventurers. The issue was less the contracts' language or terms than that they were negotiated individually. Workers often earned different wages, salaries, or piece rates and different benefits, such as housing, pasture rights for livestock, or firewood. William Sullivan has attributed the relative tranquility of labor relations in Pennsylvania's early-nineteenth-century iron industry largely to such a bargaining system. Few workers who possessed identical deals with ironmasters meant few workers who bore the same grievances against them. Sullivan notes that ironworkers managed to unite against adventurers only around common concerns, such as the quality and variety of provisions or the form that payment should take—the situation that Barnard confronted when he became manager.[65]

Barnard also won because he exploited other divisions between ironworkers. By the early nineteenth century, some skilled ironworkers had developed a craft consciousness—especially moulders and forgemen, the men who handled iron and literally shaped it. The moulders at Delaware

Furnace believed that they shared interests which superseded those of the carpenters. Barnard needed the moulders more than he needed the carpenters. Hiring them to build a new storehouse thwarted the carpenters and it kept the moulders happy because they remained together and gainfully employed. Barnard's gesture served him well when the furnace hands struck—moulders' support would have bolstered their cause. Unlike Robert Downs, the moulders apparently saw no problem with the way Barnard ran the furnace. The most they would do was scribble their names on a petition to get Thomas Adams rehired.

Barnard gloated too soon. Within a few months the local labor pool had shriveled. In July 1823, Barnard dreamed of sending off the founder, a keeper, and several moulders for their poor work habits. "Could I replace them," he vowed, "I would discharge them every day before I would put up with it, but I am troubled with too many raw hands already." By the next summer the surplus of hands had evaporated. Barnard pleaded with Wright to "come down and regulate prices of wages & C. The wages are too high, but I do not see as they can be lowered before next blast." He could not replace a collier because he "did not wish to do any thing or have any thing done to offend him if it could be possibly avoided. He has considerable influence with the Petersons and some other good hands." The following January, colliers won another concession. After Barnard directed them to stop making charcoal, so many objected "that I fear I shall have to countermand the orders. . . . The hands all declare they will go to the city, but I will not settle with them. I have offered them wood chopping." Barnard also counseled compromise when he advised Wright to give founder Jesse Peterson a small raise. He had heard that Peterson had left the furnace and did not plan to return. Paying him more would be "far preferable to getting *any* new founder" because Peterson "understands our ore, and manner of working." To his manager's relief, the founder stayed on.[66]

The moulders were the most challenging workers for Barnard and Wright. They, along with forgemen, proved the most eager and the best able to contest ironmasters' power. They possessed scarce knowledge vital to ironmaking, which made their services valuable and sought after. Moulders demanded, and usually enjoyed, a degree of workplace autonomy that other ironworkers envied. They formed and participated in communication networks which enabled them to find jobs. Moulders could demand security from adventurers and usually get it.

Recruiting reliable moulders for Delaware Furnace was a constant struggle. In October 1824, Barnard reported that Jesse Peterson had engaged "three good moulders" at New Jersey's Batsto Iron Works. Three years later, Delaware hired two in the District of Columbia. Wright and his agents also turned to former employees. Barnard urged Wright to consult Joseph Baughman, who had made castings at Delaware, should he fail to hire enough moulders in New Jersey. Wright probably took his manager's advice to heart, since he had received word that Baughman had recommended Joseph Williams, John Shifer, and Charles Evans as "Sober Industrious young men."[67]

Moulders and other furnace hands at times solicited jobs from Samuel Wright. Batsto founder Samuel McAnnine wrote Wright to express interest in working for him and offered recommendations from Jesse Richards at Batsto and John Richards at the Weymouth Iron Works. Waples, after mentioning to Wright that he wanted two moulders because "it is a pitty that Such Iron as is now making should be run into pigs," noted that "Samuel Laning a Moulder at Etna Furnace Sent us word if you wanted he would come. He is a good Moulder." In 1824, Delaware employed three moulders "from Buckley's furnace in Pa." after it went out of blast. They stayed until replacements arrived.[68]

Adventurers had to resort to a variety of measures to engage moulders. In 1822, John King and Joseph Klutz promised Henry B. Grubb that they would make castings at Mt. Hope Furnace during the next blast. Their contract tied their compensation to what nearby furnaces paid; King and Klutz would earn what moulders made at Joanna Furnace for each ton of hollowware that they cast. Grubb pledged to match the price that Hopewell Furnace paid for making stove plates. King and Klutz would get one-third of the value of what they produced in cash, two-thirds in castings and merchandise. If the moulders preferred to be paid in castings, Grubb agreed to deliver them anywhere not exceeding the distance to Lancaster from Mt. Hope "at the Wholesale price he may be selling at, provided the quantity is not less than one waggon Load."[69]

Such detailed contracts are no longer extant for Delaware Furnace, but records of cash advances to moulders are. Samuel Wright advanced Robert Larre four dollars in 1825. William M. Matthews received $8.50 before he headed to Delaware in 1827. John Headrick obtained cash advances for himself and his son. His four daughters, who lived in Philadelphia, drew a total of twenty-six dollars on his account during May and June 1827. Two years later, Headrick received another twenty dollar advance on wages for himself and his son. Wright also used cash to lure moulders from farther away. In November 1823, Wright sent Bela Kingman twenty dollars to come to Delaware Furnace "with men from Massachusetts."[70]

The payment culminated an effort to recruit Kingman that took over a year. Kingman first offered his services to Wright in August 1822. He asked to know what types of castings Wright would expect him and his associates to make and he reminded Wright that it "has been a uniform practice" that ironmasters pay the travel expenses of moulders who journeyed long distances. Should Wright decline, he should then advise them when they should go to Philadelphia and he should support them until they could begin work. Kingman also warned that if he and his peers "should be under necessity of spending time in procuring sand, or any other labor than moulding we shall expect to be paid wages." Some of what Kingman said sounded better to Wright. "I should be as unwilling to introduce an Intemperate Man, as you would be to employ such a Character" answered Wright's demand for sober workers, as did assurance that "we shall not go so far to spend our time in Idleness or dissipation, or our wages in Whiskey or peach Brandy. We are all middle aged Men, [and] have worked in various different Furnaces." Kingman believed that "it will not be best to contract for a stipulated time, but to reserve to ourselves the privilige of quiting at any time by giving you seasonable notice, & you to be at liberty to discharge us when you please." Wright liked what he read. On the back of Kingman's letter he scribbled "does Col. Waples know this man—write me." Waples knew Kingman. He remembered him as capable and honest, though it worried him that the moulder did not wish to commit to Delaware.[71]

Kingman was getting impatient. Two weeks had passed and Wright had not told him what types of castings the moulders would make, whether he would assume travel expenses, when they would get paid, or how many men he would hire. He chastised:

> It is not the business of a day, to make the necessary arrangements for leaving home for 8 Months. If you are disposed to answer these Queries, I presume you will do it Immediately, for by your answers we shall determine whither to go, or not. If you do not answer them, we will drop the correspondence. I can find business nearer home. If *good* workmen cannot make *good wages* in your furnace by doing 3 days work in 2 days, we do not wish to go.

A few weeks later Kingman informed Wright that he and his peers would "comply with your terms so far as to go to the Furnace, & if we find the prospect (on seeing the flasks, patterns, sand, &c.) such as to suit us, we shall commence mouldings," though they insisted on "the privilige of quiting when we please after having given you or your Manager 20 days notice." The deal fell through. The moulders decided to engage elsewhere for the season, mostly because they could not reach an agreement with Wright.[72]

The next fall, Kingman and three companions began work at Delaware Furnace. Barnard soon fired all four. Two, he claimed, cast shoddy stove plates. The other two worked so slowly "that it never would answer to have a sand heap occupied by them." Barnard also complained that the four men lost "more plates by bad pouring" than the rest of the moulders combined. Kingman refused to go quietly. Neither he nor his associates had heard a complaint until Barnard discharged them. Kingman admitted to difficulties; they were not accustomed to the moulding sand used at Delaware. Wright sided with his manager.[73]

Kingman was in many ways no different from other moulders. He tried to negotiate the best deal that he could for himself and his men. He conveyed clearly what he would brook and what he valued. What he most wanted was that Wright and Barnard respect him and his craft by taking his queries seriously and allowing him a wide berth to determine how and how long he would work.

Yet Bela Kingman was unlike other moulders in that he did not demand job security. A few months after Kingman left, Barnard argued that Wright could not stay in business unless he addressed moulders' desire for secure employment and steady income. Delaware Furnace had to be "connected with a furnace in Jersey to be carried on to advantage" so that Wright could "then have a picked sett of hands. Till that happens this furnace must depend too much on temporary hands who have no ties of interest to bind them." Five days later, Barnard advised Wright to consider producing only pig iron because Delaware might never have

> a full set of good moulders. The *climate* is a very *serious objection*, for men of families—and even young men cannot be retained in any subordination after the Jersey furnaces go in blast, even though they agree to stay the blast. They are, after that time wavering and unsettled, ready to go at any moment, and in fact, ready to do any act, that may lead to their discharge. This is at present the situation of almost every man at the furnace, and I believe it impossible for any manager to get good work out of hands in that frame of mind. Make it the interest or necessity of hands to stay and they are under command. But the moment that bond is loosed, the moment they see a prospect of lengthening their job, or changing their situation perhaps for a better [one]. The man who has the management of them need not be envied, where men have to be coaxed and threatened from day to day like children 'tis no job to be craved. This is the most serious difficulty under which we labour, and I fear it is comparatively speaking insurmountable.

Unless he could provide steadier work, Wright might as well shut down the furnace.[74]

Wright acquired two furnaces in southern New Jersey. Wright's moulders soon considered it their right to work at them. In 1827, Barnard asked Wright to come to Delaware to negotiate agreements with moulders and warned him that several threatened to leave "unless they make a *positive agreement* with you to go to Dover." Five years later, a rumor circulated among moulders at Wright's other New Jersey furnace—they might not get berths at Dover as they expected. All got drunk and four of six stopped working. To Gardiner Wright's astonishment, the moulders "went so far as to say they would leave the furnace unless they thought they could get work at Dover when they returned," especially since "they had worked in father's employ for 7 and 8 years and thought it hard to be turn'd off now." Perhaps the moulders bluffed by equating the prospect of no jobs at Dover with being "turn'd off." Even if they were, their assertion underscores their belief that their employer owed them security.[75]

It is nearly impossible to imagine moulders, or any other ironworkers, issuing a similar demand sixty years earlier. It would have fallen on deaf ears. Their craft was their property, though that property was worth little without a job. Moulders and other skilled ironworkers were acutely aware that they depended on adventurers in a society that celebrated independence. They could not afford to buy their own enterprises. What better way to preserve their livelihood and to mask their dependence than to demand the right to employment? Curiously, Delaware's moulders couched their grievance, at least as Gardiner Wright remembered it, not as an assertion of manly defiance but as a complaint of an emotional bond violated—after years of loyal service, it was "hard" to be treated this way. Did the Wrights have no shame? This too would not have been uttered sixty years earlier in the iron business.

Similar sentiments echo in words that James McFadden scratched out to Samuel Wright. In 1833, McFadden, a filler and keeper at Delaware Furnace, begged help to secure a $200 settlement that he believed the furnace owed him. He reminded Wright that he had "being a faithful servant" and that "now i am cripeled and have not worked a day Sense the second of may last and the have settleed with all the rest of the furnas men that worked at the same time at furnase but me." McFadden pleaded that he had

> Don more hard work than any other man that Ever worked at Delaware furnace and your books will show it. . . . I never flenched at keeping or filling and now my work is Eight hundred Dollars since i had a settlement

> and last winter and it very cold wether and my child telling his mother that we shall frese. I got a load of wood that was fifty cents trade and Cornel W. D. Waples would not let the Clark in furnace store pay for it.

Waples would not show him mercy. What could McFadden do but appeal to Wright? He might have pursued the matter in court, but winter loomed. He had no time for legal maneuvers.[76]

McFadden made the best move that he could. He asked for no more than what the ledger said was his, though a "faithful servant" deserved better, especially one who was "cripeled." McFadden went further in his letter. He invited Wright into his home to hear his son condemn them both. Maybe the boy said that he feared the winter, maybe he did not. That the boy confessed to his mother and not to McFadden directly only made the scene more poignant, and more humiliating to his father. McFadden's point was that he had failed to provide for his family, despite his best efforts, because Wright had not honored his duty to McFadden. Wright had unmanned him before his wife and boy—only he could set things aright. To do otherwise would expose Wright as a heartless ingrate who fled his obligations. Here was free labor plotting while in check.

Storming Hell, Storming Washington

The bargain and the sentiment that James McFadden invoked arose within the iron industry during the revolutionary era. Colonial adventurers emphasized coercion—open or concealed—in their relations with ironworkers. They seldom pretended that they or their enterprises should or could lift up those whom they employed. The Revolution, the efforts of evangelical Christian missionaries, and the industry's struggle against foreign-made iron prompted Pennsylvania and New Jersey ironmasters to reimagine their relationship to ironworkers. Concern for the souls and for the civic identities of white ironworkers became, by 1800, part and parcel of the industrious revolution that ironmasters tried to nurture. They reaped most of what they sowed. Ironworkers responded favorably—to evangelical Christianity, to their employers' political initiatives, or to both—mostly because ironworkers concluded that doing so served their own needs.

In 1778, John Lesher, owner of the Oley Iron Works, protested to the Executive Council of Pennsylvania that foraging Continental soldiers had destroyed his pastures, burned his fences, and forcibly requisitioned hay, cattle, and hogs. Lesher wielded the language of property rights—the troops and the government had unfairly seized what they took—and contended that the Army had cut off its nose to spite its face. He would have to idle his forge and his furnace, which "will be a loss to the public as well as myself as there is so great a call for iron at present for public use." Lesher knew he had a strong argument—ironmaking was a matter of public concern and the Oley Iron Works was a strategic asset. It helped too that he had impeccable credentials as a patriot.[77]

Lesher did not think that his case was strong enough, so he fortified it by charging that the Army had rendered him unable to meet his responsibilities to his workers. Lesher considered many of them "my Family, which I am necessiated to maintain," who numbered "near 30 Persons, not reckoning Colliers, Wood Cutters, and other day Labourers." In a sense, they were Lesher's family. He fed them; and he, like other adventurers, shielded them from military service by employing them. If ironmaking had become patriotic by arming the United States, then Lesher's relationship to Oley's hands was a matter of public concern. Pennsylvania authorities had to acknowledge and honor his commitment to them. They depended on Lesher, and the commonwealth's cause depended on his ironworks. As Lesher reckoned it, by thwarting his ability to fulfill his obligations to his "family" of ironworkers, the Army threatened to undermine the nation that was struggling to emerge.[78]

Lesher and most other adventurers hitched the future of their enterprises to that of the new republic. They did not believe that they would ever reap the benefits of national independence without a strong national government. An American iron industry needed political stability, a dependable credit system, a truly national market in which they could sell their iron across state lines, and, above all, protection of the domestic market from imported iron. All prompted most adventurers to support ratification of the Constitution. They led some to become ardent Federalists. In 1789, Batsto Furnace shareholder John Cox urged William Richards to line up employees to vote for Cox's favored ticket in New Jersey's first election under the constitution.[79] Several ironmasters, including Robert Coleman, attended the 1790 convention which wrote a far more conservative constitution for Pennsylvania. Coleman and Robert Jenkins joined troops that suppressed the Whiskey Rebellion in 1794. George Ege was elected to Congress as a Federalist in 1795 and in 1796. Two other Pennsylvania ironmasters, Jenkins and Daniel Udree, served in Congress as Republicans—Jenkins in 1807–1811, and Udree in 1813–1815 and 1822–1825. All supported high tariffs on imported iron.[80]

Adventurers' campaigns for tariffs shaped how they related to ironworkers.[81] When they pressed for government protection, they argued that their employees were freemen whose toil supported thousands of dependents and the republic. Ironworkers could not simply serve as rhetorical props—they had to raise their voices in chorus with their employers if adventurers were to win more attentive ears in state legislatures and in Congress. Free white ironworkers' signatures and votes underscored ironmasters' claim that they were "freemen" who often elected sympathetic candidates to state legislatures and to Congress. This was especially true in Pennsylvania, where all adult taxpaying residents who were freemen and citizens had possessed the right to vote since 1776. Certainly many of the nearly 800 men who signed Pennsylvania forgeowners' 1785 petition for duties on bar iron were ironworkers. Ironworkers, like urban artisans, may well have supported passage of the Constitution on the grounds that it would empower a national government to impose duties on imported manufactures. Adventurers may also have been able to call upon British traditions which viewed the iron trade as an entity in which ironmasters and ironworkers shared interests and to which both owed allegiance in order to mobilize employees' support for their employer's political initiatives.[82] The gradual emergence of near universal white manhood suffrage throughout most of the North during the early nineteenth century raised adventurers' stake in ironworkers' political activities.[83]

Adventurers particularly needed their employees' help after the War of 1812, when foreign iron began to flood into the United States. Pennsylvania manufacturers, led by ironmasters, inundated Congress with pleas for relief in 1817. To make Washington heed their cries, adventurers had to document how much their industry mattered to the nation. In 1817, a convention of ironmasters held in Philadelphia directed John Weidman to ask Jacob M. Haldeman to report: how much capital furnaces and forges in Cumberland County employed, how much iron they made annually, how much production might increase if demand grew, and the "Number of Persons employed at each works, also the Number of women and children supported by the laborers and who are dependant upon the works for their subsistance." Weidman also enclosed "Coppies of two Memorials, . . . the one to be signed by those immediately interested in Iron Works, and the other by those Citizens of all denominations friendly to these establishments" which Haldeman was "to give as extensive a circulation as possible" and return to Philadelphia. The Convention and Haldeman needed his employees' help. They provided the information on how many people actually depended on Cumberland. They would also surely furnish many of the signatures on at least one of the memorials that Haldeman was to circulate.[84]

The Philadelphia convention's efforts bore fruit. In 1818, Congress increased duties on hammered bar iron and for the first time levied taxes specifically on imported pig iron. The ironmasters' victory ushered in two decades of increasingly heavy protection of the domestic iron market. Adventurers won higher duties in 1824 and in 1828; they turned back attempts to lower them for another ten years. It paid to employ freemen, especially when it came time to do battle in Washington.[85]

By 1800, ironworkers also joined their employers in combating their enemies from the underworld. Ironworks in Pennsylvania and New Jersey became regular stops for evangelists during the revolutionary era. Few itinerant preachers dared to visit them before. Legend has it that George Whitfield, the British Atlantic world's most famous traveling evangelist, narrowly escaped death at the hands of Warwick Furnace workers. Rebecca Grace supposedly saved him. The story made Methodist Benjamin Abbott seem braver when he accepted her invitation to preach at Warwick in 1780. Abbott recalled that Warwick

> for wickedness was next door to hell; here they swore that they would shoot me. Mrs. Grace, hearing of their threats, and being herself unwell, and not able to attend, sent a person to moderate the furnace-men and colliers. . . . I went into the house, and preached with great liberty: several of the colliers' faces were all in streaks where the tears ran down their cheeks.

Abbott then called on Rebecca Grace, who "took me by the hand, and said, 'I was never so glad to see a man in the world, for I was afraid some of the furnace-men had killed you, for they swore bitterly that they would shoot you.' " He preached the next day at the furnace "and the Lord was very precious, many wept and sighed." Grace later donated a building for Methodists to hold services, and she willed land for a meeting house to her township's Methodists.[86]

Rebecca Grace was but the first of several adventurers who aspired to serve God by promoting evangelical Christianity to ironworkers. Methodist bishop Francis Asbury remembered Batsto Iron Works owner William Richards as "a friend of ours for many years," though he considered Sarah Richards and other ironmasters' wives even greater allies. "Where are my sisters Richards, Vanleer, Potts, Rutter, Patrick, North?" he lamented after preaching at Coventry in 1812: "At rest in Jesus; and I am left to pain and toil: courage, my soul, we shall overtake them when our task is done!" Moral suasion had long been the work of wives of ironmasters and their agents, a role reinforced in the early nineteenth century by evangelical Christianity and by an ethos of "republican motherhood" which cast white women as the nurturers of the nation's future. This would not seem amiss to most ironworkers, and it might have served to deflect away from ironmasters any resentment or discomfort over such sentimentality.[87]

Indeed, an ironmistress could confront hired men in ways that her husband could not. In 1856, Rev. John B. Laman eulogized Catherine Carmichael Jenkins, wife of Robert Jenkins of Windsor Forges. He recalled that Jenkins had tried to curb his employees' drinking and failed. His wife took over, first by attempting to persuade them of "the injury they were doing to themselves, and the great sin they were committing against that kind Being whom she adored as her God." When that plea fell on deaf ears, she ordered a servant to gather bottles of rum that belonged to the workmen and bring them to the dining table. At meal time, she invited the men to step forward to claim their liquor. None did. Jenkins grabbed the bottles, carried them to an open window, and smashed them against the outside wall. She then turned to the shocked men and warned, "If they be replaced by others, they shall share the same fate." Robert Jenkins could not have done what she did. Stunned silence was the only socially acceptable response the workers could have mustered against Catherine Carmichael Jenkins. To do otherwise would have been unmanly.[88]

As Francis Asbury noted of William Richards, adventurers hardly remained content to stand behind their wives as they campaigned to save ironworkers. They involved themselves personally in spreading the Word. In 1791, Asbury preached at Batsto and "advised the people to build a house for the benefit of those men so busily employed day and night, Sabbaths not excepted, in the manufacture of iron—rude and rough, and strangely ignorant of God." William Richards and his family complied. By 1796, Batsto had a meetinghouse at Pleasant Mills. Twelve years later, Joseph Ball, William Richards's nephew and his son Samuel's partner in Martha Furnace and the Weymouth Iron Works, deeded the grounds on which the rebuilt Pleasant Mills church and its cemetery stood to its trustees, who included William Richards and his son Jesse. They were, Ball instructed, to offer the church first to Methodist ministers, though any Christian denomination could use it to propagate the gospel. Ball and Samuel Richards had meetinghouses constructed at Martha and Weymouth. Perhaps no ironworks endorsed Methodism more dramatically than did Colebrook Furnace, where on October 19, 1809, the clerk noted "Camp Meeting. Hell Stormd, a number of the hands assisting at the siege."[89]

Why did adventurers want their hired hands to storm Hell, especially as soldiers of the Methodist Episcopal Church? Why did so many ironwork-

Figure 14. "Seely Bunn, the black smith, at a camp meeting near Georgetown," 1809. Sketch in pencil, pen, and ink by Benjamin Henry Latrobe. Courtesy of the Maryland Historical Society, Baltimore, Maryland. Latrobe drew this while at a Methodist camp meeting in Virginia sponsored primarily by Henry Foxall, an ironmaster. A camp meeting in southern New Jersey or Pennsylvania would have looked quite similar.

ers enlist to fight that battle? The answers depend on which historians you ask. Labor historians, following the lead of English historian E. P. Thompson, have argued that entrepreneurs promoted evangelical Christianity in general and Methodism in particular because doing so made for better and more obedient employees.[90] Although Christian teaching did seem to promote better work habits and submissiveness, I would emphasize that most ironmasters believed in what they said and did; theirs was neither false faith nor feigned concern for ironworkers' souls. There were exceptions. John Jacob Faesch thought religion was good because it kept "the lower classes in subordination." David Potts, Springton Forge's owner, was horrified by a camp meeting that he attended in 1826.[91]

Adventurers indeed knew that evangelical Christianity, especially Methodism, would advance the industrious revolution that they wanted to forge. Some observers with ties to the industry said as much. Roland Curtin reported in 1803 that "the major part of Dunlop's [forge] hands are becoming Methodists, which prevents the rapid sale of whiskey I have had in November and December." Abraham Sharpless marveled at how industrious and sober the employees at his rolling mill had become after they formed a Methodist society in 1807.[92]

Methodism asked converts to discipline themselves by internalizing a set of codes on how they should behave. This promised ironmasters a way to remake the culture of ironworkers and do away with the messy drama of broken bottles and rote inscription of fines in ledgers to punish ironworkers who failed or refused to honor their duties. Evangelical Christianity offered a relatively nonconfrontational path to reform, especially when dealing with tradesmen who valued their autonomy and could market their skills easily. Methodism's evocation of a family of believers buttressed adventurers' claims that they and their workers were a harmonious community that shared the same interests and should act accordingly. Such metaphors dovetailed with the sentiments that adventurers evoked when they lobbied government. The promotion of evangelical Christianity also served workers, ironmasters could claim, by connecting them to an evolving network of sacred and secular institutions that knit the new nation together.[93]

Labor historians have traditionally had more difficulty explaining why so many working people agreed to "besiege Hell." Conversion might win them favor with their employers. Evangelical Christianity's emphasis on self-improvement appealed to those who already were sober, upright, and industrious or who aspired to be.[94] The enthusiasm of evangelical preaching and ritual offered cathartic release to men and women who wanted assurance.

Historians who focus on the study of religion, as well as an increasing number of labor historians, have stressed that working people converted to Methodism and to evangelical Protestantism because it answered their needs. Amidst the political, social, and cultural tumult of the new nation, evangelical sects promised ordinary people self-empowerment, an enhanced sense of community, and equality in this life and the next. Converts had the power to effect their own salvation. They could elect to participate in rituals and join new groups that would bind them to others with similar beliefs and goals near their homes and across the country. Methodism particularly resonated with those who sought to reinforce their conviction that they were the equals of anyone. How converts acted on that ideal sometimes confounded what their proselytizing employers intended; they sometimes invoked their faith to press claims for greater equality at work as well as at worship.[95]

Samuel Richards and the men whom he employed at his South Jersey ironworks embraced Methodism mostly for their own reasons. For Richards, it played a key role in a campaign to bring greater order to life and work at Martha Furnace and the Weymouth Iron Works. Clerks at Martha took a keen interest in what workers did off the job; they recorded brawls between employees and incidents in which ironworkers beat their wives. They also noted important milestones in the lives of workers and their families: when workers' children were born, when workers or members of their families died, and where they were buried—which was usually at the Pleasant Mills cemetery near Batsto. Martha's clerks kept the timebook partly as an account of community life for Richards and Ball to read when they visited. It documented and reinforced their view that Martha hands were family.[96]

By the 1820s, Methodism and paternalism had become institutionalized at Richards's ironworks. Sunday School classes began at Weymouth in 1821. Within six years, as many as "70 Schollars" attended. In 1825, Samuel Richards and his wife went to a local camp meeting. The following year the furnace prepared for Christmas with a "Watch Night meeting" and a "Love feast & sacrament." They were likely well attended, especially by some of the thirty-five people who "Joined the M. Society" on one day the previous summer. Care for Weymouth's children accompanied care for their parents' souls. In 1830, a doctor came to vaccinate them against smallpox.[97]

The fatherly concern of the Richards men had limits. In 1813, Martha Furnace's stack exploded while John Craig was filling it. The accident left him "very much burnt" and made him disappear from furnace records. Two years later John Potter "got burn't" at Washington Furnace. A few weeks later the clerk "notified the overseer of the poor to take John Potter away," which he soon did. For the Richards family, paternalism did not cover ironworkers' bodies. They apparently believed, as industrial employers and judges increasingly concluded during the nineteenth century, that their workers knowingly assumed whatever risk their jobs entailed. If they were badly injured on the job, that was their problem.[98]

Little overt or organized resistance met Richards's campaign to storm the culture of Martha Furnace and the Weymouth Iron Works on several fronts. Indeed, many hands signed on enthusiastically. Over a dozen paid to send their children to Weymouth's school when it first opened. No one objected loudly enough to create an echo in Martha's timebook after Asa Lanning's boys were expelled from school for fighting, a common pastime for their fathers. The battle for the souls of ironworkers was also lopsided. To be sure, the side that planned the attack told the story of the battle and so held the power to declare victory. Against camp meetings, preachers, and exhortations to come to God that could net two dozen new recruits at a time, James McEntire, a carpenter and Martha Furnace's most notorious absentee drunk, in 1809 mustered a hammer to nail a set of deer antlers "over the door of the House where the people assembled on the 7th day of the week & sprinked the Blood round about." McEntire's protest may have poked fun at his employers' efforts to institutionalize their beliefs and their vision of an industrious revolution. But he probably acted alone and the antlers came down almost immediately.[99]

What made Richards's efforts appear so successful? Something in what he offered appealed to ironworkers. Schooling and health care for their children were tangible benefits. So was the network that Richards created. He owned several ironworks and connected his employees to wider opportunities for employment. By sponsoring preachers, Richards offered his employees a chance to expand their community. Public embrace of Methodism provided a way to gain Richards's favor, no small consideration since he and his family dominated South Jersey's iron industry. Jacob Trusty, an African American who was one of four converts to Methodism named in the Martha Diary, perhaps considered it. Methodism also offered Richards's workers, among them Trusty, a chance to forge their own spiritual destinies and to stake a claim to equality in a nation that celebrated it in the abstract but often violated it in reality. Other new Methodists at Martha interpreted its principles more freely than Richards would have liked. Edward Rutter converted, but soon resumed drinking heavily. Methodism might have comforted him; he fell, but what mattered more was that he tried to get up again. Rutter may have also heard Methodism's celebration of equality as support for his decision to set his agendas, whether they agreed with those of Samuel Richards or not.[100]

Then again, Rutter may have been opting out of the reformation that Richards was promoting. Many others did. Most could leave if they found the climate that Richards had created too hostile or threatening, and some certainly exercised that right. Others simply paid it little mind. It helped that Richards did not force his beliefs on employees, especially not on tradesmen. Martha's owners never punished McEntire for hanging the antlers, or for getting drunk and missing work so often. Besides, Richards could not push his employees too hard. He, after all, needed them in more than one way. Theirs were the hands and heads that made what he sold; theirs were also the voices and the votes that enabled him to command attention in Washington. One of Richards's most crucial campaigns was against imported iron; he sat on the Committee of Correspondence of the convention that convened in Philadelphia in 1817 and solicited information and signatures from the region's ironmasters for presentation to Congress.[101]

Mid-Atlantic adventurers and ironworkers had created an industrious revolution which superficially resembled that of New England in the seventeenth century. Industrial work had an avowedly moral purpose; it was to reform those who did it and to save the society in which they lived. Adventurers after the Revolution also tried to connect their employees to the emerging religious and political institutions of their day to mold them into men fit to make iron and to save the iron industry in a new nation.

There were key differences. For the men who had placed such hope in the Lynn and New Haven ironworks, reform was a means to stave off damnation that might result from inviting such unruly and blasphemous men into their midst. Pennsylvania and New Jersey adventurers of the early republic had more mundane concerns—they needed to control labor costs and they needed help in manning the ramparts against incursions from foreign competitors. They employed free men, most of whom shared their desire to keep tariffs as high as possible. Before Congress and the nation adventurers and ironworkers stood as one, knit together by mutual dependence and by politics. Together they forged a democratic industrial capitalism, the likes of which the world had never seen before.[102]

We must remember that this was the story that adventurers wanted to tell. Samuel and Jesse Richards convened with other "Friends of Domestic Industry" in New York in late 1831. There the assembled manufacturers and their allies proclaimed to the nation that to them labor was not "the mere instrument of capital, the mere handmaid to furnish the profits of the capitalist," but rather "an intelligent active principle,—the partner and sharer in the increase of wealth produced by the united action of both. We have no class in America corresponding with the operatives,—the human machines of Europe." The United States never would so long as it maintained high tariffs, which, they claimed, benefited workingmen most of all. What they made, "protected by our own free government" was "in effect, the government that holds us together, and makes us one people."[103]

Manufacturing and "free" labor, the latter largely the rhetorical handiwork of ironmasters, had bound the nation together. Adventurers and ironworkers had indeed created a new order. The industrious revolution and the American Revolution together were making a relatively harmonious industrial revolution. Ironmasters' paternalism, their nationalism—and their ability to harvest workers' votes and sometimes their souls in the service of both—saw to that.[104] All made it possible to pretend that there was no "class" in the United States, at least not one that was analogous to industrializing Europe. But the nation would soon dissolve into sectional squabbling and civil war. So would its industrial revolution.

CONCLUSION
LEGACIES OF ANGLO AMERICA'S INDUSTRIOUS REVOLUTION

In 1799, Julian Ursyn Niemcewicz visited New Jersey's Mount Hope Furnace. "From one hundred to two hundred souls, counting women and children, belong to this tremendous undertaking," but "only about 40 are truly workers." Most of them, he thought, hardly worked. "Though the pay is good," he noted,

> there are no instances of one of these workers having built up a fortune. Attached as they are to the place on which a large number of them were born, and above all subject to the force of habit, this strong lord of human nature, these people keep their place. Above all their trade or life, which at first glance appears so heavy and so industrious, when at more closely is seen as only a constant alternation of heavy work with complete inactivity. During an 18 or 20 hour period there is a half an hour of heavy devilishly hot work when they take the molten iron from the furnace and pour it into the forms. For the rest of the time, whether by day or by night, there is needed only the adding of charcoal and flux and light supervision. The largest part of the time is spent in talking, cards or sleeping. Furthermore, these people do not know the cares and troubles of domestic life. The owner keeps a store, from which he provides food, drink, and clothing, etc., taking the money from their pay and reserving 10 per cent for himself. It is not strange that in this situation, where the sum of inactivity so pleasant to man, exceeds work, and

> far distant from the sight of better conditions, they are content with their lot.

"They are born and die charcoal burners and iron workers," Niemcewicz concluded from his observations.[1]

In Niemcewicz's story, early Anglo America's industrious revolution had passed Mount Hope's workers by. They lacked ambition; they had no desire to seek the economic independence that signified full manhood to most men of their time. Mostly born and bred to ironmaking, they remained content to enjoy comfortable, secure lives without having to exert themselves unduly, thanks to the nearly self-contained world that Mount Hope's adventurer John Jacob Faesch had created for them. Indeed, Niemcewicz implicitly contrasted their habits with those of Faesch and of their German parents who had accompanied him to build the furnace. Faesch, "as every European in America who knows a craft or a trade useful here, earned himself an estate." The men who Niemcewicz observed, had not and never would, even though they knew useful trades. In becoming American ironworkers, they became lazy. America, the Faesch family, and the iron business had spoiled them. Niemcewicz echoed what was already a common refrain concerning first-generation Americans: they were unworthy of their immigrant parents.[2]

Niemcewicz may have portrayed Mount Hope's staff unsympathetically, but he observed clearly how adventurers and ironworkers merged early Anglo America's industrious and industrial revolutions. The iron industry acculturated immigrants—be they Europeans or Africans—and their children into dominant ways of life, thought, and belief, though it had to recast the relationship between work, masculinity, and independence so that early Anglo America's industrious and industrial revolutions did not collide. The ultimate purpose of work to a man was to become free and remain free, to be his own boss and to position his sons to do the same. Obeying that fundamental principle of early Anglo America's industrious revolution fully would have hampered the development of manufacturing. By conjoining the industrial and industrious revolutions, ironmaking forced adventurers and ironworkers to reckon with expectations of independence that neither could ever meet. Slavery, a state of perpetual dependence, provided the principal means by which many colonial adventurers and most southern ironmasters resolved that dilemma, though they had to modify bondage to encourage and reward industry. Industrial slavery offered slave men a wider range of opportunities to participate in the industrious revolution, and it made their bonds more supple and even harder to shatter.

There were no slave ironworkers at Mount Hope or at many other iron-works in Pennsylvania and New Jersey when Niemcewicz visited. Adventurers and free white ironworkers had by then begun to reconcile the tension between the industrious and industrial revolutions by stressing their interdependence—by casting their relationship in terms of reciprocity. Capital and labor ideally shared the same interests and the same goals, including preservation of a link between work, manhood, and the achievement of material well-being. Conflict occurred principally to restore balance within a system that most could live with and that some agreed served the interests of all directly concerned.

Consider the situation from the perspective of the moulders who Niemcewicz criticized. Their craft earned them respect, a comfortable living, and considerable autonomy on the job. Moreover, they knew that leaving the employ of men like Faesch meant leaving their trade and their identities. They had few hopes of becoming adventurers—that took too much capital and was too great a risk in what was an intensely competitive business. Their lives, Mount Hope's ironworkers might have told Niemcewicz, reflected the best choice available to them. It was the best way for them to be American men. What Niemcewicz interpreted as surrender of ambition and lazy dependence on an indulgent employer reflected a rough consensus among white men that permitted the industrious and industrial revolutions to proceed together, especially when slavery was absent.

This was especially true in a new nation that celebrated white masculine independence and had fought a successful revolution to defend it. Therein lay much of the allure to northern industrial entrepreneurs and to many of their employees of paternalism, evangelical Christianity, and reform movements. All provided language and intellectual frameworks that proprietors and workers found empowering. All helped mask employees' dependence on their employers, largely because they permitted both parties to unite around a shared definition of masculinity which stressed interdependence and rested on their ability to participate directly in electoral politics and to shape the institutions that structured and defined community. Paternalism, as practiced in the North on ironworks and within other industries, helped to give workers the sense that they were actors within a larger, even national, community.[3]

To be sure, some white northern workers attacked the notion that industrial capitalism and their own independence could coexist. For them, and for the majority of labor historians who have sympathized with them, paternalism concealed the true costs of industrial capitalism from workers and so misled them. It only sustained the early nineteenth-century myth, embedded in legal and political thought, that entrepreneurs and laborers negotiated and executed consensual agreements as equals and that employers owed employees at least whatever compensation a free market would bear. In other words, free labor was not really "free." The attempts of paternalism and of most reform movements to reconcile the industrious, industrial, and American revolutions hid industrial capitalism's "true" face from most of its "natural" adversaries. Indeed, by the antebellum era, southern apologists for black slavery joined many northern urban artisans and millhands in denouncing the "wage slavery" that industrial entrepreneurs had imposed upon employees by rendering them so dependent. Such notions registered more powerfully with more workers after labor activists in the 1840s began to characterize "wage slavery" as "white slavery," a change which buttressed workers' claims to whiteness and dramatized the evil of a system which perpetuated and worsened such inequality between putative equals. After all, how could white men "enslave" other white men in a republic? Far more northern workers, however, scarcely heeded such claims. They believed, or at least acted as if they believed, that employment was a temporary means to independence and that they and their employers shared common goals.[4]

Slavery blocked the path to independence, masters and slaves knew. This hurdle, many masters asserted, only harmonized their interests and those of slaves better. Antebellum defenders of slavery rejoiced that much of the South's industrial workforce had no political voice and celebrated it as a safeguard of democracy. That dictated that the paternalism that southern ironmasters and slave masters espoused would differ drastically in its ends from that practiced to the north. Northern and southern ironmasters alike saw in paternalism a means to make workers more industrious. But the paternalism practiced in antebellum southern industries, plantations, and households was essentially private and privatized—its principal public purpose being to shield slavery and masters from abolitionist attack. Southern paternalists never envisioned a place in civil society for those they claimed to benefit. Paternalism never sought to imbue slaves with a sense that they belonged to something greater than that which their masters directly controlled. They remained, as Walter Johnson puts it in his work on the antebellum slave market, "people with a price." Law, politics, culture, and booming demand for slaves ensured that they would remain so until civil war terminated slavery within the United States.[5]

The regional differences that resulted from the industrious, industrial, and American revolutions emerged most starkly during the Civil War. It exposed the weakness of industrial slavery, perhaps nowhere more vividly than in iron manufacturing. The South could not muster the expertise and the flexibility necessary to arm the Confederacy well. The North's superior industrial capacity powered the conquest of the Confederacy and forced its reabsorption into the Union. So did the support of northern ironmasters. George Dawson Coleman, member of one of the most prominent ironmaking families of the early republic, personally raised and gave financial support to the 93rd Pennsylvania Volunteers. David Potts, an heir to the most prominent family of Pennsylvania ironmasters, contributed substantial sums to raise volunteers for the Union Army. But perhaps equally important was the bond that industrial paternalism had helped to forge between manhood and nation. Industrial workers, fired by nationalist causes and moral crusades that they shared with men like Potts and Coleman, volunteered to fight to restore the Union. Their enslaved southern counterparts, forced to manufacture goods to defend a nation whose founding principles declared their perpetual inferiority, did not meet them on the battlefield, at least not as adversaries.[6]

Though the North won the Civil War, its idealized version of the industrious revolution, which linked work, manhood, and civic identity, was in retreat throughout the United States by 1877. Changes in the North undermined it from within. Riots and violent strikes rewrote the script that the United States was a classless nation which had escaped the conflicts between capital and labor that plagued Europe. The millions who poured into the nation's fields and factories from East Asia, Mexico, and southern and eastern Europe effectively had to reinvent free labor, partly because they seemed to uneasy elites at best indigestible, at worst a contagion that might consume the nation and its values. It took a generation before entrepreneurs and Progressive reformers began again to see and to portray working people as potential Americans.[7]

The antebellum North's idealized vision of the industrious revolution also foundered against the legacies of slavery. After all, the industrial revolution had never extended to free black men in the antebellum North, who white workers and white entrepreneurs together excluded from nearly all factories. Slave men in the early national and antebellum South participated more fully in the industrious revolution than they did during the colonial era, but only in carefully circumscribed ways which encouraged their industry but denied them membership in society. Once emancipated, they eagerly seized opportunities to work toward independence and to exercise their newly won right to vote and hold elected office. Their aspirations and their actions testified to the degree to which slavery and the industrious revolution had made them American.[8]

By seeking to become independent and equal economic and political actors, freedmen threatened the reintegration of the South into a national market. Southern elites, local proprietors, and northern investors alike believed that neither the restoration of commercial agriculture nor postbellum industrialization could proceed with working men, especially freedmen, who pursued and often attained independence. Like colonial adventurers before them, they believed that economic development demanded coercion and labor's disempowerment. The New South's industrial paternalism, its disenfranchisement of working people, its erection of a comprehensive system of racial segregation, and its suppression of labor unions all undergirded the belief that economic development and social stability required that working people be kept outside the bounds of civil society—to render their voices, especially when expressed collectively, as inaudible as possible. This represented the legacy of the industrious revolution that arose during the colonial era and was secured by the American Revolution. Because it took slaves to make Anglo America's industrious revolution, entrepreneurs and politicians were certain that it would take coerced and excluded workers to modernize the South.[9]

The conjunction of the American Revolution with the industrious and industrial revolutions echoed well into the twentieth century. The struggle for independence and American manhood that skilled white ironworkers were among the first to wage presaged the development of American trade unionism in the late nineteenth century, particularly in the emphasis of unions on controlling the work process to enhance job security and on preserving members' dignity as masters of their craft and as providers for their families. Industrial employment became the vehicle through which entrepreneurs often self-consciously strove to assimilate immigrants. The welfare capitalism that many industrial employers designed in the early twentieth century tried to wean immigrants and their children from their own ethnically centered organizations and to dissuade them from joining unions. It also unintentionally raised workers' expectations of their employers and encouraged workers to participate in electoral politics and in the unions that the Congress of Industrial Organizations began to organize during the 1930s. Their embrace of politics and unions prompted a backlash which impeded workers' ability to organize, partly because industries migrated to avail themselves of lower labor costs that the South's hostility to unions permitted.[10]

What significance do Anglo America's industrious revolution and the adventurers and ironworkers who participated in it hold for us today? Perhaps they matter most to poor women. After all, our politicians celebrate the virtue of paid employment and promote it as the means by which welfare mothers, who have been deemed most dependent on others and therefore most in need of moral uplift, may develop a work ethic and rise to dignity and independence. Otherwise, the industrious revolution and my interpretation of it seem to be of little import, at least at first glance. Slavery ended generations ago, though campaigns seeking reparations for slavery have brought to wider attention the roles that bondage played in economic development in the United States.[11] Relatively few Americans today aspire to be their own boss, so the dilemma that perpetual employment posed for Anglo American men of the past appears insignificant. Moreover, the rusting hulks that fill so many cities and towns bearing silent witness to deindustrialization make even the industrial revolution seem remote to millions of Americans—an anachronism rendered obsolete by electronic technology and cheap imports from faraway places. To them, the Saugus Iron Works, Hopewell Furnace, or another ironworks reconstructed, maintained, and interpreted by the National Park Service or by state historical agencies represent less the quaint foundation of their nation's industrial might than nostalgic and pastoral memorials to it.

But if the issues that this story raises about adventurers, ironworkers, and the industrious revolution that they mostly made together seem detached from the lives of most who will likely read it, they might resonate far more powerfully outside the United States. The principal audience for early Anglo America's three revolutions lives within what economists and politicians call "the developing world" or "emerging markets." Having triumphed in the Cold War and having embarked upon a global campaign against terrorism, the United States argues to the world that only its model of mostly unregulated capitalism and representative democratic government, both legacies of its Revolution, can deliver billions in Africa, Asia, and Latin America from grinding poverty and tyranny. Open your markets to the world and only then will prosperity and freedom follow, our leaders proclaim. Labor rights, environmental protection, and public investment in health and education lie outside of the realm of "trade" and so must be rejected as well-intentioned but misguided concerns, at least for the foreseeable future.[12]

To dissenters and even to many international proponents of free trade, this smacks of hypocrisy and willful ignorance of U.S. history. After all, hasn't the United States protected its industries with tariffs for more than two centuries? Perhaps not surprisingly, I would agree with these dissenters, though I would add that what the United States preaches partly reflects its history as the only developed nation which depended heavily and directly on slave labor to launch its industrial revolution and its revolution for political independence. It took slaves to help resolve the tension between capitalism, manhood, and democracy in early Anglo America; it took slavery to forge America. One result has been a nation which has often relied on coercion and the effective exclusion of working people from civil society, one which has frequently seen humans as little more than economic assets motivated primarily by the desire to enrich themselves and to consume. I think that it should come as little surprise that most leaders of the United States today prescribe the same vision as the blueprint that the majority of humanity should follow in the twenty-first century.

ABBREVIATIONS FOR SELECTED ARCHIVES, MANUSCRIPT COLLECTIONS, AND SERIAL PUBLICATIONS

AHR	*American Historical Review*
AISIC	American Iron and Steel Institute Collection, Acc. 1631, Box 229
ATL	William Allen & Joseph Turner Letterbook, 1755–1774, Library Company of Philadelphia Collections
BM	British Museum—Additional Manuscripts 29600
Case Papers, Weaver *v.* Mayburry	Case Papers, Weaver *v.* Mayburry, Superior Court of Chancery, Augusta County Court House, Staunton, Virginia
Chancery	Chancery Court (Chancery Papers, Exhibits), MSA S 528
CCCFP	*Charles Carroll of Carrollton Family Papers,* Ronald Hoffman, ed., Eleanor Darcy, associate ed., 38 reels (Annapolis, Md., 1989, microfilm), followed by reel number
CRPa	*Colonial Records of Pennsylvania,* 16 vols. (Philadelphia, 1851–1853)
Drinker	Henry Drinker Business Papers, Acc. # 176
DRL	David Ross Letterbook, 1812–1813
Duke	Rare Book, Manuscript, and Special Collections Library, Duke University, Durham, N.C.
ERF	*Records and Files of the Quarterly Courts of Essex County, Massachusetts,* 8 vols. (Salem, Mass., 1911–1921)
FFR	Forges and Furnaces Records, Acc. #212
GFC	Grubb Family Collection, Acc. 1948
GFP	Grubb Family Papers, Acc. #1967A

HDL	Henry Drinker Letterbook
HIWC	Hibernia Iron Works Collection (typescripts of originals at Morristown, N.J., National Historic Site)
HML	Manuscripts and Archives Department, Hagley Museum and Library, Greenville, Delaware
HP	Haldeman Papers, Acc. 840, No. 9
HSD	Research Library, Historical Society of Delaware, Wilmington
HSP	Historical Society of Pennsylvania, Philadelphia
IZP	Isaac Zane Papers, Coates and Reynell Papers, Acc. #140A
JAH	*Journal of American History*
JEconHist	*Journal of Economic History*
JER	*Journal of the Early Republic*
LC	Manuscripts Division, Library of Congress, Washington, D.C.
LCHS	Lancaster County Historical Society, Lancaster, Pa.
LH	*Labor History*
LIWC	Lynn Iron Works Collection, Baker Library, Harvard Business School, followed by page number
Martha	Henry H. Bisbee and Rebecca Bisbee Colesar, eds., *Martha 1808–1815: The complete Martha Furnace Diary and Daybook* (Burlington, N.J., 1976)
Mass Records	*Records of the Governor and Company of the Massachusetts Bay in New England*, Nathaniel B. Shurtleff, ed., 5 vols. (Boston, Mass. 1853–1854)
MG	*Maryland Gazette*
MHS	Manuscripts Department, Maryland Historical Society Library, Baltimore
MHM	*Maryland Historical Magazine*
MSA	Maryland State Archives, Annapolis
NHTR	*New Haven Town Records*, ed. Franklin Dexter Bowditch, 2 vols. (New Haven, 1917–1919)
NJSA	New Jersey State Archives, Trenton
PA	*Pennsylvania Archives,* 9 series, 119 vols. (Philadelphia, 1852–1935)
PA8	*Pennsylvania Archives, Eighth Series: Votes and Proceedings of the House of Representatives of the Province of Pennsylvania, Dec. 4, 1682–Sept. 26, 1776*, ed. Gertrude MacKinney and Charles F. Hoban, 8 vols. (Harrisburg, 1931–1935)
PCP	Principio Company Papers, MS 669
PG	*Pennsylvania Gazette*
PH	*Pennsylvania History: A Journal of Mid-Atlantic Studies*
PIWC	Principio Iron Works Collection
PMHB	*Pennsylvania Magazine of History and Biography*
PSA	Pennsylvania State Archives, Harrisburg

RAB	Ridgely Account Books, MS 691
RABSP:F:3	Kenneth M. Stampp, gen. ed., *Records of Ante-Bellum Southern Plantations from the Revolution to the Civil War: Series F: Selections from the Manuscript Department, Duke University Library; Part 3: North Carolina, Maryland, and Virginia*, 45 reels (Frederick, Md., 1987), followed by reel number
RMP	Ringwood Manor Papers
RP692	Ridgely Papers, MS 692
RP692.1	Ridgely Papers, MS 692.1
RUL	Rutgers University Libraries, Special Collections and University Archives, New Brunswick, N.J.
RVCL	*The Records of the Virginia Company of London*, Susan Myra Kingsbury, ed., 4 vols. (1906–1935)
SABSI:A	John H. Bracey, Jr. and August Meier, gen. eds., Charles B. Dew, editorial adviser, *Slavery in Ante-Bellum Southern Industries: Series A: Selections from the Duke University Library.* 28 reels (Bethesda, Md., 1991, microfilm), followed by reel number
SABSI:B	John H. Bracey, Jr. and August Meier, gen. eds., Charles B. Dew, editorial adviser, *Slavery in Ante-Bellum Southern Industries: Series B: Selections from the Southern Historical Collection, University of North Carolina, Chapel Hill.* 38 reels (Bethesda, Md., 1993, microfilm), followed by reel number
SHC	Southern Historical Collection
UNC	Wilson Library, University of North Carolina, Chapel Hill
UVA	The Albert and Shirley Small Special Collections Library, University of Virginia Library, Charlottesville
VHS	Virginia Historical Society, Richmond
VMHB	*Virginia Magazine of History and Biography*
WAL	William Allen Letterbook, 1753–1770, Shippen Family Papers, Acc. #595C
WBP	William Bolling Papers
WFP	Wright Family Papers, Acc. #1665
WMQ	*The William and Mary Quarterly*, 3d. Series
WP	*Winthrop Papers*, 6 vols. (Boston, 1929–)
WWP	William Weaver Papers

NOTES

Introduction

1. I borrow "industrious revolution" from Jan de Vries's studies of early modern Europe's economy, best articulated in "The Industrial Revolution and the Industrious Revolution," *JEconHist* 54 (1994): 249–70.

2. For a biography of Udree, see George Winterhalter Schultz, "Major General Daniel Udree: Oley Ironmaster, Soldier, Statesman," *Historical Review of Berks County* 1 (1936): 66–70.

3. On self-fashioning and early American nationalism, see Laura Rigal, *The American Manufactory: Art, Labor, and the World of Things in the Early Republic* (Princeton, 1998); David Waldstreicher, *In the Midst of Perpetual Fetes: The Making of American Nationalism, 1776–1820* (Chapel Hill, 1997); and Joyce Appleby, *Inheriting the Revolution: The First Generation of Americans* (Cambridge, Mass., 2000).

4. Daniel Vickers, "Competency and Competition: Economic Culture in Early America," *WMQ* 47 (1990): 3–29; and Jack P. Greene, *Pursuits of Happiness: The Social Development of Early Modern British Colonies and the Formation of American Culture* (Chapel Hill, 1988).

5. "Record of the Male and Female Slaves recorded in the County of Berks in the Commonwealth of Pennsylvania, according to an Act of General Assembly entitled an Act 'For the Gradual Abolition of Slavery' " [typescript copy], Mary Owen Steinmetz Collection, Acc. #1690, HSP.

6. Many scholars have shaped my understanding of work, especially Patrick Joyce, "The Historical Meanings of Work: An Introduction," and John Rule, "The Property of Skill in the Period of Manufacture," in *The Historical Meanings of Work*, ed. Patrick Joyce (Cambridge, 1987), 1–30, 99–118; Camilla Townsend, *Tales of Two Cities: Race and Economic Culture in Early Republican North and South America* (Austin, Tex., 2000); and Jacqueline Jones, *American Work: Four Centuries of Black and White Labor* (New York, 1998).

7. Orlando Patterson, *Slavery and Social Death: A Comparative Study* (Cambridge, Mass., 1982). See also Walter Johnson, *Soul by Soul: Life inside the Antebellum Slave Market* (Cambridge, Mass., 1999); and David Eltis, *The Rise of African Slavery in the Americas* (New York,

2000). Peter Kolchin argues that historians should consider slavery within the history of work in "The Big Picture: A Comment on David Brion Davis's 'Looking at Slavery from Broader Perspectives,'" *AHR* 105 (2000): 467–71.

8. Among the best studies of work in early Anglo America are: Stephen Innes, ed., *Work and Labor in Early America* (Chapel Hill, 1988); Ira Berlin and Philip D. Morgan, eds., *Cultivation and Culture: Labor and the Shaping of Slave Life in the Americas* (Charlottesville, Va., 1993); Marcus Rediker, *Between the Devil and the Deep Blue Sea: Merchant Seamen, Pirates, and the Anglo-American Maritime World, 1700–1750* (New York, 1987); Laurel Thatcher Ulrich, *A Midwife's Tale: The Life of Martha Ballard, Based on Her Diary, 1785–1812* (New York, 1990); *The Age of Homespun: Objects and Stories in the Creation of an American Myth* (New York, 2001); Daniel Vickers, *Farmers and Fishermen: Two Centuries of Work in Essex County, Massachusetts, 1630–1830* (Chapel Hill, 1994); W. Jeffrey Bolster, *Black Jacks: African American Seamen in the Age of Sail* (Cambridge, Mass., 1997); and Jones, *American Work*.

9. Greene, *Pursuits of Happiness;* Mary Beth Norton, *Founding Mothers and Fathers: Gendered Power and the Forming of American Society* (New York, 1996); and Kathleen M. Brown, *Good Wives, Nasty Wenches, and Anxious Patriarchs: Gender, Race, and Power in Colonial Virginia* (Chapel Hill, 1996).

10. Richard S. Dunn, "Servants and Slaves: The Recruitment and Deployment of Labor," in *Colonial British America: Essays in the New History of the Early Modern Era*, ed. Jack P. Greene and J. R. Pole (Baltimore, 1984), 157–94; Aaron S. Fogelman, "From Slaves, Convicts, and Servants to Free Passengers: The Transformation of Immigration in the Era of the American Revolution," *JAH* 85 (1998): 43–76. On race, slavery, and divisions of labor, see Jones, *American Work*, 23–168.

11. Linda K. Kerber, *Women of the Republic: Intellect and Ideology in Revolutionary America* (Chapel Hill, 1980), and "'No Political Relation to the State': Conflicting Obligations in the Revolutionary Era," in *No Constitutional Right to Be Ladies: Women and the Obligations of Citizenship* (New York, 1998), 3–46; and Jeanne Boydston, *Home and Work: Housework, Wages, and Ideology in the Early Republic* (New York, 1990).

12. Jones, *American Work*, 165–272; and Ira Berlin, *Many Thousands Gone: The First Two Centuries of Slavery in North America* (Cambridge, Mass., 1998), 217–365.

13. Sean Wilentz, *Chants Democratic: New York City and the Rise of the American Working Class, 1788–1850* (New York, 1984); Christine Stansell, *City of Women: Sex and Class in New York, 1789–1860* (Urbana, Ill., 1987); Ronald Schultz, *The Republic of Labor: Philadelphia Artisans and the Politics of Class, 1720–1830* (New York, 1993); Thomas Dublin, *Women at Work: The Transformation of Work and Community in Lowell, Massachusetts, 1826–1860* (New York, 1979); Eric Foner, *Free Soil, Free Labor, Free Men: The Ideology of the Republican Party before the Civil War* (New York, 1970); and W.E.B. DuBois, *The Souls of Black Folk*, ed. David W. Blight and Robert Gooding-Williams (Boston, 1997), 34.

14. Several books focus on work within the iron industry before the Civil War. I have found the most valuable of these to be: Arthur Cecil Bining, *Pennsylvania Iron Manufacture in the Eighteenth Century*, 2d ed. (Harrisburg, 1987); E. N. Hartley, *Ironworks on the Saugus: The Lynn and Braintree Ventures of the Company of Undertakers of the Ironworks in New England* (Norman, Okla., 1957); Joseph E. Walker, *Hopewell Village: The Dynamics of a Nineteenth Century Iron-Making Community* (Philadelphia, 1966); Ronald L. Lewis, *Coal, Iron, and Slaves: Industrial Slavery in Maryland and Virginia, 1715–1865* (Westport, Conn., 1979); and Charles B. Dew, *Bond of Iron: Master and Slave at Buffalo Forge* (New York, 1994). On assimilation and industrial capitalism, see Herbert G. Gutman, "Work, Culture, and Society in Industrializing America, 1815–1919," *AHR* 78 (1973): 531–88.

15. My case that ironworks were factories and examples of early industrialization draws from Jonathan Prude, "Capitalism, Industrialization, and the Factory in Post-Revolutionary America," in *Wages of Independence: Capitalism in the Early American Republic*, ed. Paul A. Gilje (Madison, Wis., 1997), esp. 86–87; and Milan Myska, "Pre-Industrial Iron-Making in the Czech Lands: The Labour Force and Production Relations circa 1350–circa 1840," *Past and Present* 82 (1979): 47–49.

16. David Eltis argues that the development of freedom in Europe structured slavery in the Americas in *Rise of African Slavery*, esp. 273–80.

17. Wilentz, *Chants Democratic;* Dublin, *Women at Work;* Anthony F.C. Wallace, *Rockdale: The Growth of an American Village in the Early Industrial Revolution . . .* (New York, 1978); Jonathan Prude, *The Coming of Industrial Order: Town and Factory Life in Rural Massachusetts, 1810–1860* (New York, 1983); Philip Scranton, *Proprietary Capitalism: The Textile Manufacture at Philadelphia, 1800–1885* (New York, 1983); and Jones, *American Work,* 191–272.

18. For arguments that slavery, economic development, and modernity were compatible, see Joyce E. Chaplin, *An Anxious Pursuit: Agricultural Innovation and Modernity in the Lower South, 1730–1815* (Chapel Hill, 1993); Mark M. Smith, *Mastered by the Clock: Time, Slavery, and Freedom in the American South* (Chapel Hill, 1997); and Robin Blackburn, *The Making of New World Slavery: From the Baroque to the Modern, 1492–1800* (London, 1997). On slavery's role in stimulating development in the North Atlantic world, see Blackburn, *Making of New World Slavery;* Eric Williams, *Capitalism and Slavery* (Chapel Hill, 1944); Sidney W. Mintz, *Sweetness and Power: The Place of Sugar in Modern History* (New York, 1985); and Charles Bergquist, "The Paradox of American Development," in *Labor and the Course of American Democracy: U.S. History in Latin American Perspective* (London, 1996), 8–42.

19. Arthur Cecil Bining, *British Regulation of the Colonial Iron Industry* (Philadelphia, 1933), 3–4, 24–31, 122, 134.

20. Northampton Furnace Journal, 1790–1796, Box 5, RAB, MHS. Alfred Crosby presents double-entry accounting as one technique that enabled Europe's rise to global power in *The Measure of Reality: Quantification and Western Society, 1250–1600* (New York, 1997), 199–223. On double-entry accounting and its power to shape knowledge and discourse in early modern England, see Mary Poovey, *A History of the Modern Fact: Problems of Knowledge in the Sciences of Wealth and Society* (Chicago, 1998), 29–91. I thank my colleague Dick Pearce for directing me to Poovey's work. On women's virtual anonymity within account books, see Lisa Norling, *Captain Ahab Had a Wife: New England Women and the Whalefishery, 1720–1830* (Chapel Hill, 2000), 29–39; and Jeanne Boydston, "The Woman Who Wasn't There: Women's Market Labor and the Transition to Capitalism in the United States," *JER* 16 (1996): 183–206.

21. *NHTR,* 2:149.

22. By "Middle Colonies," I mean New Jersey, Pennsylvania, and Delaware.

Chapter 1. Mastered by the Furnace

1. Joshua Hempstead, "Cecil County in 1749," *MHM* 49 (1954): 348.

2. Leonard W. Labaree et al, eds., *The Papers of Benjamin Franklin,* 36 vols. (New Haven, 1959–) 1:310. See also Michael W. Robbins, *The Principio Company: Iron-Making in Colonial Maryland 1720–1781* (New York, 1986).

3. Henry Drinker to Robert Bowne & Co., May 11, 1790, *HDL,* 1790–1793, Drinker, HSP. See also Joseph E. Walker, *Hopewell Village: The Dynamics of a Nineteenth Century Iron-Making Community* (Philadelphia, 1966), 32–37.

4. On bloomery forges, see Frederick Overman, *The Manufacture of Iron in All Its Various Branches,* 3d ed., revised (Philadelphia, 1854), 245–48; Samuel Gustaf Hermelin, *Report about the Mines in the United States of America, 1783,* trans. Amandus Johnson (Philadelphia, 1931), 55–58; and Robert B. Gordon, *American Iron, 1607–1900* (Baltimore, 1996), 90–100.

5. E.N. Hartley, *Ironworks on the Saugus: The Lynn and Braintree Ventures of the Company of Undertakers of the Ironworks in New England* (Norman, Okla., 1957), 272–79; Ann B. Markell, "Solid Statements: Architecture, Manufacturing, and Social Change in Seventeenth-Century Virginia," in *Historical Archeology of the Chesapeake,* ed. Paul A. Shackel and Barbara J. Little (Washington, 1994), 56–58; Robert Erskine to Richard Willis and the American Iron Company, Mar. 27, 1772 [typescript], 1:50, Box 1, RMP, NJSA; and Overman, *Manufacture of Iron,* 383.

6. David Harvey, "Reconstructing the American Bloomery Process," *The Colonial Williamsburg Historic Trades Annual* 1 (1988): 29–37.

7. On the development of blast furnaces and refinery forges, see R. F. Tylecote, *The Early History of Metallurgy in Europe* (New York, 1987), 325–48; and William Rostoker and Bennet Bronson, *Pre-Industrial Iron: Its Technology and Ethnology* (Philadelphia, 1990), 153–65. On the spread of the indirect process across Britain, see H. R. Schubert, *History of the British Iron and Steel Industry* (London, 1957), 157–94.

8. Schubert, *History of the British Iron and Steel Industry*, 152–53; Rostoker and Bronson, *Pre-Industrial Iron*, 157–59, 162–65.

9. Charles Carroll to Charles Carroll of Carrollton, Jan. 9, 1764, *MHM* 12 (1917): 26–28; Julian Ursyn Niemcewicz, *Under Their Vine and Fig Tree: Travels through America in 1797–1799, 1805 with some further account of life in New Jersey*, trans. and ed. Metchie J. E. Budka (Elizabeth, N.J., 1965), 223–24; Joseph M. Paul Daybook, 1800–1821, Joseph M. Paul Papers, Acc. #192, HSP. Although many ironworks' accounts have survived, few permit complete reconstruction of a firm's financial history for more than a few years. More problematic are the limitations of the accounts themselves, which often determined profit and loss from cash flow only and which never systematically attempted to address concerns such as depreciation of plant and equipment. See Judith A. McGaw, "Accounting for Innovation: Technological Change and Business Practice in the Berkshire Paper Industry," *Technology and Culture* 26 (1985): 703–25.

10. Drinker to Richard Blackledge, Oct. 4, 1786, HDL, 1786–1790, Drinker, HSP, in Thomas M. Doerflinger, "How to Run an Ironworks," *PMHB* 108 (1984): 366; and Jeffrey F. Zabler, "A Microeconomic Study of Iron Manufacture, 1800–1830" (Ph.D. diss., University of Pennsylvania, 1970), 53–77.

11. Charles E. Hatch, Jr., and Thurlow Gates Gregory, "The First American Blast Furnace, 1619–1622: The Birth of a Mighty Industry on Falling Creek in Virginia," *VMHB* 70 (1962): 266–77; Hartley, *Ironworks on the Saugus*, 59–83, 117–42; Principio Company to John England, Sept. 25, 1723, Principio Company Papers, MS 669, MHS; G. MacLaren Brydon, "The Bristol Iron Works in King George County," *VMHB* 42 (1934): 97–102; Robbins, *Principio Company*, 303.

12. Arthur C. Bining, *British Regulation of the Colonial Iron Industry* (Philadelphia, 1933), 36–38.

13. *Archives of Maryland*, 72 vols. (Baltimore, 1883–1972) 33:467–69; 37:540–41; 46:469–70; William Waller Hening, *The Statutes at Large; Being a Collection of all the Laws of Virginia, From the First Session of the Legislature in the Year 1619*, 18 vols. (Richmond, 1809–1823), 4:228–30; 296–300; and Thomas Penn to Rev. Mr. Barton, Apr. 11, 1764, Letter Book VIII, 48–51, Penn Family Papers, Acc. #485A, HSP. On the General Loan Office, see Mary M. Schweitzer, *Custom and Contract: Household, Government, and the Economy in Colonial Pennsylvania* (New York, 1987), 79, 115–39, 162.

14. Brendan McConville, *These Daring Disturbers of the Public Peace: The Struggle for Property and Power in Early New Jersey* (Ithaca, 1999), 94–104; Johann David Schoepf, *Travels in the Confederation [1783–1784]*, trans. and ed. Alfred J. Morrison, 2 vols. (Philadelphia, 1911; reprint, New York, 1968), 1:36–37; and Board of Proprietors of the Eastern Division of New Jersey, *Minutes*, 3 vols. (Perth Amboy, N.J., 1960) 3:224, 346. For more on crowd violence directed at adventurers and ironworks, see Irene D. Neu, "The Iron Plantations of Colonial New York," *New York History* 33 (1952): 6–7.

15. Bining, *British Regulation*, 38–80; and Benjamin Franklin, "Observations Concerning the Increase of Mankind," in *Papers of Benjamin Franklin* 4:225–34.

16. Harry Scrivenor, *History of the Iron Trade: From the Earliest Records to the Present Period*, 2d ed. (London, 1854; reprint, 1967), 58, 81; Bining, *British Regulation*, 128–32.

17. Principio Iron Works Ledger, 1727, PIWC, HSD; John Tayloe (1687–1747) Account Book, 1740–1741, and John Tayloe (1721–1779) Account Book, 1749–1768, Tayloe Family Papers, VHS; and Dr. Charles Carroll to William Black, Sept. 14, 1750, *MHM* 23 (1928): 382–83. See also Keach Johnson, "The Baltimore Company Seeks English Markets: A Study

of the Anglo-American Iron Trade, 1731–1755," *WMQ* 16 (1959): 37–60; and Robbins, *Principio Company*, esp. 189–237. I discuss the social origins of colonial Chesapeake adventurers in "Forging a New Order: Slavery, Free Labor, and Sectional Differentiation in the Mid-Atlantic Charcoal Iron Industry, 1715–1840" (Ph.D. diss., University of Pennsylvania, 1995), 91–98.

18. Bezís-Selfa, "Forging a New Order," 91–98, 225–29; Thomas M. Doerflinger, *A Vigorous Spirit of Enterprise: Merchants and Economic Development in Revolutionary Philadelphia* (Chapel Hill,1986), 151–54.

19. I have compiled these data in "Forging a New Order," 79, from Arthur C. Bining, *Pennsylvania Iron Manufacture in the Eighteenth Century*, 2d ed. (Harrisburg, 1987), 171–76; Kathleen Bruce, *Virginia Iron Manufacture in the Slave Era* (New York, 1931), 16–79, 454–55; Joseph T. Singewald, Jr., *The Iron Ores of Maryland, with an Account of the Iron Industry* (Baltimore, 1911), 139–77; and Charles S. Boyer, *Early Forges and Furnaces in New Jersey* (Philadelphia, 1931) 12–25, 154–90. The figures I have used denote only forges and furnaces built during each period and not those then in operation. Figures for forges include bloomeries. Where an enterprise went by the term "ironworks," I assumed one furnace and one forge unless I had data to indicate otherwise.

20. Boyer, *Early Forges and Furnaces in New Jersey*, 12–25, 159–90.

21. Principio Company to Nathaniel Chapman, Sept. 11, 1758, Thomas Ruston Papers, LC.

22. Paul F. Paskoff, *Industrial Evolution: Organization, Structure, and Growth of the Pennsylvania Iron Industry, 1750–1860* (Baltimore, 1983), 39–67; Robbins, *Principio Company*, 230–37; and Johnson, "Baltimore Company Seeks English Markets." See also Dr. Charles Carroll to William Black, July 4, 1743, *MHM* 20 (1925): 272; Charles Carroll, barrister, to Messrs. Gale and Pensonby, Aug. 20, 1758, *MHM* 32 (1937): 184; and Carroll, barrister, to Williams Anderson, Sept. 25, 1759, in ibid., 359.

23. James Old to Curtis Grubb, Mar. 27, 1774, Box 1, GFP, HSP; James & Drinker to Lancelot Cowper & Co., Aug. 4, 1774, James & Drinker Foreign Letters, Drinker, HSP; Paskoff, *Industrial Evolution*, 67–69.

24. Peter Hasenclever, *The Remarkable Case of Peter Hasenclever* (London, 1773; reprint, Newfoundland, N.J., 1970), 5–8; David Curtis Skaggs, "John Semple and the Development of the Potomac Valley," *VMHB* 92 (1984): 284–302; Alan L. Karras, *Sojourners in the Sun: Scottish Migrants in Jamaica and the Chesapeake, 1740–1800* (Ithaca, 1992), 93–99; John Keppely to Jasper Yeates, Dec. 14, 1774, Box 1, Jasper Yeates Family Papers, MG-137, PSA; and Rev. Thompson P. Ege, *History and Genealogy of the Ege Family in the United States 1738–1911* (Harrisburg, 1911), 146–54.

25. Bruce, *Virginia Iron Manufacture*, 24–79; Bining, *Pennsylvania Iron Manufacture*, 163; Boyer, *Early Forges and Furnaces*, 93–97, 181–84; Northampton Furnace Timebook, 1775–1776, Box 15, RAB, MHS; and Isaac Zane to Sarah Zane, May 19, 1777, Box A-115, IZP, HSP.

26. Zane to John Pemberton, May 29, 1775; Zane to Sarah Zane, May 19, 1777, Box A-115, IZP, HSP. On inflation, see Paskoff, *Industrial Evolution*, 69–71. On the impact of military campaigns, see Bining, *British Regulation*, 112–13; Bruce, *Virginia Iron Manufacture*, 60–63.

27. Bining, *British Regulation*, 111, and *Pennsylvania Iron Manufacture*, 124–25.

28. Hermelin, *Report*, 74; Schoepf, *Travels*, 1:36–37; Latrobe to Erick Bollmann, July 1, 1809, in *The Correspondence and Miscellaneous Papers of Benjamin Henry Latrobe*, ed. John C. Van Horne et al., 3 vols. (New Haven, 1984–1988), 2:743; and Gordon, *American Iron*, 40–44. Hermelin estimated that one clearcut acre yielded sixteen to twenty cords of wood and that it took twenty-five to forty years for new trees to grow large enough to harvest. Hopewell Furnace consumed 5,000 to 6,000 cords of wood a year; its woodcutters cleared over 200 acres annually. See *Report*, 47–48; and Walker, *Hopewell Village*, 141.

29. On iron exports, see Charles H. Evans, *Exports, Domestic and Foreign from the American Colonies to Great Britain from 1697 to 1789, Inclusive. Exports, Domestic from the United States to All*

Countries from 1789 to 1883, Inclusive. (Washington, D.C., 1884; reprint, New York, 1976), 36–37.

30. [Committee on Historical Research, Pennsylvania Society of the Colonial Dames of America], *Forges and Furnaces in the Province of Pennsylvania* (Philadelphia, 1914), 55–56; Aubrey C. Land, *The Dulanys of Maryland: A Biographical Study of Daniel Dulany, the Elder (1685–1753) and Daniel Dulany, the Younger (1722–1797)* (Baltimore, 1955), 326–27; Commissioners to Preserve Confiscated British Property (Ledger and Journal), 1781–1782, S132; Commissioners to Preserve Confiscated British Property (Sales Book), 1781–1782, S134, MSA; and *MG*, Aug. 30, 1781, Jan. 17, 1782, Feb. 27, 1783. I discuss confiscation and slave ironworkers in chapter 5.

31. I devote more attention to this issue in chapter 6. See also my "Forging a New Order," 220–29.

32. For 1784 to 1800 I used Bining, *Pennsylvania Iron Manufacture*, 171–76; Bruce, *Virginia Iron Manufacture*, 16–79, 128–48, 454–55; Singewald, Jr., *Iron Ores of Maryland*, 139–77; and Boyer, *Early Forges and Furnaces in New Jersey*. I used the same counting method as I did for the colonial era.

33. *Compendium of the Enumeration of the Inhabitants and Statistics of the United States as Obtained from the Returns of the Sixth Census* (Washington, D.C., 1841; reprint, New York, 1976).

34. Ibid.; Arthur D. Pierce, *Iron in the Pines: The Story of New Jersey's Ghost Towns and Bog Iron* (New Brunswick, N.J., 1957), 16–17; Bining, *Pennsylvania Iron Manufacture*, 50–54, 173–76; Paskoff, *Industrial Evolution*, 73–74; Samuel Sydney Bradford, "The Ante-Bellum Charcoal Iron Industry of Virginia" (Ph.D. diss., Columbia University, 1958), 12–16.

35. Glenn Porter and Harold C. Livesay, *Merchants and Manufacturers: Studies in the Changing Structure of Nineteenth-Century Marketing* (Baltimore, 1971), 37–61.

36. Peter Temin, *Iron and Steel in Nineteenth-Century America: An Economic Inquiry* (Cambridge, Mass., 1964), 38–44; Walker, *Hopewell Village*, 153–54; "Agreement between Samuel Richards and the City Council of New Orleans," Sept. 21, 1822; "Agreement [of Philadelphia's 'Watering Committee'] with Samuel Richards for castings and cast Iron pipes to be delivered in 1825 and 1826," Dec. 3, 1825; and "Agreement between Samuel Richards and Albert Stein . . . on behalf of . . . Richmond," Sept. 16, 1830, Box 2, Samuel Richards Papers, Acc. #547, HSP.

37. Paul Paskoff argues that the Constitution stimulated Pennsylvania's iron industry in *Industrial Evolution*, 72–73.

38. [Louis McLane], *Documents Relative to the Manufactures in the United States*, 4 vols. (Washington, D.C., 1833; reprint [in 3 vols.], New York, 1969), 2:205, 219–20, 255–393; Edward B. Grubb to Henry C. Grubb, Jan. 18, 1833, Box 6, GFP, HSP. On ironmasters' collection of data, see John Weidman to Jacob M. Haldeman, Oct. 25, 1817, Box 13, HP, HML; and Benjamin Reeves to William Weaver, Aug. 17, 1831, WWP, Duke, *SABSI:A*, 23. On tariffs, see Frank W. Taussig, *The Tariff History of the United States*, 8th ed., revised (New York, 1964), 46–59; and Paskoff, *Industrial Evolution*, 75–79. I discuss lobbying for tariffs and its impact on how adventurers viewed ironworkers in chapter 6.

39. According to Jeffrey Zabler, the Pennsylvania iron industry's health was inversely related to that of the national economy for most of the early nineteenth century. In other words, when the U.S. economy fared relatively poorly, the iron business did well. See Zabler, "Microeconomic Study," 220–22.

40. David Kizer & Co. to Jacob M. Haldeman, Mar. 23, 1820, Feb. 13, 1821, and J. G. Lowrey to Jacob M. Haldeman, Feb. 12, 1821, Box 13, HP, HML; Anderson to Haldeman, Jan. 31, 1822, Apr. 1, 1822, Box 14, HP, HML. On the Panic of 1819, see Charles Sellers, *The Market Revolution: Jacksonian America, 1815–1846* (New York, 1991), 137–71.

41. Lewis Webb & Co. to William Weaver, Dec. 1, 1825, WWP, Duke, *SABSI:A*, 22; Webb & Co. to Weaver, Dec. 6, 1828, Box 1, Letters, Weaver-Brady Papers (#38–98), UVA; John Donihoe to Weaver, Oct. 8, 1830; and John Schoolfield to Weaver, Apr. 30, 1829, WWP, Duke *SABSI:A*, 23.

42. Bining, *Pennsylvania Iron Manufacture*, 19–38; and Charles B. Dew, *Bond of Iron: Master and Slave at Buffalo Forge* (New York, 1994), 63–66.

43. Alfred D. Chandler, Jr., "Anthracite Coal and the Beginnings of the Industrial Revolution in the United States," *Business History Review* 46 (1972): 146–65; Temin, *Iron and Steel*, 51–80; Louis C. Hunter, "The Influence of the Market upon Techniques in the Iron Industry in Western Pennsylvania up to 1860," *Journal of Economic and Business History* 1 (1929; reprint, New York, 1964), 241–81; and Charles K. Hyde, *Technological Change and the British Iron Industry, 1700–1870* (Princeton, 1977).

44. Hasenclever, *Remarkable Case*, 17; John Potts to Robert E. Hobart, Dec. 16, 1794, Keeptryst Furnace Collection, Acc. 31, HML; and Louis C. Hunter, *A History of Industrial Power in the United States, 1780–1930: Volume One: Waterpower in the Century of the Steam Engine* (Charlottesville, Va., 1979), 110–11.

45. Walker, *Hopewell Village*, 238; and Martha Furnace Diary and Timebook, 1808–1815, Acc. 339, HML. Jacqueline Jones notes the centrality of clearing forests to male colonists' labor routines in *American Work: Four Centuries of Black and White Labor* (New York, 1998), 27–29.

46. Overman, *Manufacture of Iron*, 84; Jacob L. Bunnell, ed., *Sussex County Sesquicentennial* (Newton, N.J., 1903), 29, quoted in Peter O. Wacker, *The Musconetcong Valley of New Jersey: A Historical Geography* (New Brunswick, N.J., 1968), 111. On colliers paying for wood, see Cornwall Furnace Journal, 1768–1769, FFR, HSP. Ironmasters usually expected a chopper to cut one and one-half to two cords a day. A stack of hewn logs four feet wide, four feet high, and eight feet long made a cord.

47. Overman, *Manufacture of Iron*, 80–81, 115–17; and Walker, *Hopewell Village*, 245.

48. Clement Brooke to Robert Carter, Apr. 14, 1778, Robert Carter Papers, MS 1228, MHS.

49. Deposition of John Doyle, May 16, 1826, Case Papers, Weaver *v.* Mayburry; Colebrookdale Furnace Ledger, 1740–1743, Durham Forge Ledger, 1744–1749, FFR, HSP; and Michael V. Kennedy, "Working Agreements: The Use of Subcontracting in the Pennsylvania Iron Industry, 1725–1789," *PH* 65 (1998): 492–508.

50. John Bezís-Selfa, "A Tale of Two Ironworks: Slavery, Free Labor, Work, and Resistance in the Early Republic," *WMQ* 56 (1999): 688–89. I discuss fines in chapter 4.

51. Andover Furnace Day Book/Journal, May 19, 1773–November 24, 1777, Taylor Family (of Hunterdon County, N.J.) Papers, MC 885, RUL; and Barnard to Samuel G. Wright, July 5, 1824, Box 1, WFP, HML.

52. John Evelyn provides an excellent description of making charcoal to fuel ironworks in *Sylva, or a Discourse of Forest-Trees and the Propagation of Timber in His Majesties Dominions* (London, 1664), 100–2. See also Overman, *Manufacture of Iron*, 104–7; and Walker, *Hopewell Village*, 242–45.

53. *MG*, Apr. 30, 1767; Benjamin Abbott, *The Experience and Gospel Labours of the Rev. Benjamin Abbott: To Which is Annexed, A Narrative of His Life and Death*, 4th ed. (New York, 1813), 131; Charles Ridgely & Co. Account Book, 1774–1780, LC; and Kingsbury Furnace Timebook, 1767–1769, PIWC, HSD. Ironworks sometimes transferred risks to colliers. Wood burned by accident or neglect, small yields of charcoal, or excessive amounts of uncharred wood all often resulted in fines. I discuss this issue in chapters 4 and 6.

54. Overman, *Manufacture of Iron*, 106–7.

55. Peter Kalm, *Travels in North America*, 1770 ed., ed. Adolph B. Benson (New York, 1966), 157; Schoepf, *Travels*, 2:7; Hermelin, *Report*, 15, 21–25; William Byrd, "A Progress to the Mines," in *The Writings of Colonel William Byrd of Westover in Virginia Esqr.*, ed. John Spencer Bassett (New York, 1901, reprint, New York, 1970), 352–53; Joseph Hoff to Lord Stirling, Feb. 4, 1776; Hoff to Ichabod Barnet, Apr. 4, 1776; and Hoff to Murray Samson & Co., May 27, 1776, HIWC, RUL.

56. Niemcewicz, *Under Their Vine and Fig Tree*, 228; and Schoepf, *Travels*, 1:197–98.

57. Doerflinger, "How to Run an Ironworks," 365; Schoepf, *Travels*, 1:198, 202; and Accokeek Furnace Journal, 1749–1760, Kingsbury Furnace Timebook, 1767–1769, PIWC, HSD.

58. David Ross to John Staples, June 2, 1813, DRL, 1812–1813, VHS; *PG*, Dec. 13, 1739; and William Eddis, *Letters from America*, ed. Aubrey C. Land (Cambridge, Mass., 1969), 43. I discuss these issues more in chapter 3.

59. Bining, *Pennsylvania Iron Manufacture,* 82; and Robbins, *Principio Company,* 169–73. On blast furnace technology, see R. Bruce Council, Nicholas Honerkamp, and M. Elizabeth Will, *Industry and Technology in Antebellum Tennessee: The Archeology of Bluff Furnace* (Knoxville, Tenn., 1992), 20–37.

60. Overman, *Manufacture of Iron,* 204–5. On furnace work, see Walker, *Hopewell Village,* 231–37.

61. Barnard to Samuel G. Wright, Aug. 19, 1823, Nov. 4, 1822, Box 1, WFP, HML.

62. John Fuller, Jr., to the Prince of San Severino, Oct. 24, 1754, in *The Fuller Letters, 1728–1755: Guns, Slaves, and Finance,* ed. David Crossley and Richard Saville (Lewes, England, 1991), 282; and Overman, *Manufacture of Iron,* 204–5.

63. John D. Tyler, "Technological Development: Agent of Change in Style and Form of Domestic Iron Castings," in *Technological Innovation and the Decorative Arts,* ed. Ian M. G. Quimby and Polly Anne Earl (Charlottesville, Va., 1974), 141–42.

64. Ibid., 144–48, 150; Deborah Ducoff-Barone, "Marketing and Manufacturing: A Study of Domestic Cast Iron Articles Produced at Colebrookdale Furnace, Berks County, Pennsylvania, 1735–1751," *PH* 50 (1983): 30–31.

65. For examples of such work, see Henry C. Mercer, *The Bible in Iron: Pictured Stoves and Stoveplates of the Pennsylvania Germans,* 3d ed. (Doylestown, Pa., 1961), 23–48. On how moulders cast stove plates, see Donald Allen Crownover, *Manufacturing and Marketing of Iron Stoves at Hopewell Furnace, 1835–1844* (Washington, D.C., 1970), 54–65.

66. John G. Smith to Wright, Feb. 24, 1822; and Derick Barnard to Wright, June 2, 1823, Mar. 24, 1827, Box 1, WFP, HML.

67. *Martha,* June 2, 1813; Barnard to Wright, Nov. 4, 1822, Box 1, WFP, HML.

68. Erskine to Richard Atkinson, Sept. 19, 1770, Vol. 2, Box 1, RMP, NJSA. For an excellent discussion of forge work, see Dew, *Bond of Iron,* 10–11, 187.

69. Erskine to Richard Atkinson, Sept. 21, 1770, Vol. 1, Box 1, RMP, NJSA. Excess sulfur made pig iron "red short," while too much phosphorus yielded "cold short" iron. Ironmasters largely avoided the former by sticking to charcoal as a fuel and by not smelting ores composed largely of sulfides and sulfates. Controlling phosphorus content was far more difficult. See Gordon, *American Iron,* 7–8, 153.

70. *Virginia Gazette* (Purdie), Mar. 7, 1766; Peter Coale to Capt. Charles Ridgely, June 21, 1783, Box 1, RP692, MHS; William Davis to William Weaver, July 7, 1832, WWP, Duke, *SABSI:A,* 23; Dew, *Bond of Iron,* 173.

71. Deposition of William Norcross, May 23–24, 1826, Case Papers, Weaver *v.* Mayburry; Drinker to Pearsall & Pell, Mar. 23, 1787, HDL, 1786–1790; Drinker to Isaac Stoutenbough & Son, July 15, 1790, HDL, 1790–1793; Drinker to Thomas & James Stevenson, June 11, 1801, HDL, 1800–1802, Drinker, HSP; "A letter from some blacksmiths complaining of the quality of the iron made at Spring forge to Robert Coleman," May 27, 1801, Robert Coleman #2, Box 2, Folder 3, Coleman Papers Collection, MG-275, LCHS. On appearance and marketing of bar iron, see A. Allyene to Dr. Charles Carroll, Apr. 2, 1750, *MHM* 23 (1928): 379.

72. Doerflinger, "How to Run an Ironworks," 363, 364.

73. Daniel Vickers, "Competency and Competition: Economic Culture in Early America," *WMQ* 47 (1990): 3–29; Thomas M. Doerflinger, "Rural Capitalism in Iron Country: Staffing a Forest Factory, 1808–1815," *WMQ* 59 (2002): 3–38; Michael V. Kennedy, "Working Agreements," and "An Alternate Independence: Craft Workers in the Pennsylvania Iron Industry, 1725–1775," *Essays in Economic and Business History* 16 (1998): 113–25.

74. Clement Brooke to Robert Carter, Oct. 20, 1776, Robert Carter Papers, MS 1228, MHS. On staple crops and slave labor, see Peter Kolchin, *Unfree Labor: American Slavery and Russian Serfdom* (Cambridge, Mass., 1987), 17–26; Carville E. Earle, "A Staple Interpretation of Slavery and Free Labor," *Geographical Review* 68 (1978), 51–65; and Todd H. Barnett, "The Evolution of 'North' and 'South': Settlement and Slavery on America's Sectional Border, 1650–1810" (Ph.D. diss., University of Pennsylvania, 1993), 84–121, 178–204.

75. *CRPa* 3:392, 584. On coercion and the growth of eastern Europe's iron industry, see Maria Ågren, ed., *Iron-Making Societies: Early Industrial Development in Sweden and Russia,*

1600–1900 (Oxford, 1998); Hugh D. Hudson, Jr., Bruce J. DeHart, and David M. Griffiths, "Proletarians by Fiat: The Compulsory Ural Metallurgical Labor Force, 1630–1861," *International Labor and Working-Class History* 48 (1995): 94–111; and Milan Myska, "Pre-Industrial Iron-Making in the Czech Lands: The Labour Force and Production Relations circa 1350–circa 1840," *Past and Present* 82 (1979): 44–72.

Chapter 2. Molding Men

1. *NHTR* 2:275–76. See also Jane Kamensky, *Governing the Tongue: The Politics of Speech in Early New England* (New York, 1997).

2. Edmund S. Morgan, *American Slavery, American Freedom: The Ordeal of Colonial Virginia* (New York, 1975), 215–49; and Kathleen M. Brown, *Good Wives, Nasty Wenches, and Anxious Patriarchs: Gender, Race, and Power in Colonial Virginia* (Chapel Hill, 1996), 149–54.

3. H. R. Schubert, *History of the British Iron and Steel Industry* (London, 1957), 188–94; T. C. Barnard, "Sir William Petty as Kerry Ironmaster," *Proceedings of the Royal Irish Academy. Section C [Ireland]* 82 (1982): 1–32.

4. *RVCL* 1:352, 472; 3:278; 4:141.

5. Ibid. 3:475–76. On Falling Creek, see Charles E. Hatch, Jr., and Thurlow Gates Gregory, "The First American Blast Furnace, 1619–1622: The Birth of a Mighty Industry in Virginia," *VMHB* 70 (1966): 259–96.

6. *RVCL* 3:278; and Morgan, *American Slavery*, esp. 108–30.

7. Thomas Hariot, *A briefe and true report of the new found land of Virginia* (London, 1588), B3, in *Virginia: Four Personal Narratives* (New York, 1972); and *RVCL* 3:308–9. On deforestation, see Schubert, *History of the British Iron and Steel Industry*, 219–22.

8. Morgan, *American Slavery*, 28–30.

9. George Thorpe and John Pory to Edwin Sandys, May 15/16, 1621, *RVCL* 3:446; and George Yeardley to [Sandys], [1619], *RVCL* 3:128–29. On attempts to anglicize Virginia Algonkians, see Brown, *Good Wives*, 42–74. On English and Native assessments of each other, see Karen Ordahl Kupperman, *Indians and English: Facing Off in Early America* (Ithaca, 2000).

10. *RVCL* 3:165–66; and *RVCL* 1:307–8, 335.

11. *RVCL* 1:585–88. Francis Jennings argues that the school was a ruse to attract investors to the Virginia Company in *The Invasion of America: Indians, Colonialism, and the Cant of Conquest* (Chapel Hill, 1975), 53–56.

12. *RVCL* 1:588–89. On Sandys and his overseas ventures, see Theodore K. Rabb, *Jacobean Gentleman: Sir Edwin Sandys, 1561–1629* (Princeton, 1998), 319–85.

13. *RVCL* 1:640, 565. On reorganization of Virginia's government, see Wesley Frank Craven, *The Southern Colonies in the Seventeenth Century* (Baton Rouge, La., 1949), 147–55.

14. *RVCL* 3:558–59. On English views of Natives and their development of racialized views of them, see Joyce E. Chaplin, *Subject Matter: Technology, the Body, and Science on the Anglo-American Frontier* (Cambridge, Mass., 2001).

15. *RVCL* 3:670–71; *RVCL* 2:384; and David Harvey, "Reconstructing the American Bloomery Process," *The Colonial Williamsburg Historic Trades Annual* 1 (1988): 19–38.

16. Edward Johnson, *Wonder-Working Providence of Sions Saviour in New-England* (London, 1654), reprinted in *Johnson's Wonder-Working Providence 1628–1651*, ed. J. Franklin Jameson (New York, 1910), 245–48. For arguments that Puritan ideas of commonwealth and economic development reinforced one another, see Stephen Innes, *Creating the Commonwealth: The Economic Culture of Puritan New England* (New York, 1995); Margaret Ellen Newell, *From Dependency to Independence: Economic Revolution in Colonial New England* (Ithaca, 1998); and John Frederick Martin, *Profits in the Wilderness: Entrepreneurship and the Founding of New England Towns in the Seventeenth Century* (Chapel Hill, 1991).

17. Bernard Bailyn, *The New England Merchants in the Seventeenth Century* (Cambridge, Mass., 1955), 45–86.

18. Johnson, *Wonder-Working Providence*, 245–47.

19. William Hubbard, *A General History of New England from the Discovery to MDCLXXX* in *Collections of the Massachusetts Historical Society*, 2d. series, 10 vols. (Boston, 1814–1823; reprint, New York, 1968), 5:374–75; and E. N. Hartley, *Ironworks on the Saugus: The Lynn and Braintree Ventures of the Company of Undertakers of the Ironworks in New England* (Norman, Okla., 1957), 215–305.

20. Emmanuel Downing to John Winthrop, ca. Aug. 1645, *WP* 5:38, 48–49. On enslavement of Natives, see Jill Lepore, *The Name of War: King Philip's War and the Origins of American Identity* (New York, 1998), 150–70; and Michael C. Fickes, "'They Could Not Endure That Yoke': The Captivity of Pequot Women and Children after the War of 1637," *New England Quarterly* 73 (2000): 58–81. On complaints regarding wages, see Richard Saltonstall to Emmanuel Downing, Feb. 4, 1632, in Everett Emerson, ed., *Letters from New England: The Massachusetts Bay Colony, 1629–1638* (Amherst, Mass., 1976), 92. On regulation of wages in seventeenth-century New England, see Richard B. Morris, *Government and Labor in Early America* (New York, 1946; reprint, Boston, 1981), 55–84.

21. Alison Games, *Migration and the Origins of the English Atlantic World* (Cambridge, Mass., 1999); David Cressy, *Coming Over: Migration and Communication between England and New England in the Seventeenth Century* (New York, 1987), 52–63.

22. Robert Child to John Winthrop, Jr., June 27, 1643, *WP* 4:395–96; and Joshua Foote to John Winthrop, Jr., Sept. 20, 1643, *WP* 4:415. On the Irish rebellion, see Schubert, *History of the British Iron and Steel Industry*, 190.

23. Foote to John Winthrop, Jr., May 20, 1643, *WP* 4:379–80; and "Petition of John Winthrop, Jr., to Parliament," *WP* 4:424–25.

24. Foote to Winthrop, Jr., Sept. 20, 1643, *WP* 4:415; and Promoters of the Ironworks to Winthrop, Jr., Mar. 13, 1648, *WP* 5:209.

25. *WP* 5:209; Richard S. Dunn, *Puritans and Yankees: The Winthrop Dynasty of New England, 1630–1717* (Princeton, 1962; paperback ed., New York, 1971), 87–96. See also Robert C. Black, III, *The Younger John Winthrop* (New York, 1966).

26. *Mass Records* 2:103–4, 125–28, 185–86; Hartley, *Ironworks on the Saugus*, 90–96; Newell, *From Dependency to Independence*, 55–66.

27. *Mass Records* 2:127–28, 185–86.

28. *Mass Records* 3:142; and Hartley, *Ironworks on the Saugus*, 97.

29. B. G. Awty, "The Continental Origins of Wealden Ironworkers, 1451–1544," *Economic History Review* 34 (1981): 524–39; Daniel Vickers, "Work and Life on the Fishing Periphery of Essex County, Massachusetts, 1630–1675," in *Seventeenth-Century New England*, ed. David Grayson Allen and David D. Hall (Boston, 1984), 83–117; and Vickers, *Farmers and Fishermen: Two Centuries of Work in Essex County, Massachusetts, 1630–1830* (Chapel Hill, 1994), 91–100, 129–41.

30. On manhood and communal expectations in New England, see Mary Beth Norton, *Founding Mothers and Fathers: Gendered Power and the Forming of American Society* (New York, 1996); Ann M. Little, "A 'Wel Ordered Commonwealth': Gender and Politics in New Haven Colony, 1636–1690" (Ph.D. diss., University of Pennsylvania, 1996); and Lisa Wilson, *Ye Heart of A Man: The Domestic Life of Men in Colonial New England* (New Haven, 1999).

31. *ERF* 1:245, 271; and Wilson, *Ye Heart of A Man*, 99–114.

32. Here I largely agree with Innes, *Creating the Commonwealth*, 254–63, and with Hartley, *Ironworks on the Saugus*, 202–9.

33. *ERF* 1:130, 198–200.

34. *ERF* 1:134–35, 138, 173–74; and *Mass Records* 2:243; 3:127.

35. *ERF* 1:136, 153, 159.

36. Ibid., 1:181, 184.

37. See also David Thomas Konig, *Law and Society in Puritan Massachusetts: Essex County, 1629–1692* (Chapel Hill, 1979).

38. *Mass Records* 3:227–28, 257; Kamensky, *Governing the Tongue*, 145; Margaret E. Newell, "Robert Child and the Entrepreneurial Vision: Economy and Ideology in Early New England," *New England Quarterly* 68 (1995): 223–56; and Hartley, *Ironworks on the Saugus*, 117–21, 134–40, 158–59.

39. LIWC, Baker Library, Harvard Business School, Harvard University, Boston, Mass., followed by page number, 273, 409. On litigation and Lynn, see Hartley, *Ironworks on the Saugus,* 215–43.

40. *ERF* 1:289–93; LIWC, 409, Baker Library, Harvard Business School, Harvard University, Boston, Mass.

41. *NHTR* 1:330–31, 349.

42. Ibid., 2:117–23, 134, 140–41.

43. *Ibid.,* 134. On New Haven's legal culture, see Gail Sussman Marcus, "'Due Execution of the Generall Rules of Righteousnesse': Criminal Procedure in New Haven Town and Colony, 1638–1658," 99–137, and John M. Murrin, "Magistrates, Sinners, and a Precarious Liberty: Trial by Jury in Seventeenth-Century New England," 170–82, both in *Saints and Revolutionaries: Essays on Early American History,* ed. David D. Hall, John M. Murrin, and Thad W. Tate (New York, 1984); Cornelia Hughes Dayton, *Women before the Bar: Gender, Law, and Society in Connecticut, 1639–1789* (Chapel Hill, 1995), 27–34.

44. This paragraph and the next draw from *NHTR* 2:117–23. Mary Beth Norton proclaims the Pinions seventeenth-century New England's "most dysfunctional family" and uses them to illustrate the importance attached to family governance in *Founding Mothers,* 27–38. On prosecution of consensual sex and sexual assault in New Haven, see Dayton, *Women before the Bar,* 173–81, 234–43.

45. *NHTR* 2:122–23. Dayton notes that Morran was the only man not convicted in a sexual assault case in New Haven before 1700. See *Women before the Bar,* 237.

46. *NHTR* 2:146.

47. Ibid., 159–62, 215.

48. This paragraph and the next three come from *NHTR* 2:148–51.

49. Ibid., 178, 199, 222–23, 240, 251, 267, 297–98, 307, 345, 355, 369, 379, 388.

50. Ibid., 181–83, 201, 204, 211, 215, 222–23; Norton, *Founding Mothers,* 36–37; and Dayton, *Women before the Bar,* 166.

51. Karen Ordahl Kupperman, *Providence Island, 1630–1641: The Other Puritan Colony* (New York, 1993); and Downing to John Winthrop, ca. Aug. 1645, *WP* 5:48–49.

52. LIWC, 148–49, 151, Baker Library, Harvard Business School, Harvard University, Boston, Mass. On Lynn's Scottish servants, see Hartley, *Ironworks on the Saugus,* 146–47, 154–55, 198–202.

53. Company of Undertakers to Gifford, Apr. 26, 1652, LIWC, 53; LIWC, 67; "Deposition of Samuel Hart," LIWC, 229; *ERF* 2:96; and Bond and Company of Undertakers to Gifford, Apr. 26, 1652, LIWC, 49, Baker Library, Harvard Business School, Harvard University, Boston, Mass.

54. LIWC, 50; LIWC, 175–76, 295, 331, 391, Baker Library, Harvard Business School, Harvard University, Boston, Mass.

55. "Colonel Lewis Morris & Comp.y Pattent for ye Land Belonging to ye Iron Works Called the Mannor of Tinton," Oct. 25, 1676, and "sir George Carteret's Grant of Privileges for the Tinton Iron Works," May 1, 1677, Box 1, Robert Morris Papers, MC 588, RUL; and Dean Freiday, "Tinton Manor: The Iron Works," *Proceedings of the New Jersey Historical Society* 70 (1952): 250–61.

56. Freiday, "Tinton Manor," 253–61; George Scot, *The Model of the Government of the Province of East Jersey* (Edinburgh, 1685), 128–29; and Graham Russell Hodges, *Slavery and Freedom in the Rural North: African Americans in Monmouth County, New Jersey, 1665–1865* (Madison, Wis., 1997), 9–10. Most of Lewis Morris's will and inventory is published in Robert Bolton, *The History of the Several Towns, Manors, and Patents of the County of Westchester from its First Settlement to the Present Time,* rev. ed., ed. C. W. Bolton, 2 vols. (New York, 1881), 2:464–69.

Chapter 3. Passages through the Ledgers

1. Kingsbury Furnace Journal, 1768–1770, PIWC, HSD; Provincial Court (Judgment Record), April 1770, DD 16, MSA, S551, MdHR 808–2, 1/17/2/23, MSA.

2. Provincial Court (Judgment Record); Commissioners to Preserve Confiscated British Property (Ledger and Journal), 1781–1782, MSA S132, MSA.

3. López was likely one of the first Chesapeake slaves to sue for freedom by claiming that his bondage violated his right to liberty under British custom. Most Maryland slaves who filed manumission suits based their claims upon descent from a white woman. See T. Stephen Whitman, *The Price of Freedom: Slavery and Manumission in Baltimore and Early National Maryland* (Lexington, Ky., 1997), 63–67.

4. Bernard Bailyn, *Voyagers to the West: A Passage in the Peopling of America on the Eve of the Revolution* (New York, 1987), 247–54. See also Christine Daniels, "Gresham's Laws: Labor Management on an Early Eighteenth-Century Chesapeake Plantation," *Journal of Southern History* 62 (1996): 205–38.

5. Charles Carroll of Carrollton to Charles Carroll & Co., Dec. 8, 1773; Clement Brooke to Charles Carroll & Company, Feb. 4, 1774, *CCCFP*, M–4214. See also Ronald Hoffman, *Princes of Ireland, Planters of Maryland: A Carroll Saga, 1500–1782* (Chapel Hill, 2000), 229–32.

6. "Account of Persons Employed at the Baltimore Iron Works, April 30th, 1734," *CCCFP*, M–4215; Charles Carroll of Annapolis to Charles Carroll of Carrollton, Jan. 9, 1764, *MHM* 12 (1917): 27; Charles G. Steffen, *From Gentlemen to Townsmen: The Gentry of Baltimore County, Maryland, 1660–1776* (Lexington, Ky., 1993), 52–53; Commissioners to Preserve Confiscated British Property (Ledger and Journal), 1781–1782, MSA S132, MSA. Ronald L. Lewis discusses the scale of slaveholding in the colonial Chesapeake's iron industry in *Coal, Iron, and Slaves: Industrial Slavery in Maryland and Virginia, 1715–1865* (Westport, Conn., 1979), 7, 21–26.

7. John England to Joseph Farmer, Apr. 25, 1723; England to John Ruston, Apr. 27, 1723, PCP, MHS.

8. Principio Company to England, Sept. 11, 1723, and Sept. 25, 1723, PCP, MHS.

9. England to Joshua Gee, July 12, 1723, PCP, MHS; Lois Green Carr and Lorena S. Walsh, "Economic Diversification and Labor Organization in the Chesapeake, 1650–1820," in *Work and Labor in Early America*, ed. Stephen Innes (Chapel Hill, 1987), 158–59; and Robert J. Steinfeld, *The Invention of Free Labor: The Employment Relation in English and American Law and Culture, 1350–1870* (Chapel Hill, 1991), 66–116.

10. John D. Cushing, ed., *The Laws of the Province of Maryland* (Wilmington, Del., 1978), 119–20.

11. England to Joseph Farmer, Jan. 4, 1724, and England to Joshua Gee, Jan. 13, 1724, PCP, MHS.

12. Principio Company to England, Sept. 11, 1723, and England to Joseph Farmer, Jan. 1724, PCP, MHS.

13. Principio Iron Works Ledger, 1723–1725, PIWC, HSD; and Principio Iron Works Day Book, 1732, Principio Forge Accounts, MS 2472, MHS. On the development of clock time consciousness in the South, see Mark M. Smith, *Mastered by the Clock: Time, Slavery, and Freedom in the American South* (Chapel Hill, 1997), 20–22, 40–50, 63–67.

14. Principio Iron Works Ledger, 1723–1725, and Principio Iron Works Ledger, 1724–1728, PIWC, HSD. In 1724, Durham's annual salary was £24.

15. England to Farmer, Jan. 1724, and "Indenture of Joseph Reading," PCP, MHS.

16. On British forgemen, see Chris Evans and Göran Rydén, "Kinship and the Transmission of Skills: Bar Iron Production in Britain and Sweden, 1500–1800," in *Technological Revolutions in Europe: Historical Perspectives*, ed. Maxine Berg and Kristine Bruland (Cheltenham, UK, 1998), 188–206. On company partners recruiting workers, see William Russell to William Chetwynd, Feb. 27, 1726, and Russell to Chetwynd, Apr. 17, 1725, BM, LC.

17. PCP, MHS; England to [ ], 1723, PCP, MHS.

18. England to Joseph Farmer, Jan. 1724, and Chetwynd to England, Sept. 19, 1725, PCP, MHS.

19. Chetwynd to England, Aug. 19, 1726, BM, LC. At the time, the company's thirteen

slaves may have comprised about one-quarter of its employees. See "A List of Men's Names Belonging To the Iron Works of Principio," [no date, probably mid 1720s], PCP, MHS.

20. Fayrer Hall, *The Importance of the British Plantations in America to This Kingdom; with the State of their Trade, and Methods for Improving it; as also a Description of the several Colonies there* (London, 1731), 76–77.

21. William Russell to Chetwynd, April 17, 1725, BM, LC; Principio Company to England, Sept. 25, 1725, and Chetwynd to England, Oct. 5, 1725, PCP, MHS.

22. William Russell to William Chetwynd, April 17, 1725, BM, LC.

23. On slavery and Afro-Britons, see Philip D. Morgan, "British Encounters with Africans and African-Americans, circa 1600–1780," in *Strangers within the Realm: Cultural Margins of the First British Empire,* ed. Bernard Bailyn and Philip D. Morgan (Chapel Hill, 1991), 159–60, 165–67, 191, 193–95; and F. O. Shyllon, *Black Slaves in Britain* (London, 1974), 11–16. On British forgemen see Evans and Rydén, "Kinship and the Transmission of Skills." On opposition of colonial white artisans to slaves in their trades, see Richard B. Morris, *Government and Labor in Early America* (New York, 1946; reprint, Boston, 1981), 182–86.

24. Principio Forge Ledger, 1726, Principio Forge Accounts, MS 2472, MHS; Principio Iron Works Ledger, 1727, and Principio Iron Works Ledger, 1728, PIWC, HSD.

25. Principio Iron Works Ledger, 1727, PIWC, HSD.

26. "Indenture of Joseph Reading," PCP, MHS.

27. Principio Iron Works Ledger, 1728, PIWC, HSD. See also Christine Daniels, "'Without Any Limitacon of Time': Debt Servitude in Colonial America," *LH* 36 (1995): 232–50.

28. John Wrightwick to John England, Oct. 2, 1730, PCP, MHS; and Commissioners to Preserve Confiscated British Property (Ledger and Journal), 1781–1782, MSA.

29. Elizabeth Chapman Denny Vann and Margaret Collins Denny Dixon, *Virginia's First German Colony* (Richmond, 1961), 9–26, 28, 33–46; and William L. Sanders, ed., *The Colonial Records of North Carolina,* 10 vols. (Raleigh, 1886–1890), 1:973–75.

30. William Byrd, "A Progress to the Mines," in *The Writings of Colonel William Byrd of Westover in Virginia Esqr,* ed. John Spencer Bassett (New York, 1901; reprint, New York, 1970), 354.

31. Ibid., 345, 351, 360.

32. Ibid., 362; Lester J. Cappon, ed., *Iron works at Tuball: terms and conditions for their lease as stated by Alexander Spotswood on the twentieth day of July 1739* (Charlottesville, Va., 1945), 17, 19–20.

33. Byrd, "A Progress to the Mines," 362.

34. Stephen Onion to Charles Carroll of Annapolis, Dec. 15, 1735, and Carroll of Annapolis to Edmund Jenings, Dec. 21, 1735, *CCCFP,* M–4213.

35. Edmund Jenings to Charles Carroll of Annapolis, Dec. 15, 1735, Richard Snowden to the Baltimore Company, Dec. 19, 1735, and Dr. Charles Carroll to Carroll of Annapolis, Dec. 22, 1735, *CCCFP,* M–4213.

36. Cappon, *Iron works at Tuball,* 20.

37. On adventurers who traded in servants or slaves, see *MG,* May 20, 1729, June 21, 1753, Nov. 8, 1753; and Aubrey C. Land, *The Dulanys of Maryland: A Biographical Study of Daniel Dulany, the Elder (1685–1753) and Daniel Dulany, the Younger (1722–1797)* (Baltimore, 1955), 105–6. For examples of planter-ironmasters contributing slaves to their ironworks, see Baltimore Company Minute Book, 1731–1774, *CCCFP,* M–4214; "Charles Carroll of Annapolis and Daniel Carroll of Duddington: Account of Time Worked By Negroes, July 31, 1733," *CCCFP,* M–4215; and John Tayloe Account Book, 1749–1768, Tayloe Family Papers, VHS.

38. A. Roger Ekirch, *Bound for America: The Transportation of British Convicts to the Colonies, 1718–1775* (New York, 1987), 111–16, 124–25. See also Farley Grubb, "The Transatlantic Market for British Convict Labor," *JEconHist* 60 (2000): 94–122.

39. Stevenson, Randolph & Cheston Factorage Book, 1767–74, 1781, Box 5, Cheston-Galloway Papers, MS 1994, MHS. See also Stevenson, Randolph & Cheston Account of Servants 1774–75, Box 6, Samuel Dorsey, Jr., to James Cheston, Aug. 10, 1772, Box 11, James Cheston to Thomas, Samuel, and John Snowden, Jan. 7, 1771, Cheston to Corbin Lee, Jan. 7,

1771, James Cheston Letterbook, 1768–1771, Box 8, Cheston-Galloway Papers, MS 1994, MHS.

40. Byrd, "A Progress to the Mines," 351; "A List of Men's Names Belonging To the Iron Works of Principio," PCP, MHS; "Account of Time Worked by Negroes" and "Account of Persons Employed at the Baltimore Iron Works, April 30th, 1734," *CCCFP*, M–4215, MHS; Kingsbury Furnace Timebook, 1767–1769, PIWC, HSD; and Caleb Dorsey & Co., Elk Ridge Furnace Ledger AA, 1761–1762, #1497, Elk Ridge Furnace Ledger BB, 1762–1764, #1498, Chancery, MSA. On the ethnic and regional origins of Africans whom the slave trade brought to the Chesapeake, see Philip D. Morgan, *Slave Counterpoint: Black Culture in the Eighteenth-Century Chesapeake and Lowcountry* (Chapel Hill, 1998), 58–79; and Lorena S. Walsh, "The Chesapeake Slave Trade: Regional Patterns, African Origins, and Some Implications," *WMQ* 58 (2001): 139–70.

41. Byrd, "A Progress to the Mines," 345; and Chetwynd to England, Aug. 19, 1726, BM, LC. See also the Baltimore Company's directive in the winter of 1766–67 permitting its clerk to "purchase African Negroes, provided they are well season'd to the Country," in Baltimore Company Minute Book, 1731–1774, *CCCFP*, M–4214, MHS. On mortality rates for Africans, see Morgan, *Slave Counterpoint*, 444–45; and Allan Kulikoff, *Tobacco and Slaves: The Development of Southern Cultures in the Chesapeake, 1680–1800* (Chapel Hill, 1986), 326–27.

42. Morgan, *Slave Counterpoint*, 81–82, 84–95; Allan Kulikoff, "A 'Prolifick' People: Black Population Growth in the Chesapeake Colonies, 1700–1790," *Southern Studies* 16 (1977): 391–428.

43. Paul G. E. Clemens, *The Atlantic Economy and Colonial Maryland's Eastern Shore: From Tobacco to Grain* (Ithaca, 1980); and *MG*, July 27 and Nov. 30, 1748, Aug. 8, 1765. See also Carville B. Earle, "A Staple Interpretation of Slavery and Free Labor," *Geographical Review* 68 (1978): 51–65.

44. On diversification and teaching slaves trades, see Carville V. Earle, *The Evolution of a Tidewater Settlement System: All Hallow's Parish, Maryland, 1650–1783* (Chicago, 1975), 101–31; Gerald W. Mullin, *Flight and Rebellion: Slave Resistance in Eighteenth-Century Virginia* (New York, 1972), 7–12, 19, 32–33; Kulikoff, *Tobacco and Slaves*, 396–401; and Morgan, *Slave Counterpoint*, 204–54. On the tendency of Chesapeake planters to denigrate Africans' ability to learn trades, see idem., 232.

45. *Virginia Gazette* (Purdie), March 7, 1766. On recruitment of Africans and rice cultivation, see Peter Wood, *Black Majority: Negroes in Colonial South Carolina from 1670 through the Stono Rebellion* (New York, 1974), 56–62; Daniel C. Littlefield, *Rice and Slaves: Ethnicity and the Slave Trade in Colonial South Carolina* (Baton Rouge, La., 1981), 74–114; and Judith A. Carney, *Black Rice: The African Origins of Rice Cultivation in the Americas* (Cambridge, Mass., 2001), 89–90.

46. "Indenture of Joseph Reading," PCP, MHS; Stephen Onion to Charles Carroll of Annapolis, Feb. 11, 1737, *CCCFP*, M–4213; "Proposal of Charles Carroll," *MHM* 25 (1930): 299; "Committee Report to the House of Commons," April 1738, Penn Mss. Papers Relating to Iron, Peltries, Trade, Etc. 1712–1817, 13:51, Penn Family Papers, Acc. #485A, HSP.

47. Mullin, *Flight and Rebellion*, 84–85.

48. Principio Forge Daybook, 1732, Principio Forge Accounts, MS 2472, MHS; Patuxent Iron Works Journal A&B, 1767–1794, #1504, Chancery, MSA; and Kingsbury Furnace Journal, 1745–1746, and Accokeek Furnace Journal, 1749–1760, PIWC, HSD.

49. On slave hiring in Maryland, see Kulikoff, *Tobacco and Slaves*, 405–7. I discuss slave hiring in more detail in chapter 5.

50. Byrd, "A Progress to the Mines," 351–54; and Accokeek Furnace Journal, 1749–1760, and Kingsbury Furnace Journal, 1768–1770, PIWC, HSD. See also North East Forge Journal, 1753, North East Forge Journal, 1757–1758, [Kingsbury Furnace] Waste Book/Journal, 1765, PIWC, HSD.

51. Morgan, *Slave Counterpoint*, 136.

52. *Archives of Maryland,* 72 vols. (Baltimore, 1883–1972) 37:486, 541; Cushing, ed., *Laws of the Province of Maryland,* 63, 191; and Peter Linebaugh and Marcus Rediker, *The Many-Headed Hydra: Sailors, Slaves, Commoners, and the Hidden History of the Revolutionary Atlantic* (Boston, 2000).

53. *PG,* Apr. 11, 1754; Brooke to Charles Carroll & Company, Feb. 4, 1774; and Baltimore Company Minute Book, 1731–1774, *CCCFP,* M–4214. See also Kenneth Morgan, "English and American Attitudes towards Convict Transportation 1718–1775," *History* 72 (1987): 416–31; and Ekirch, *Bound for America,* 134–40.

54. "Description of White Servants Taken January 1772," Box 14, RAB, MHS; *MG,* Oct. 15, 1761; and Caleb Dorsey & Co., Elk Ridge Furnace Journal AA, 1761–1762, #1494, Chancery, MSA. I counted 221 incidents of a slave or servant running away from a Chesapeake ironworks before 1776: 122 of convicts, 52 of servants (which may include some convicts), 46 of slaves, 1 unknown; of which 15 convicts, 7 slaves, and 2 servants tried to run away more than once. I compiled these figures from *MG, PG, Virginia Gazette* (all editions), and Lathan A. Windley, ed., *Runaway Slave Advertisements: A Documentary History from the 1730s to 1790,* 4 vols. (Westport, Conn., 1983), v. 1–2.

55. Charles Ridgely & Co. Account Book, 1774–1780, LC; and Northampton Furnace Journal, 1775–1778, Box 3, RAB, MHS. See also R. Kent Lancaster, "Almost Chattel: The Lives of Indentured Servants at Hampton-Northampton, Baltimore County," *MHM* 94 (1999): 341–62.

56. Northampton Furnace Journal, 1775–1778, Box 3, RAB, MHS; Randolph Hulse to Ridgely, Feb. 22, 1777, Box 2, RP692.1, MHS.

57. *MG,* Sept. 17, 1767.

58. *Diary and autobiography of John Adams,* L. H. Butterfield, ed., Leonard C. Faber and Wendell D. Garrett, asst. eds., 4 vols. (Cambridge, Mass., 1961), 2:261; and William Eddis, *Letters from America,* ed. Aubrey C. Land (Cambridge, Mass., 1969), 38. For examples of convicts and slaves running away together, see *MG,* May 2, 1754; and *Virginia Gazette* (Pinkney), Nov. 23, 1775. Ronald Hoffman contends that the Carrolls punished white servants as harshly as slaves in *Princes of Ireland, Planters of Maryland,* 258–59.

59. On tobacco, see Lorena S. Walsh, *From Calabar to Carter's Grove: The History of a Virginia Slave Community* (Charlottesville, Va., 1997), 63–65. On rice, empowerment, and the development of slave economies, see Joyce E. Chaplin, "Tidal Rice Cultivation and the Problem of Slavery in South Carolina and Georgia, 1760–1815," *WMQ* 49 (1992): 29–61; Carney, *Black Rice,* 69–106; and Betty Wood, *Women's Work, Men's Work: The Informal Slave Economies of Lowcountry Georgia* (Athens, Ga., 1995), 12–79.

60. Ian Fowler, "Babungo: A Story of Iron Production, Trade and Power in a Nineteenth-Century Ndop Plains Chiefdom (Cameroons)," (Ph.D. diss., University of London, 1990), 344, quoted in Eugenia W. Herbert, *Iron, Gender, and Power: Rituals of Transformation in African Societies* (Bloomington, Ind., 1993), 127. On iron's significance to Africa's past, see Jan Vansina, *Paths in the Rainforests: Toward a History of Political Tradition in Equatorial Africa* (Madison, Wis., 1990), 58–61. On iron as currency and on its role within the slave trade, see Elizabeth Donnan, ed., *Documents Illustrative of the History of the Slave Trade to America,* 4 vols. (Washington, 1930–1935), 4:77–80; and Joseph C. Miller, *The Way of Death: Merchant Capitalism and the Angolan Slave Trade* (Madison, Wis., 1988), 71–104.

61. Herbert, *Iron, Gender, and Power,* 25–39, 78–89; and Candice L. Goucher and Eugenia Herbert, "The Blooms of Banjeli: Technology and Gender in West African Iron Making," 40–57, in *The Culture and Technology of African Iron Production,* ed. Peter R. Schmidt (Gainesville, Fla., 1996).

62. Herbert, *Iron, Gender, and Power,* 26–27. See also Walter Cline, *Mining and Metallurgy in Negro Africa* (Menasha, Wis., 1937), 114–40.

63. Herbert, *Iron, Gender, and Power,* 65–75.

64. I discuss this last point in more detail in chapter 5. On furnaces gendered as female in Anglo American traditions, see John Fuller, Jr., to the Prince of San Severino, Oct. 24, 1754, in *The Fuller Letters 1728–1755: Guns, Slaves, and Finance,* ed. David Crossley and Richard Saville (Lewes, England, 1991), 282.

65. Jon Butler has asserted that Africans in colonial British North America endured a "spiritual holocaust" in which their religious beliefs as systems suffered irreparable harm. While Butler overstated the extent to which slavery degraded African spiritualities, his argument is helpful in trying to understand and imagine how Africans might have comprehended that which was both familiar and strange, like the world that I have tried to describe here. See his *Awash in a Sea of Faith: Christianizing the American People* (Cambridge, Mass., 1990), 151–63. W. Jeffrey Bolster notes how European seafaring technology at first seemed strange and baffling to Africans in *Black Jacks: African American Seamen in the Age of Sail* (Cambridge, Mass., 1997), 57.

66. Bolster, *Black Jacks*, 24–26, 44–67.

67. See Michael Mullin, *Africans in America: Slave Acculturation and Resistance in the American South and the British Caribbean, 1736–1831* (Urbana and Chicago, 1992), 71.

68. On gold mining in Africa and slave labor, see Paul E. Lovejoy, *Transformations in Slavery: A History of Slavery in Africa* (New York, 1983), 116–17; Ivor Wilks, *Asante in the Nineteenth Century: The Structure and Evolution of a Political Order* (London, 1975), 176–77; and Goucher and Herbert, "Blooms of Banjeli," 49. On Akan-Asante society and meanings of slavery within it, see A. Norman Klein, "Slavery in the Context of Asante History," in *The Ideology of Slavery in Africa*, ed. Paul E. Lovejoy (London, 1981), 149–67.

69. Kathleen J. Higgins, "Masters and Slaves in a Mining Society: A Study of Eighteenth-Century Sabará, Minas Gerais," *Slavery & Abolition: A Journal of Slave and Post-Slave Studies* 11 (1990): 58–73, and *"Licentious Liberty" in a Brazilian Gold-Mining Region: Slavery, Gender, and Social Control in Eighteenth-Century Sabará, Minas Gerais* (University Park, Pa., 1999); and María Elena Díaz, *The Virgin, the King, and the Royal Slaves of El Cobre: Negotiating Freedom in Colonial Cuba, 1670–1780* (Stanford, 2000), 199–223.

70. Judith Carney argues that growing rice for a global market in colonial British North America destroyed many African meanings and practices involved in its cultivation. See *Black Rice*, 124–40. On the "Africanization" of North American slavery from the late seventeenth to the mid-eighteenth centuries, see Ira Berlin, *Many Thousands Gone: The First Two Centuries of Slavery in North America* (Cambridge, Mass., 1998), 103–5, 110–26.

71. Lewis, *Coal, Iron, and Slaves*, 111–13; Charles B. Dew, *Bond of Iron: Master and Slave at Buffalo Forge* (New York, 1994), 107–8; and Robert S. Starobin, *Industrial Slavery in the Old South* (New York, 1970), 77–91. For a provocative perspective on these issues, see Stephen Fenoaltea, "Slavery and Supervision in Comparative Perspective: A Model," *JEconHist* 44 (1984): 635–68.

72. Principio Iron Works Ledger, 1728, Principio Iron Works Ledger, 1731, Principio Iron Works Ledger, 1734, and Accounts of Cash, 1741, PIWC, HSD.

73. Caleb Dorsey & Co., Elk Ridge Furnace Journal, 1758/10/1–1761/08/31, #1493, Elk Ridge Furnace Journal, Journal BB, 1762–1764, #1495, Chancery, MSA. On slaves selling produce to planters, see Morgan, *Slave Counterpoint*, 358–64

74. Principio Iron Works Ledger, 1737–1738; North East Forge Journal, 1745–1746; North East Forge Journal, 1748–1750; North East Forge Cash Book, 1751–1757; North East Forge Journal, 1754–1755; North East Forge Journal, 1755; North East Forge Waste Book, 1757; North East Forge Journal, 1757–1758; North East Forge Waste Book, 1759, PIWC, HSD.

75. Charles Dew provides an excellent discussion of overwork in *Bond of Iron*, 108–21. See also Lewis, *Coal, Iron, and Slaves*, 119–27. On task systems, see Philip D. Morgan, "Work and Culture: The Task System and the World of Lowcountry Blacks, 1700 to 1880," *WMQ* 39 (1982): 537–74, and "Task and Gang Systems: The Organization of Labor on New World Plantations," in *Work and Labor*, ed. Innes, 189–220.

76. On sugar harvest and processing, see Roderick A. McDonald, "Independent Economic Production by Slaves on Antebellum Louisiana Sugar Plantations," in *Cultivation and Culture: Labor and the Shaping of Slave Life in the Americas*, ed. Ira Berlin and Philip D. Morgan (Charlottesville, Va., 1994), 283–85. On credit for cutting wood, see [Kingsbury Furnace] Waste Book, 1754, North East Forge Journal, 1754–1755, PIWC, HSD; and Caleb Dorsey &

Co., Elk Ridge Furnace Journal, 1758/10/1–1761/08/31, #1493, Chancery, MSA. See also Lewis, *Coal, Iron, and Slaves*, 120–21.

77. On task systems and slave economies, see Morgan, "Work and Culture"; Wood, *Women's Work, Men's Work*, 53–79, 101–21; and Robert Olwell, *Masters, Slaves, and Subjects: The Culture of Power in the South Carolina Low Country, 1740–1790* (Ithaca, 1998), 141–80.

78. Caleb Dorsey & Co., Elk Ridge Furnace Journal BB, 1762–1764, #1495, Chancery, MSA. On overwork and its advantages for masters, see O. Nigel Bolland, "Proto-Proletarians? Slave Wages in the Americas: Between Slave Labour and Free Labour," in *From Chattel Slaves to Wage Slaves: The Dynamics of Labour Bargaining in the Americas*, ed. Mary Turner (Bloomington, Ind., 1995), 123–47.

79. Lewis, *Coal, Iron, and Slaves*, 121; and Patuxent Iron Works Journal A&B, 1767–1794, #1504, Chancery, MSA. Regarding the impact of commerce between masters and their slaves on their relationship, see Robert Olwell, "'A Reckoning of Accounts': Patriarchy, Market Relations, and Control on Henry Laurens's Lowcountry Plantations, 1762–1785," in *Working Toward Freedom: Slave Society and Domestic Economy in the American South*, ed. Larry E. Hudson (Rochester, N.Y., 1994), 33–52.

80. Elk Ridge Furnace Journal, 1758/10/1–1761/08/31, #1493, Chancery, MSA; Patuxent Iron Works Journal A&B, 1767–1794, #1504, Chancery, MSA; and Principio Company Store Sales Book, 1764, PIWC, HSD; North East Forge Cash Book, 1751–1757, PIWC, HSD; Caleb Dorsey & Co., Elk Ridge Furnace Journal BB, 1762–1764, #1495, Chancery, MSA. On slaves buying cloth, garments, and apparel, see Wood, *Women's Work, Men's Work*, 66–67. On clothing's meanings to slaves see idem., 57–61; and Shane White and Graham White, "Slave Clothing and African-American Culture in the Eighteenth and Nineteenth Centuries," *Past and Present* 148 (1995): 149–86.

81. Michael A. Gomez, *Exchanging Our Country Marks: The Transformation of African Identities in the Colonial and Antebellum South* (Chapel Hill, 1998), 110, 130–31; T. C. McCaskie, *State and Society in Pre-Colonial Asante* (Cambridge, 1995), 75.

82. On the relatively quick erosion of African cultures in the Chesapeake colonies, see Morgan, *Slave Counterpoint*, 451–52; and Mullin, *Africa in America*, 24–27. On scholarly debates concerning the difficulties of defining African ethnicity during the era of the Atlantic slave trade, see Philip D. Morgan, "The Cultural Implications of the Atlantic Slave Trade: African Regional Origins, American Destinations and New World Developments," *Slavery & Abolition: A Journal of Slave and Post-Slave Studies* 18 (1997): 122–45.

83. North East Forge Journal, 1754–1755, Kingsbury Furnace Journal, 1768–1770, PIWC, HSD. On consumption in the eighteenth-century Anglo American world, see Neil McKendrick et al., *The Birth of a Consumer Society: The Commercialization of Eighteenth-Century England* (Bloomington, Ind., 1985); and T. H. Breen, "An Empire of Goods: The Anglicization of Colonial America, 1690–1776," *Journal of British History* 25 (1986): 467–99, and "Narrative of Commercial Life: Consumption, Ideology, and Community on the Eve of the American Revolution," *WMQ* 50 (1993): 471–501. On slave artisans and acculturation, see Mullin, *Flight and Rebellion*, 83–98, 105–23, and *Africa in America*, 230–40.

84. On overwork and unequal opportunity, see Lewis, *Coal, Iron, and Slaves*, 120–22. On the gendered dimensions of diversification, see Carr and Walsh, "Economic Diversification and Labor Organization," 176–82; and Lorena S. Walsh, "Slave Life, Slave Society, and Tobacco Production in the Tidewater Chesapeake, 1620–1820," in *Cultivation and Culture*, ed. Berlin and Morgan, 199.

85. This is a key point in nearly every essay in Berlin and Morgan, eds., *Cultivation and Culture.*

86. Commissioners to Preserve Confiscated British Property (Ledger and Journal), 1781–1782, MSA; Steffen, *From Gentlemen to Townsmen*, 52–53. On slave family life, see Morgan, *Slave Counterpoint*, 501–19, 530–58; and Kulikoff, *Tobacco and Slaves*, 352–80.

87. See especially Morgan, *Slave Counterpoint*, 258–96. I discuss these issues extensively in chapter 5.

88. See Hoffman, *Princes of Ireland, Planters of Maryland*, 250–58. On masters' sense of

slaves as human principally because they were workers and economic actors, see Joyce E. Chaplin, *An Anxious Pursuit: Agricultural Innovation and Modernity in the Lower South, 1730–1815* (Chapel Hill, 1993), and "Slavery and the Principle of Humanity: A Modern Idea in the Early Lower South," *Journal of Social History* 24 (1990–1991): 299–316.

89. Stephen Onion to Charles Carroll of Annapolis, Jan. 24, 1735, Jan. 30, 1735, and Onion to [unknown recipient], June 18, 1735, *CCCFP,* M–4213, MHS.

90. Kingsbury Furnace Timebook, 1767–1769, PIWC, HSD; *Virginia Gazette* (Pinkney), Nov. 23, 1775; and *Maryland Journal & Baltimore Advertiser,* Mar. 20, 1781.

91. *MG,* June 10, 1762; Caleb Dorsey & Co., Elk Ridge Furnace Journal AA, 1761–1762, #1494; Elk Ridge Furnace Journal, 1758/10/1–1761/08/31, #1493, Chancery, MSA; and *MG,* Aug. 2, 1759. On differences in the behavior of African-born and American-born fugitive slaves, see Morgan, *Slave Counterpoint,* 446–52, 464–68.

92. Caleb Dorsey & Co., Elk Ridge Furnace Journal AA, 1761–1762, #1494, and Elk Ridge Furnace Journal BB, 1762–1764, #1495, Chancery, MSA.

93. Elk Ridge Furnace Journal BB, 1762–1764, #1495, Chancery, MSA.

94. On slavery as entrapment whether in accommodation or resistance, see Alex Bontemps, *The Punished Self: Surviving Slavery in the Colonial South* (Ithaca, 2001).

Chapter 4. The Best Poor Man's Country

1. William Allen and Joseph Turner to John Griffiths, Jr., Oct. 6, 1760; Allen and Turner to John Griffiths, July 25, 1760, ATL, HSP.

2. Allen and Turner to John Griffiths, Jr., Oct. 6, 1760, WAL, HSP; Susan E. Klepp and Billy G. Smith, eds., *The Infortunate: The Voyage and Adventures of William Moraley, an Indentured Servant* (University Park, Pa., 1992), 87, 89; and Allen and Turner to John Allen, Oct. 21, 1761, Turner to Henry Keppely, Oct. 30, 1761, and Turner to Lynford Lardner, Dec. 13, 1762, ATL, HSP.

3. Allen and Turner to John Griffiths, July 25, 1760, and Turner to Keppely, Oct. 30, 1761, ATL, HSP; and *PG,* Oct. 4, 1770.

4. James Logan to Nehemiah Champion, Nov. 6, 1728, James Logan Letter Book [Parchment], 1717–1731, Logan Family Papers, Acc. #379, HSP.

5. *PA8,* 3:1846, 1959–60; and Darold D. Wax, "Negro Import Duties in Colonial Pennsylvania," *PMHB* 97 (1973): 24, 26–27.

6. Marianne S. Wokeck, *Trade in Strangers: The Beginnings of Mass Migration to North America* (University Park, Pa., 1999), 70–71, 74–75, 94–95; Darold D. Wax, "Negro Imports into Pennsylvania, 1720–1766," *PH* 32 (1965): 261–81; and Michael V. Kennedy, "Furnace to Farm: Capital, Labor, and Markets in the Pennsylvania Iron Industry, 1716–1789" (Ph.D. diss., Lehigh University, 1996), 89–93.

7. David Galenson, *White Servitude in Colonial America: An Economic Analysis* (New York, 1981), 24–28, 86–93; Wokeck, *Trade in Strangers,* and "The Flow and Composition of German Immigration to Philadelphia, 1727–1775," *PMHB* 105 (1981): 249–78.

8. Coventry Forge Ledger, 1727–1730, FFR, HSP; and Moraley, *The Infortunate,* 109–10. On servants' prospects once free, see Sharon V. Salinger, *"To serve well and faithfully": Labor and indentured servants in Pennsylvania, 1682–1800* (New York, 1987), 115–36.

9. *PA8,* 3:2679–80; Allen & Turner to Anthony Simpson, Oct. 28, 1756, ATL, HSP; "A List of Servants Belonging to the Inhabitants of Pennsylvania and Taken into His Majesty's Service, 1757" [photocopy], HSP; and Allen and Turner to John Griffiths, July 25, 1760, ATL, HSP. On enlistment, see Salinger, *"To serve well and faithfully,"* 105–7; and Matthew C. Ward, "An Army of Servants: The Pennsylvania Regiment during the Seven Years' War," *PMHB* 109 (1995): 75–94. On servitude, the threat of war, and colonial labor systems north of the Mason-Dixon line, see Jacqueline Jones, *American Work: Four Centuries of Black and White Labor* (New York, 1998), 109–68.

10. *CRPa,* 4:437; and Allen & Turner to Anthony Simpson, Oct. 28, 1756, ATL, HSP.

11. Allen and Turner to John Allen, Nov. 3, 1761, ATL, HSP. See also Richard B. Morris, *Government and Labor in Early America* (New York, 1946; reprint paperback ed., Boston, 1981), 416.

12. Charles Read (IV) to James Pemberton, Jan. 23, 1771, 22:94, Pemberton Family Papers, Acc. #484A, HSP.

13. Allen and Turner to John Perks, Dec. 20, 1756, Allen and Turner to John Griffiths, Jr., Aug. 22, 1760, June 1, 1761, and Allen and Turner to Perks, Nov. 11, 1760, ATL, HSP. Slitting mills cut flattened sheets of bar iron into strips, which were often used to make nails.

14. Christopher Hanes, "Turnover Cost and the Distribution of Slave Labor in Anglo-America," *JEconHist* 56 (1996): 307–29.

15. Kennedy, "Furnace to Farm," 93–98. Copies of inventories for Thomas and John Potts are in Mrs. Thomas [Isabella] Potts James, *Memorial of Thomas Potts, Junior* (Cambridge, Mass., 1874), 86–88, 113–15; Carl D. Oblinger, "New Freedoms, Old Miseries: The Emergence and Disruption of Black Communities in Southeastern Pennsylvania, 1780–1860" (Ph.D. diss., Lehigh University, 1988), 31–32; and Paul N. Schaeffer, "Slavery in Berks County," *Historical Review of Berks County* 6 (1941): 112.

16. Mary Schweitzer, *Custom and Contract: Household, Government, and the Economy in Colonial Pennsylvania* (New York, 1987), 49–54; Kennedy, "Furnace to Farm," 3–8; and Thomas M. Doerflinger, "Rural Capitalism in Iron Country: Staffing a Forest Factory, 1808–1815," *WMQ* 59 (2002): 3–38.

17. Benjamin Welsh to Curtis Grubb, June 4, 1767, Box 1, GFP, HSP; Cornwall Furnace Journal, 1769–1770, Cornwall Furnace Journal, 1770–1772, Cornwall Furnace Journal, 1772–1774, and Cornwall Furnace Journal, 1774–1776, FFR, HSP; and Ferguson McElwaine to Peter Grubb, May 9, 1769, Box 1, GFP, HSP.

18. Coventry Forge Ledger C, 1732–1733, and Coventry Forge Ledger, 1756–1759, FFR, HSP; *PG*, Oct. 4, 1770; *Archives of the State of New Jersey. First Series, Documents Relative to the Colonial History of the State of New Jersey*, 33 vols. (Newark and Paterson, N.J., 1880–1928), 26:319; and *PG*, Feb. 2, 1769.

19. Allen and Turner to John Griffiths, July 25, 1760, and Allen and Turner to John Griffiths, May 24, 1758, ATL, HSP; and Hanes, "Turnover Cost." On slave artisans in northern cities, see Ira Berlin, *Many Thousands Gone: The First Two Centuries of Slavery in North America* (Cambridge, Mass., 1998), 179–80.

20. Quoted in J. Smith Futhey and Gilbert Cope, *History of Chester County, Pennsylvania, with Genealogical and Biographical Sketches* (Philadelphia, 1881), 347.

21. *PG*, Aug. 2, 1764; Lynford Lardner Journal D, 1752–1760, Box 3, Lardner Family Papers, Acc. #2171, HSP.

22. Allen and Turner to John Griffiths, Jr., Nov. 15, 1759, ATL, HSP; and Chris Evans, *"The Labyrinth of Flames": Work and Social Conflict in Early Industrial Merthyr Tydfil* (Cardiff, Wales, 1993), 71–88. On elite colonists' notions of gentility, see Gordon S. Wood, *The Radicalism of the American Revolution* (New York, 1992), 24–42.

23. "Meeting of the Owners of the Carlisle Iron Works, Minutes," June 29, 1765 [photostat], Carlisle Iron Works Papers, Iron Industry and Trade Folder, Society Collection, HSP.

24. Coventry Forge Ledger C, 1732–1733, Coventry Forge Ledger, 1742–1748, New Pine Forge Timebook, 1760–1763, and New Pine Forge Journal A, 1760–1762, FFR, HSP.

25. Allen and Turner to John Perks, July 9,1762, ATL, HSP. White ironworkers who trained or hired slaves may also have seen such arrangements as a part of a subcontracting system that became common during the eighteenth century. See Michael V. Kennedy, "Working Agreements: The Use of Subcontracting in the Pennsylvania Iron Industry, 1725–1789," *PH* 65 (1998): 492–508.

26. Pine Forge Ledger, 1744–1751, New Pine Forge Journal A, 1760–1762, Charming Forge Journal A, 1772–1775, Cornwall Furnace Journal, 1774–1776, and Hopewell Forge Journal F, 1771–1774, FFR, HSP.

27. Pine Forge Ledger, 1744–1748, Pine Forge Ledger, 1748–1758, Hopewell Forge Journal F, 1771–1774, Berkshire Furnace Daybook, 1767–1769, and Cornwall Furnace Journal,

1769–1770, FFR, HSP; and Andover Furnace Journal, May 19, 1773–Nov. 24, 1777, Taylor Family of Hunterdon County, N.J. Papers, 1769–1882, MC 885, RUL.

28. *PG*, June 13, 1751, July 12, 1744, July 17, 1746, and Dec. 2, 1746. On urban tradesmen and slave ownership, see Gary B. Nash and Jean R. Soderlund, *Freedom by Degrees: Emancipation in Pennsylvania and Its Aftermath* (New York, 1991), 20; and Shane White, *Somewhat More Independent: The End of Slavery in New York City, 1770–1810* (Athens, Ga., 1991), 7–8.

29. Coventry Forge Ledger, E, 1734–1740, Coventry Forge Ledger, 1742–1748, Coventry Forge Ledger, 1756–1759, and Hopewell Forge Timebook, 1768–1775, FFR, HSP.

30. *PG*, Nov. 10, 1763, Mar. 14, 1765.

31. Society Collection (under "Branson, William"), HSP. Adam had worked for Branson at Coventry Forge. See Coventry Forge Ledger, 1742–1748, FFR, HSP.

32. Israel Acrelius, *A History of New Sweden*, trans. William M. Reynolds (Philadelphia, 1874; reprint, New York, 1972), 168.

33. Coventry Forge Ledger, 1736–1741, Anna Nutt Ledger A, 1742–1748, Coventry Forge Daybook, 1746–1754, New Pine Forge Journal A, 1760–1762, and Pine Forge Ledger, 1748–1758, FFR, HSP.

34. Ferguson McElwaine to Peter Grubb, May 9, 1769, Box 1, GFP, HSP; Cornwall Furnace Journal, 1769–1770, and Cornwall Furnace Journal, 1772–1774, FFR, HSP.

35. Charming Forge Journal, 1763–1767, FFR, HSP.

36. Hopewell Forge Journal D.E., 1769–1771, and Hopewell Forge Timebook, 1768–1775, FFR, HSP.

37. Hopewell Forge Timebook, 1768–1775, and Hopewell Forge Journal D.E., 1769–1771, FFR, HSP.

38. Hopewell Forge Journal A, 1765–1767, FFR, HSP; and "List of Negroes Returned into the Clerk's Office by Peter Grubb," October 21, 1780 [copy], Box 3, and "Return of Negro Servants in Pursuance of the Act Passed 29 March 1788 of the Negro Children of the Minors," Sept. 22, 1788, Box 2, GFP, HSP.

39. "Sketches of Several Northampton County, Pennsylvania, Slaves," *PMHB* 22 (1898): 503. On Moravians and slavery, see John F. Sensbach, *A Separate Canaan: The Making of an Afro-Moravian World in North Carolina, 1763–1840* (Chapel Hill, 1998), esp. 52–53, 130–34. See also Merle G. Brouwer, "Marriage and Family Life among Blacks in Colonial Pennsylvania," *PMHB* 99 (1975): 368–72.

40. *PG*, Apr. 26, 1764; and Lynford Lardner Journal D, 1752–1760, Box 3, Lardner Family Papers, Acc. #2171, HSP.

41. Nash and Soderlund, *Freedom By Degrees*, 32–40; Ira Berlin, "Time, Space, and the Evolution of Afro-American Society on British Mainland North America," *AHR* 85 (1980): 45–54, and *Many Thousands Gone*, 185–86; Graham Russell Hodges, *Slavery and Freedom in the Rural North: African Americans in Monmouth County, New Jersey, 1665–1865* (Madison, Wis., 1997), 16–20; Chester County Register of Slaves, 1780 [copy], Misc. Box 2, Pennsylvania Abolition Society Papers, Acc. #490, HSP; "Record of the Male and Female Slaves recorded in the County of Berks in the Commonwealth of Pennsylvania, according to an Act of General Assembly entitled 'an Act For the Gradual Abolition of Slavery &c.' . . ." [typescript copy], Mary Owen Steinmetz Collection, Acc. #1690, HSP; and Slave Register, Lancaster County, Slave Records of Lancaster County Collection, MG–240, LCHS.

42. Coventry Forge Daybook, 1746–1754, FFR, HSP. On burial practices, see Sylvia R. Frey and Betty Wood, *Come Shouting to Zion: African American Protestantism in the American South and British Caribbean to 1830* (Chapel Hill, 1998), 23–26, 51–53.

43. *PG*, Oct. 2, 1746; Charming Forge Journal, 1763–1767, FFR, HSP; and *PG*, Nov. 3, 1763. By "Dutch," Joe's masters meant German, which was probably the most common language spoken at Charming Forge. On the origins of slaves imported into the mid-Atlantic colonies, see Berlin, *Many Thousands Gone*, 182–83.

44. On Pennsylvania and the convict servant trade, see A. Roger Ekirch, *Bound for America: The Transportation of British Convicts to the Colonies, 1718–1775* (New York, 1987), 137–40.

45. Eliza Cope Harrison, ed., *Philadelphia Merchant: The Diary of Thomas P. Cope* (South Bend, Ind., 1978), 63–64.

46. Berkshire Furnace Daybook, 1767–1769, FFR, HSP. On clothing's significance, see Shane White and Graham White, "Slave Clothing and African-American Culture in the Eighteenth and Nineteenth Centuries," *Past and Present*, 148 (August 1995): 149–86.

47. *PG*, Nov. 3, 1763; Charming Forge Journal, 1764–1767, FFR, HSP; and *PG*, Sept. 15, 1763, Oct. 11, 1770. On runaways and reinvented identities, see David Waldstreicher, "Reading the Runaways: Self-fashioning, Print Culture, and Confidence in Slavery in the Eighteenth Century Mid-Atlantic," *WMQ* 56 (1999): 243–72. See also Jonathan Prude, "To Look upon the 'Lower Sort': Runaway Ads and the Appearance of Unfree Laborers in America, 1750–1800," *JAH* 78 (1991): 124–59.

48. *PG*, May 24, 1775, Oct. 11, 1775, July 17, 1776; "Record of the Male and Female Slaves recorded in the County of Berks," Mary Owen Steinmetz Collection, Acc. #1690, HSP. The most common spelling of the forge's name is Birdsboro, but in the ad for the runaway slave it was spelled "Birdsborough."

49. Nash and Soderlund, *Freedom by Degrees*, 85; and Jean R. Soderlund, *Quakers and Slavery: A Divided Spirit* (Princeton, 1985), 87–172.

50. For examples of stoves, see Henry Chapman Mercer, *The Bible in Iron: Pictured Stoves and Stoveplates of the Pennsylvania Germans*, 3d. ed. (Doylestown, Pa., 1961).

51. *Records of Individuals Bound Out as Apprentices, Servants, Etc. and of Germans and Other Redemptioners in the Office of the Mayor of the City of Philadelphia, October 3, 1771 to October 5, 1773* (Lancaster, Pa., 1907; reprint, Baltimore, 1973), 42–43; Aaron Spencer Fogleman, *Hopeful Journeys: German Immigration, Settlement, and Political Culture in Colonial America* (Philadelphia, 1996), 69–99, 127–48.

52. Tulpehocken Forge Day Book, 1754–1756, and Tulpehocken Forge Account Book, 1757–1760, FFR, HSP; Peter Hasenclever, *The Remarkable Case of Peter Hasenclever, Merchant* (London, 1773; reprint, Newfoundland, N.J., 1970), 5–7; and "Recollections of Mrs. Elizabeth Doland" in *The Newark Centinel*, April 1, 1851 [transcript], Box 2, Vol. 3, p. 25, RMP, NJSA.

53. Colebrookdale Furnace Ledger, 1740–1743, and Charming Forge Journal, 1763–1767, FFR, HSP; and Deborah Ducoff-Barone, "Marketing and Manufacturing: A Study of Domestic Cast Iron Articles Produced at Colebrookdale Furnace, Berks County, Pennsylvania, 1735–1751," *PH* 50 (1983): 30–31.

54. Allen and Turner to John Griffiths, Mar. 25, 1761, WAL; HSP; Robert Erskine to Robert Willis and the American Company, Mar. 27, 1772, Box 1, Vol. 1, RMP, NJSA; Julian Ursyn Niemcewicz, *Under Their Vine and Fig Tree: Travels through America in 1797–1799, 1805 with some further account of life in New Jersey*, trans. and ed. Metchie J. E. Budka (Elizabeth, N.J., 1965), 227.

55. Johann David Schoepf, *Travels in the Confederation [1783–1784]*, trans. and ed. Alfred J. Morrison, 2 vols. (Philadelphia, 1911; reprint, New York, 1968), 1:200.

56. Hasenclever, *Remarkable Case*, 9. See also James M. Ransom, *Vanishing Ironworks of the Ramapos: The Story of the Forges, Furnaces, and Mines of the New Jersey–New York Border Area* (New Brunswick, N.J., 1966), 17–24; and Irene D. Neu, "The Iron Plantations of Colonial New York," *New York History* 33 (1952): 11–17.

57. Erskine to [American Iron Co.], Oct. 31, 1771, Box 1, Vol. 2, RMP, NJSA.

58. Erskine to the American Iron Company, Mar. 27, 1772 [typescript of missing original], Box 1, Vol. 1, RMP, NJSA; and Niemcewicz, *Under Their Vine*, 227.

59. *PG*, July 5, 1775. On the Irish servant trade and Irish immigration, see Wokeck, *Trade in Strangers*, 167–219. On British immigration, see Bernard Bailyn, *Voyagers to the West: A Passage in the Peopling of America on the Eve of the Revolution* (New York, 1987). Thomas Doerflinger stresses immigration's significance to the development of the iron industry and the mid-Atlantic's economy in "Rural Capitalism in Iron Country," 36–38.

60. Kennedy, "Working Agreements," and "Furnace to Farm," 115–270.

61. Arthur Cecil Bining, *Pennsylvania Iron Manufacture in the Eighteenth Century*, 2d ed.

(Harrisburg, Pa., 1987), 106; and Paul F. Paskoff, *Industrial Evolution: Organization, Structure, and Growth of the Pennsylvania Iron Industry, 1750–1860* (Baltimore, 1983), 11–14. See also Arthur Cecil Bining, *British Regulation of the Colonial Iron Industry* (Philadelphia, 1933), 84–85. I discussed the cartel in chapter 1.

62. [Committee on Historical Research, Pennsylvania Society of the Colonial Dames of America], *Forges and Furnaces in the Province of Pennsylvania* (Philadelphia, 1914), 76, 119–21; Frederic S. Klein, "Robert Coleman, Millionaire Ironmaster," *Journal of the Lancaster County Historical Society* 64 (1960): 17–33; and Bining, *Pennsylvania Iron Manufacture*, 118. Historian John E. Pomfret has observed for New Jersey that "the ironmasters of Morris, Bergen, and Sussex [Counties] constituted a tight coterie." See Pomfret, *Colonial New Jersey: A History* (New York, 1973), 201.

63. See Mrs. Thomas Potts James, *Memorial of Thomas Potts, Junior* (Cambridge, Mass., 1874), 20–22, 42–43, 73–88, 91–153; and Linda McCurdy, "The Potts Family Iron Industry in the Schuylkill Valley," (Ph.D. diss., Pennsylvania State University, 1974).

64. Colonial ironmasters may have emulated the British iron industry, where most ironworks were owned by a loosely connected network of partnerships and intermarried families. See Arthur Raistrick, *Quakers in Science and Industry* (London, 1950), 89–160; and Evans, *"Labyrinth of Flames,"* 52–70, 121–37.

65. Mount Pleasant Furnace Ledger, 1737–1750, Pine Forge Ledger, 1744–1751, and Colebrookdale Furnace Ledger, 1743–1751, FFR, HSP.

66. See Neil McKendrick, John Brewer, and J. H. Plumb, eds., *The Birth of a Consumer Society: The Commercialization of Eighteenth-Century England* (Bloomington, Ind., 1982); Carole Shammas, *The Pre-Industrial Consumer in England and America* (New York, 1990); and Cary Carson, Ronald Hoffman, and Peter J. Albert, eds., *Of Consuming Interests: The Style of Life in the Eighteenth Century* (Charlottesville, Va., 1994).

67. Hasenclever, *Remarkable Case*, 83; Jacob L. Bunnell, ed., *Sussex County Sesquicentennial* (Newton, N.J., 1903), 29. Quoted in Peter O. Wacker, *The Musconetcong Valley of New Jersey: A Historical Geography* (New Brunswick, N.J., 1968), 111. Some historians have argued that ironworks' stores did not gouge workers. See Bining, *Pennsylvania Iron Manufacture*, 112; and Alfred Gemmell, "The Charcoal Iron Industry in the Perkiomen Valley" (M.A. thesis, University of Pennsylvania, 1948), 84.

68. Allen and Turner to John Griffiths, Dec. 30, 1760, Allen and Turner to John Griffiths, Jr., June 1, 1761, ATL, HSP.

69. Allen and Turner to John Griffiths, Jr., June 1, 1761, ATL, HSP; and Ruth H. Bloch, "The Gendered Meanings of Virtue in Revolutionary America," *Signs: Journal of Women in Culture and Society* 13 (1987): 37–58. On men as consumers, see Margot Finn, "Men's Things: Masculine Possession in the Consumer Revolution," *Social History* 25 (2000): 133–55. On views of women, work, and consumption, see Jeanne Boydston, *Home and Work: Housework, Wages, and the Ideology of Labor in the Early Republic* (New York, 1990), 27–29; and Cornelia Hughes Dayton, *Women before the Bar: Gender, Law, and Society in Connecticut* (Chapel Hill, 1995), 102–3.

70. Allen to Samuel Walker Sr. & Co., Dec. 15, 1765, WAL, HSP.

71. *The Statutes at Large of Pennsylvania from 1682 to 1801*, comp. James T. Mitchell and Henry Flanders, 18 vols. (Harrisburg, 1895–1915), 4:65–67. *PA8*, 3:2318; and *Statutes at Large of Pennsylvania*, 4:301–3. On licensing, see Peter Thompson, *Rum Punch and Revolution: Taverngoing and Public Life in Eighteenth-Century Philadelphia* (Philadelphia, 1998), 30–51.

72. Carl Raymond Woodward, *Ploughs and Politicks: Charles Read of New Jersey and His Notes on Agriculture, 1715–1774* (New Brunswick, N.J., 1941), 92; *The Laws of the Royal Colony of New Jersey*, comp. Bernard Rush, 4 vols. (Trenton, 1977–1986), in *Archives of the State of New Jersey: Third Series*, 5 vols., 4:593–94.

73. Colebrookdale Furnace accounts, Feb. 6, 1734, Box 2, Potts Family Papers, Acc. #520, HSP; and Mount Pleasant Furnace Ledger, 1737–1741, and Colebrookdale Furnace Ledger, 1735–1747, FFR, HSP.

74. Warwick Furnace Ledger, 1750–1756 (1760), and William Bird Ledger A, 1755–1765, FFR, HSP.

75. Andover Furnace Day Book/Journal, May 19, 1773–November 24, 1777, Taylor Family (of Hunterdon County, N.J.) Papers, 1769–1882, MC 885, RUL; and New Pine Forge Journal A, 1760–1762, and New Pine Forge Timebook, 1760–1763, FFR, HSP.

76. Cornwall Furnace Day Book, 1771–1773, Cornwall Furnace Journal, 1772–1774, and Cornwall Furnace Journal, 1774–1776, FFR, HSP.

77. Warwick Furnace Ledger, 1759–1762, Charming Forge Journal, 1763–1767, Warwick Furnace Journal, 1765–1769, Cornwall Furnace Journal, 1768–1769, Cornwall Furnace Journal, 1769–1770, Cornwall Furnace Journal, 1770–1772, Cornwall Furnace Journal, 1772–1774, and Cornwall Furnace Journal, 1774–1776, FFR, HSP.

78. Berkshire Furnace Account Book, 1768–1771, FFR, HSP. On the importance and regulation of teamsters' work, see Doerflinger, "Rural Capitalism in Iron Country," 20–23.

79. Cornwall Furnace Journal, 1768–1769, and Cornwall Furnace Journal, 1774–1776, FFR, HSP. On clock and watch ownership, see Mark M. Smith, *Mastered by the Clock: Time, Slavery, and Freedom in the American South* (Chapel Hill, 1997), 30–36.

80. Berkshire Furnace Daybook, 1767–1769, FFR, HSP; and "Meeting of the Owners of the Carlisle Iron Works, Minutes," June 29, 1765, and "Instructions given & left wth Saml Hay at ye Carlisle Iron Works 4th & 5th May 1769," [photostat], Carlisle Iron Works Papers, Iron Industry and Trade Folder, Society Collection, HSP.

81. Francis Brezina to Peter Grubb, Aug. 10, 1767, Box 1, GFP, HSP. For a reappraisal of deference, see Michael Zuckerman, "Tocqueville, Turner, and Turds: Four Stories of Manners in Early America," *JAH* 85 (1998): 13–42.

82. Curtis Grubb to Peter Grubb, Apr. 10, 1774, Samuel Jones to Peter Grubb, Apr. 12, 1774, and Curtis Grubb to Peter Grubb, Apr. 15, 1774, Box 1, GFP, HSP; and "Agreement between the Guardians of the Estate of Burd Grubb and Henry Bates Grubb and Samuel Jones," Aug. 11, 1788, Box 2, GFP, HSP.

83. "List of Negroes Returned into the Clerk's Office by Peter Grubb, October 21, 1780" [copy], Box 3, GFP, HSP; and "Agreement between the Guardians . . . and Samuel Jones," Aug. 11, 1788, Box 2, GFP, HSP.

Chapter 5. Industrial Slavery Domesticated

1. David Ross to Robert Richardson, Jan. 14, 1813, Ross to William J. Dunn, Jan. 1813, DRL, VHS.

2. "List of Slaves at the Oxford Iron Works in Families and Their Employment Taken 15 January 1811," WBP, Duke, *RABSP:F:3*, 35. On masters' ritualized distribution of clothing and food, see Charles Joyner, *Down by the Riverside: A South Carolina Slave Community* (Urbana, Ill., 1984), 106–17; and Robert Olwell, *Masters, Slaves, and Subjects: The Culture of Power in the South Carolina Low Country, 1740–1790* (Ithaca, 1998), 194–96.

3. Ross to Richardson, Jan. 14, Jan. 19, 1813, DRL, VHS. On white ideas about race and slavery in the early national South, see Winthrop D. Jordan, *White over Black: American Attitudes toward the Negro, 1550–1812* (Chapel Hill, 1968), 435–541; Ira Berlin, *Many Thousands Gone: The First Two Centuries of Slavery in North America* (Cambridge, Mass., 1998), 363–64; and Peter Kolchin, *Unfree Labor: American Slavery and Russian Serfdom* (Cambridge, Mass., 1987), 103–91.

4. On commodification and paternalism, see Walter Johnson, *Soul by Soul: Life inside the Antebellum Slave Market* (Cambridge, Mass., 1999). On master-slave relations, especially between slave men and masters, see Christopher Morris, "The Articulation of Two Worlds: The Master-Slave Relationship Reconsidered," *JAH* 85 (1998): 982–1007.

5. *Maryland Journal and Baltimore Advertiser*, Aug. 27, 1782; Latham A. Windley, ed., *Runaway Slave Advertisements: A Documentary History from the 1730s to 1790* (Westport, Conn., 1983), v. 1–2; and Carter to Thomas and Rowland Hunt, Apr. 18, 1777, Robert Carter Letterbook, 1775–1780, Robert Carter Papers, Duke, *RABSP:F:3*, 30. On slave resistance during the Revolution, see Gerald W. Mullin, *Flight and Rebellion: Slave Resistance in Eighteenth-Century*

Virginia (New York, 1972), 130–36; and Sylvia R. Frey, *Water from the Rock: Black Resistance in a Revolutionary Age* (Princeton, 1991), 55–56, 63–64, 143–71.

6. Aubrey C. Land, *The Dulanys of Maryland: A Biographical Study of Daniel Dulany, the Elder (1685–1753) and Daniel Dulany, the Younger (1722–1797)* (Baltimore, 1955), 326–27; Section 27, Carter Family Papers, VHS; and *MG*, Apr. 14, 1785.

7. *MG*, Aug. 30, 1781. Commissioners to Preserve Confiscated British Property (Ledger and Journal), 1781–1782; and Commissioners to Preserve Confiscated British Property (Sales Book), 1781–1782, MSA S 132, MSA.

8. Commissioners to Preserve Confiscated British Property (Ledger and Journal), 1781–1782; and Commissioners to Preserve Confiscated British Property (Sales Book), 1781–1782, MSA S 132.

9. On the Atlantic slave trade, see Robert McColley, *Slavery and Jeffersonian Virginia* (Urbana, Ill., 1964), 163–71; and Jeffrey R. Brackett, *The Negro in Maryland: A Study of the Institution of Slavery* (Baltimore, 1889; reprint, New York, 1969), 45–46. On manumissions during the revolutionary era, see Ira Berlin, *Slaves without Masters: The Free Negro in the Antebellum South* (New York, 1974; reprint, New York, 1981), 29–34; Richard S. Dunn, "Black Society in the Chesapeake, 1776–1810," in *Slavery and Freedom in the Age of the American Revolution*, ed. Ira Berlin and Ronald Hoffman (Charlottesville, Va., 1983), 49–82; T. Stephen Whitman, *The Price of Freedom: Slavery and Manumission in Baltimore and Early National Maryland* (Lexington, Ky., 1997), 61–118; and Christopher Phillips, *Freedom's Port: The African American Community of Baltimore, 1790–1860* (Urbana/Chicago, 1997), 35–42. On the domestic slave trade, see Brenda E. Stevenson, *Life in Black and White: Family and Community in the Slave South* (New York, 1996), 175–83; Johnson, *Soul by Soul*; and Berlin, *Many Thousands Gone*, 265–66.

10. Samuel Smith to Gov. Thomas Johnson, Apr. 1, 1778; Philip Coale to Capt. Charles Ridgely, Sept. 28, 1782, Box 2, RP692.1, MHS; George Gibson to Ridgely, Feb. 21, 1783, Box 1, RP692, MHS; and A. Roger Ekirch, *Bound for America: The Transportation of British Convicts to the Colonies, 1718–1775* (New York, 1987), 228–29, 233–36, and "Great Britain's Secret Convict Trade to America, 1783–1784," *AHR* 89 (1984): 1285–91.

11. Northampton Furnace Journals, 1785–1787, 1787–1788, Box 4, RAB, MHS; and Stewart & Plunkett to Ridgely, July 25, 1786, Box 2, RP692, MHS. See also R. Kent Lancaster, "Almost Chattel: The Lives of Indentured Servants at Hampton-Northampton, Baltimore County," *MHM* 94 (1999): 341–62.

12. Northampton Furnace Journal, 1787–1788, Box 4, RAB, MHS.

13. "Bill from Thomas Rossiter," Apr. 1786, Box 4, RP692, MHS; Ronald L. Lewis, *Coal, Iron, and Slaves: Industrial Slavery in Maryland and Virginia, 1715–1865* (Westport, Conn., 1979), 228–29; and Northampton Furnace Journal, 1787–1788, Box 4, RAB, MHS.

14. Northampton Furnace Journal, 1790–1796, Box 5, RAB, MHS. On servants' legal challenges and indentured servitude's demise, see Robert J. Steinfeld, *The Invention of Free Labor: The Employment Relation in English and American Culture, 1350–1870* (Chapel Hill, 1991), 122–72.

15. Whitman, *Price of Freedom*, 99–118; and Phillips, *Freedom's Port*, 40–56.

16. Rebecca Ridgely Account Book, 1790–1805, Box 1, and "Agreement of Charles Carnan Ridgely with Rebecca Ridgely," July 2, 1790, Box 2, Ridgely-Pue Papers, MS 693, MHS; and Northampton Furnace Journal, 1790–1796, Box 5, RAB, MHS. On slavery at Northampton, see R. Kent Lancaster, "Chattel Slavery at Hampton/Northampton, Baltimore County," *MHM* 95 (2000): 409–27. On Rebecca Ridgely, see Dee E. Andrews, *The Methodists and Revolutionary America, 1760–1800: The Shaping of an Evangelical Culture* (Princeton, 2000), 101–5. On evangelical Christianity's appeal to southern white women, see Andrews, *Methodists and Revolutionary America*, 99–122; and Christine Leigh Heyrman, *Southern Cross: The Beginnings of the Bible Belt* (New York, 1997; paperback ed., Chapel Hill, 1997), 161–205. On slave fugitives seeking refuge in Baltimore, see Whitman, *Price of Freedom*, 69–73; and Phillips, *Freedom's Port*, 65–70.

17. Louis Morton, *Robert Carter of Nomini Hall: A Virginia Tobacco Planter of the Eighteenth Century* (Williamsburg, 1941), 257–69; Isaac Zane to John Pemberton, Aug. 8, 1777, Dec. 9,

1780; "Advertisement for the Sale of the Marlboro Estate," 1794; "Inventory of the Estate of Isaac Zane," 1795; "Will of the Estate of Isaac Zane," 1795; "Bill of Sale," Nov. 21, 1795; Alex White to Sarah Zane, Sept. 12, 1795, Oct. 17, 1795; "Bill of Manumission," 1797, all in Box A–115, IZP, HSP. See also Roger W. Moss, Jr., "Isaac Zane, Jr., A 'Quaker for the Times'," *VMHB* 77 (1969): 291–306.

18. Hugh Holmes to Sarah Zane, Feb. 2, 1798, and "Will of Sarah Zane," Oct. 1, 1798, Box A–115, IZP, HSP.

19. Mackey to Sarah Zane, Oct. 4, 1802, Apr. 20, 1803, May 4, 1803, and June 15, 1803, Box A–115, IZP, HSP.

20. Daniel K. Rickter, " 'Believing that Many of the Red People Suffer Much for the Want of Food': Hunting, Agriculture, and a Quaker Construction of Indianness in the Early Republic," *JER* 19 (1999): 601–28; and Gary B. Nash, *Forging Freedom: The Formation of Philadelphia's Black Community, 1720–1840* (Cambridge, Mass., 1988), 204–10.

21. On Loudoun County Quakers, antislavery sentiment, and their alienation from other whites, see Stevenson, *Life in Black and White*, 15–17. On the impact of Gabriel's Conspiracy on Virginia politics, see Douglas R. Egerton, *Gabriel's Rebellion: The Virginia Slave Conspiracies of 1800 and 1802* (Chapel Hill, 1993). On free blacks in Loudoun County, see Stevenson, *Life in Black and White*, 258–319.

22. Unknown correspondent to Sarah Zane, June 25, 1810, Box A–115, and Fleet Smith to Sarah Zane, Dec. 10, 1816, Box A–114, IZP, HSP.

23. "Will of Sarah Zane," Nov. 23, 1812, Box A–115, IZP, HSP.

24. Sarah Zane to Jacob Rinker, June 4, 1816, Box A–115, Fleet Smith to Sarah Zane, Dec. 10, 1816, no date, Box A–114, and "Will of Sarah Zane," Mar. 24, 1819, Box A–115, IZP, HSP.

25. "Memorandum of hired Negroes at Bath Iron Works for 1829," WWP, Duke, *SABSI:A*, 22.

26. On slave hiring, see Sarah S. Hughes, "Slaves for Hire: The Allocation of Black Labor in Elizabeth City County, Virginia, 1782 to 1810," *WMQ* 35 (1978): 260–86; and James Sidbury, *Ploughshares into Swords: Race, Rebellion, and Identity in Gabriel's Virginia, 1730–1810* (New York, 1997), 187–201.

27. Ronald Lewis claims that ironmasters hired more slaves after the Revolution because they needed more flexibility. See *Coal, Iron, and Slaves*, 83. On ironmasters' social origins, see my "Forging a New Order: Slavery, Free Labor, and Sectional Differentiation in the Mid-Atlantic Charcoal Iron Industry, 1715–1840" (Ph.D. diss., University of Pennsylvania, 1995), 220–24; and Samuel Sydney Bradford, "The Ante-Bellum Charcoal Iron Industry of Virginia" (Ph.D. diss., Columbia University, 1958), 17–20.

28. Lewis, *Coal, Iron, and Slaves*, 81–103; Charles B. Dew, "Disciplining Slave Ironworkers in the Antebellum South: Coercion, Conciliation, and Accommodation," *AHR* 79 (1974): 393–418; and Egerton, *Gabriel's Rebellion*, 23–26. Midori Takagi provides an excellent overview of urban slavery's historiography in *"Rearing Wolves to Our Own Destruction": Slavery in Richmond, Virginia, 1782–1865* (Charlottesville, Va., 1999), 3–5.

29. On the role of slave hiring in discouraging manumission, see Phillips, *Freedom's Port*, 22–23. On the greater physical mobility of slave men compared to slave women (to which slave hiring contributed greatly) and its impact on gender roles and family life, see Stevenson, *Life in Black and White*, 226–57. On hiring to satisfy antebellum white aspirations to domesticity, see Keith C. Barton, " 'Good Cooks and Washers': Slave Hiring, Domestic Labor, and the Market in Bourbon County, Kentucky," *JAH* 84 (1997): 436–60.

30. "The Answer of William Weaver to a bill exhibited against him . . . by John Doyle," Dec. 15, 1831, Case Papers, Doyle v. Weaver, in Box 3, Weaver-Brady Papers (#38–98–C), UVA; Deposition of James Brawley, Apr. 18, 1826, Case Papers, Weaver *v.* Mayburry; and Jordan Davis & Co. to William Weaver, Nov. 24, 1830, WWP, Duke, *SABSI:A*, 23.

31. On self-hire, autonomy, and urban employment, see Egerton, *Gabriel's Rebellion*, 25–28; Sidbury, *Ploughshares into Swords*, 192–94, 212–13; and Takagi, *"Rearing Wolves,"* 37–54.

32. Mayburry to Weaver, Jan. 9, 1817, Box 3, William-Brady Papers (#38–98–b), UVA;

Weaver to unknown, Jan. 7, 1828, Box 3, Weaver-Brady Papers (#38–98–C), UVA; Dickinson to Weaver, Dec. 5, 1828, WWP, Duke, *SABSI:A*, 22; Wigglesworth to Weaver, Jan. 5, 1829, Box 1, Letters, Weaver-Brady Papers (#38–98), UVA; and John Chew to Weaver, Dec. 5, 1830, WWP, Duke, *SABSI:A*, 23.

33. On law and the slave hiring system, see Paul Finkelman, "Slaves as Fellow Servants: Ideology, Law, and Industrialization," *The American Journal of Legal History* 31 (1987): 269–305; and Jenny Bourne Wahl, *The Bondsman's Burden: An Economic Analysis of the Common Law of Southern Slavery* (New York, 1998), 49–77.

34. Rezin Hammond to Charles Carnan Ridgely, Feb. 22, 1794, Box 5, RP692, MHS.

35. C. Wigglesworth to William Weaver, Dec. 31, 1828, WWP, Duke, *SABSI:A*, 22.

36. R. Brooks to John Jordan, Jan. 3, 1825, WWP, Duke, *SABSI:A*, 22; and Brooks to Jordan and Irvine, Jan. 2, 1829, Jordan and Irvine Papers, State Historical Society of Wisconsin, Madison, quoted in Lewis, *Coal, Iron, and Slaves*, 100–101; and Nancy Matthews to Jordan and Irvine, Jan. 18, 1831, Jordan and Irvine Papers, quoted in Lewis, *Coal, Iron, and Slaves*, 101.

37. W. E. Dickinson to Weaver, Apr. 19, 1829, WWP, Duke, *SABSI:A*, 22.

38. "Memorandum of hired Negroes at Bath Iron Works for 1829," WWP, Duke, *SABSI:A*, 22; Charles B. Dew, *Bond of Iron: Master and Slave at Buffalo Forge* (New York, 1994), 70; Robert Crutchfield to Weaver, Jan. 10, 1830, WWP, Duke, *SABSI:A*, 22; Elizabeth Mathews to Weaver, Mar. 29, 1830, WWP, Duke, *SABSI:A*, 23; and Staples to Weaver, Jan. 4, 1830, Box 1, Letters, Weaver-Brady Papers (#38–98), UVA. On problems attracting hired slaves back to Bath and to Buffalo Forge, see Dew, *Bond of Iron*, 67–70, 75–78.

39. Robert Crutchfield to Weaver, Jan. 10, 1830, WWP, Duke, *SABSI:A*, 22.

40. Weaver to John Doyle, Oct. 17, 1827, Weaver-Brady Papers (#38–98–C), UVA; Moses McCue to Weaver, July 3, 1829, WWP, Duke, *SABSI:A*, 22; James C. Dickinson to Weaver, Jan. 21, 1833, WWP, Duke, *SABSI:A*, 23.

41. "List of Slaves," WBP, Duke, *RABSP:F:3*, 35. Lois Green Carr and Lorena S. Walsh, "Economic Diversification and Labor Organization in the Chesapeake, 1650–1820," in *Work and Labor in Early America*, ed. Stephen Innes (Chapel Hill, 1988), 176–82; and Takagi, *"Rearing Wolves,"* 28–30.

42. Jordan Davis & Co. to William Weaver, Nov. 24, 1830, WWP, Duke, *SABSI:A*, 23.

43. Northampton Furnace Journal, 1790–1796, Box 5, RAB, MHS; Ridwell Furnace Daybook, 1805–1809, #634, SHC, UNC, *SABSI:B*, 22; "Memo Cash to Negroes," WWP, Duke, *SABSI:A*, 24; and Dew, *Bond of Iron*, 114, 119.

44. On cash and economic culture, see Christopher Clark, *The Roots of Rural Capitalism: Western Massachusetts, 1780–1860* (Ithaca, 1990), 33, 67–69, 224–27. On cash's meanings to slaves, see Larry E. Hudson, Jr., " 'All That Cash': Work and Status in the Slave Quarters," in *Working toward Freedom: Slave Society and Domestic Economy in the American South*, ed. Hudson (Rochester, N.Y.,1994), 77–94. On cash and masters' attempts to regulate slave economies, see John Campbell, "As 'A Kind of Freeman'? Slaves' Market-Related Activities in the South Carolina Upcountry, 1800–1860," in *Cultivation and Culture: Labor and the Shaping of Slave Life in the Americas*, ed. Ira Berlin and Philip D. Morgan (Charlottesville, Va., 1993), 243–74.

45. On slaves and Christmas, see Shauna Bigham and Robert E. May, "The Time O' All Times? Masters, Slaves, and Christmas in the Old South," *JER* 18 (1998): 263–88.

46. On this last point, see Barton, "Good Cooks and Washers," 443–44; and Randolph B. Campbell, "Slave Hiring in Texas," *AHR* 93 (1988): 114.

47. Deposition of Ludwell Diggs, Aug. 8, 1826, Case Papers, Weaver v. Mayburry.

48. Francis Preston to John Preston, Dec. 25, 1795, Preston Family Papers, VHS.

49. Doyle to Weaver, Apr. 15, 1829, WWP, Duke, *SABSI:A*, 22.

50. Ross to William Mewburn, Aug. 1, 1813, DRL, VHS; "Bill of Complaint of William Weaver," Nov. 19, 1825, Case Papers, Weaver v. Mayburry; "Agreement between John Doyle and Dixon Hall," Box 3, Weaver-Brady Papers (#38–98–C), UVA; Bath Iron Works Ledger, 1828–1830, Box 3, Boxed Ledgers, Weaver-Brady Papers (#38–98), UVA; Doyle to Weaver, Mar. 31, 1829, WWP, Duke, *SABSI:A*, 22; and Dew, *Bond of Iron*, 106.

51. "Memorandum for Robert Richardson" Dec. 1812, and Ross to Smith, Mar. 27, 1812, DRL, VHS.

52. [Pine?] Forge Daybook, 1812–1816, Pine Forge Daybook, 1820–1825, Shenandoah County (Va.) Account Books, 1799–1838, #2934, SHC, UNC, *SABSI:B*, 25, 26.

53. Christopher L. Tomlins, "In Nat Turner's Shadow: Reflections on the Norfolk Dry Dock Affair of 1830–1831," *LH* 33 (1992): 494–518; Charles B. Dew, *Ironmaker to the Confederacy: Joseph R. Anderson and the Tredegar Iron Works* (New Haven, 1966), 22–26; and Patricia A. Schechter, "Free and Slave Labor in the Old South: The Tredegar Ironworkers' Strike of 1847," *LH* 35 (1994): 165–86. See also Jacqueline Jones, *American Work: Four Centuries of Black and White Labor* (New York, 1998), 222–32.

54. Deposition of William Norcross, May 23–24, 1826, Case Papers, Weaver *v.* Mayburry; Deposition of William Norcross, Aug. 22, 1840, Case Papers, John Alexander *v.* Sydney S. Baxter, Administrator of John Irvine & Others, Superior Court of Chancery Records, Rockbridge County Court House, Lexington, Virginia, quoted in Dew, *Bond of Iron*, 106.

55. This paragraph and the next come from Jean Libby, ed., *From Slavery to Salvation: The Autobiography of Rev. Thomas W. Henry of the A.M.E. Church* (Jackson, Miss., 1994), 26–28.

56. Ibid., 25–28. See pages xv–xxxi for a biography of Henry.

57. Libby, *From Slavery to Salvation*, 85–86; John McPherson Brien to Jonathan Meredith, June 12, 1848, Meredith Papers, MS 1367, MHS.

58. On masters' paternalism and humanitarian thought in the early republic, see Jeffrey Robert Young, *Domesticating Slavery: The Master Class in Georgia and South Carolina, 1670–1837* (Chapel Hill, 1999), esp. 123–60. My understanding of paternalism has been shaped by Morris, "The Articulation of Two Worlds"; Kolchin, *Unfree Labor*, 103–91; Johnson, *Soul by Soul;* and Eugene D. Genovese, *Roll, Jordan, Roll: The World the Slaves Made* (New York, 1972).

59. On slave families, see Herbert G. Gutman, *The Black Family in Slavery and Freedom, 1750–1925* (New York, 1976), 3–360; Ann Patton Malone, *Sweet Chariot: Slave Family and Household Structure in Nineteenth-Century Louisiana* (Chapel Hill, 1992); and Stevenson, *Life in Black and White*, 159–257.

60. Davis to Weaver, July 7, 1832, WWP, Duke, *SABSI:A*, 23.

61. Northampton Furnace Journal, 1783–1784, and Northampton Furnace Journal, 1787–1788, Box 4, RAB, MHS; and Cumberland Forge Daybook, 1802, LC.

62. Ross to James Duffield, Nov. 13, 1812, DRL, VHS.

63. Charles Dew discusses overwork, slave autonomy, and slave men as providers in *Bond of Iron*, 108–21, 162–63, 171–219. On slave men embracing provider roles and viewing their labor principally within the context of family, see Larry E. Hudson, Jr., *To Have and to Hold: Slave Work and Family Life in Antebellum South Carolina* (Athens, Ga., 1997). See also Dylan Penningroth, "Slavery, Freedom, and Social Claims to Property among African Americans in Liberty County, Georgia, 1850–1880," *JAH* 84 (1997): 405–35.

64. John Schoolfield to William Weaver, July 24, 1829, Dec. 7, 1829, WWP, Duke, *SABSI:A*, 22.

65. Schoolfield to Weaver, Sept. 26, 1829, Duke, *SABSI:A*, 22; and Dew, *Bond of Iron*, 113, 117.

66. Sidbury, *Ploughshares into Swords*, 82–86; and Ross to Dunn, Jan. 1813, DRL, VHS. See also Jean Libby, "African Ironmaking Culture among African American Ironworkers in Western Maryland, 1760–1850" (M.A. thesis, San Francisco State University, 1991).

67. "List of Slaves," WBP, Duke, *RABSP:F:3*, 35; and Dew, *Bond of Iron*, 25–27.

68. Ross to Robert Richardson, Nov. 10, 1812, Dec. 1812, Ross to Thomas Gray, Sept. 8, 1813, and Ross to Richardson, Aug. 18, 1813, DRL, VHS.

69. Ross to Thomas Evans, June 24, 1812, DRL, VHS. See also Ross to Richardson, Nov. 10, 1812, DRL, VHS.

70. Charles Dew argues that industrial slavery promoted technological conservatism in "Slavery and Technology in the Antebellum Southern Iron Industry: The Case of Buffalo Forge," in *Science and Medicine in the Old South*, ed. Ronald L. Numbers and Todd L. Savitt (Baton Rouge, La., 1989), 107–26.

71. Deposition of John Doyle, Feb. 8, 1840, Case Papers, Weaver v. Jordan, Davis & Co., Superior Court of Chancery Records, Rockbridge County Court House, Lexington, Virginia. Quoted in Dew, *Bond of Iron*, 106; Ross to Richardson, Nov. 10, 1812, DRL, VHS.

72. Historians have debated the impact of slavery on economic development and on technological innovation in the antebellum South, and most agree that slavery hampered both, though they disagree why. On slaves as absorbers of capital, see Robert E. Gallman and Ralph V. Anderson, "Slaves as Fixed Capital: Slave Labor and Southern Economic Development," *JAH* 64 (1977): 24–46. Fred Bateman and Thomas Weiss attribute slower industrialization in the South to planters' reluctance to invest in manufacturing in *A Deplorable Scarcity: The Failure of Industrialization in the Slave Economy* (Chapel Hill, 1981).

73. Langhorne & Tayloe to Weaver, Oct. 7, 1831, and Tayloe to Weaver, May 2, 1832, WWP, Duke, *SABSI:A*, 23.

74. Ira Berlin and Philip D. Morgan emphasize this point in their introduction to *Cultivation and Culture*, 21–22.

75. Dew, "Slavery and Technology."

76. Ross to Richardson, Aug. 18, 1813, DRL, VHS.

77. Ross to Richardson, Feb. 11, 1812, Feb. 19, 1813, and Sept. 25, 1813, and Ross to Dunn, July 4, 1813, DRL, VHS.

78. Ross to Richardson, Jan. 19, 1813, DRL, VHS.

79. "List of Slaves," WBP, Duke, *RABSP:F:3*, 35; and Ross to Richardson, Oct. 30, 1812, Nov. 10, 1812, "Memorandum for Richardson," [Dec. 1812], Ross to Richardson, Aug. 18, 1813, and Ross to John Duffield, Jan. 9, 1813, DRL, VHS.

80. "Memorandum for Richardson," [Dec. 1812], and Ross to William J. Dunn, Jan. 1813, DRL, VHS.

81. Arthur C. Aufderheide et al., "Lead in Bone III. Prediction of Social Correlates from Skeletal Lead Content in Four Colonial American Populations (Catoctin Furnace, College Landing, Governor's Island, and Irene Mould)," *American Journal of Physical Anthropology* 66 (1985): 353–61.

82. Charles G. Steffen, "The Pre-Industrial Iron Worker: Northampton Iron Works, 1780–1820," *LH* 20 (1979): 108–9; Bishop Francis Asbury to Ridgely, Mar. 10, 1807, Box 2, Ridgely-Pue Papers, MS 693, MHS; Elizabeth Y. Anderson, *Faith in the Furnace: A History of Harriet Chapel Catoctin Furnace, Maryland* (Thurmont, Md., 1985), 5–18; and Libby, ed., *Slavery to Salvation*, 25–26.

83. Ross to Richardson, Apr. 30, 1812, DRL, VHS; and Dew, *Bond of Iron*, 130, 178–79.

84. Diary of John Frederick Schlegel, 1799. Translated from German and quoted in Anderson, *Faith in the Furnace*, 6.

85. On Moravian evangelization, see John F. Sensbach, *A Separate Canaan: The Making of an Afro-Moravian World in North Carolina, 1763–1840* (Chapel Hill, 1998), 29–43, 192–95; Sylvia R. Frey and Betty Wood, *Come Shouting to Zion: African American Protestantism in the American South and British Caribbean to 1830* (Chapel Hill, 1998), 83–87; and Michael Mullin, *Africa in America: Slave Acculturation in the American South and the British Caribbean, 1736–1831* (Urbana/Chicago, 1992), 243–49.

86. On slave religion, see Genovese, *Roll, Jordan, Roll*, 161–284; Albert J. Raboteau, *Slave Religion: The "Invisible Institution" in the Antebellum South* (New York, 1978); and Joyner, *Down by the Riverside*, 141–71. On the compatibility of African spiritualities and evangelical Christianity, see Mechal Sobel, *Trabelin' On: The Slave Journey to an Afro-Baptist Faith* (Westport, Conn., 1979); and Frey and Wood, *Come Shouting to Zion*, esp. 118–48.

87. Sharon Ann Burnston, "The Cemetery at Catoctin Furnace, MD: The Invisible People," *Maryland Archeology* 17 (1981): 19–31; and Frey and Wood, *Come Shouting to Zion*, 23–26, 51–53.

88. Ross to Richardson, Apr. 30, 1812, DRL, VHS.

89. "List of Slaves," WBP, Duke, *RABSP:F:3*, 35; and Ross to Richardson, Apr. 30, 1812, DRL, VHS.

90. "List of Slaves," WBP, Duke, *RABSP:F:3*, 35.

91. "Memorandum for Richardson," [December 1812], Ross to Robert Richardson, Nov. 20, 1812, Jan. 5, 1813, Feb. 19, 1813, and Ross to Dunn, Jan. 1813, Apr. 1813, July 4, 1813, Aug. 1813, DRL, VHS.

92. Ross to Richardson, July 1, 1813, and Ross to John Pearce, Aug. 25, 1813, DRL, VHS.

93. Ross to Dunn, July 4, 1813 (includes copy of letter from Carpenter George to the proprietors of the Oxford Iron Works dated June 24, 1813), and Ross to Richardson, July 5, 1813, DRL, VHS.

94. Ross to Richardson, Oct. 30, Nov. 10, Dec. 1812, "Memorandum for Richardson," [Dec. 1812], Ross to Richardson, July 5, Aug. 15, 1813, and Ross to Dunn, August 28–29, 1813, DRL, VHS.

95. Ross to Richardson, Dec. 1812, DRL, VHS.

96. "Memorandum for Richardson," [Dec. 1812], and Ross to Hopkins, Aug. 25, 1813, DRL, VHS. For more on Oxford's financial problems, see Charles B. Dew, "David Ross and the Oxford Iron Works: A Study of Industrial Slavery in the Nineteenth-Century South," *WMQ* 31 (1974): 215–20; and Lewis, *Coal, Iron, and Slaves,* 187–91.

97. "List of Slaves," WBP, Duke, *RABSP:F:3,* 35; Ross to Dunn, Jan. 1813, DRL, VHS; "Deposition of William Mewburn," Oct. 26, 1812, WBP, Duke, *RABSP:F:3,* 35; Ross to Sherman, July 21, 1813; and Ross to Thomas Hopkins, July 22, 1813, DRL, VHS.

98. Dew, "David Ross and the Oxford Iron Works," 222–24.

99. David S. Garland to Weaver, May 23, 1828, John W. Schoolfield to Weaver, Feb. 28, 1828, and William C. McAllister to Weaver, Feb. 22, 1830, WWP, Duke, *SABSI:A,* 22.

100. Pennybacker to Weaver, Nov. 4, 1831, WWP, Duke, *SABSI:A,* 23. See also Benjamin Reeves to Weaver, Aug. 17, 1831, WWP, Duke, *SABSI:A,* 23.

101. *Records of the 1820 Census of Manufactures,* 27 reels (Washington, D.C., 1964–65), reel 16, 18.

Chapter 6. Manufacturing Free Labor

1. "Petition of the Manufacturers of Bar Iron within the Commonwealth of Pennsylvania to the Representatives of the Freemen of the Commonwealth of Pennsylvania," Nov. 30, 1785, Frank B. Nead Collection, Acc. #447, HSP.

2. Arthur Cecil Bining, *Pennsylvania Iron Manufacture in the Eighteenth Century,* 2d ed. (Harrisburg, 1987), 146. I borrow "disowning slavery" from Joanne Pope Melish, *Disowning Slavery: Gradual Emancipation and "Race" in New England, 1780–1860* (Ithaca, 1998).

3. Arthur C. Bining, *British Regulation of the Colonial Iron Industry* (Philadelphia, 1933), 112–13.

4. Joseph Hoff to Murray Samson & Co., Mar. 4, 1776, May 31, 1776, and June 7, 1776, HIWC, RUL.

5. *Minutes of the Council of Safety of the Province of Pennsylvania* (Harrisburg, 1852), 10:662; Charles Hoff, Jr., to Governor Livingston, July 27, 1777, and Charles Hoff, Jr., to Lord Stirling, Mar. 20, 1778, HIWC, RUL.

6. Charles Hoff, Jr., to Lord Stirling, Mar. 20, 1778, and Charles Hoff, Jr., to Lord Stirling, July 10, 1778, HIWC, RUL. See also Thomas M. Doerflinger, "Hibernia Furnace during the Revolution," *New Jersey History* 90 (1972): 97–114.

7. *CRPa,* 10:636; Bining, *Pennsylvania Iron Manufacture,* 99–100; Rev. Thompson P. Ege, *History and Genealogy of the Ege Family in the United States, 1738–1911* (Harrisburg, 1911), 153; and Hopewell Forge Timebook, 1776–1785, Cornwall Furnace Journal, 1776–1782, Cornwall Furnace Journal, 1782–1785, FFR, HSP.

8. William W. H. Davis, *History of Bucks County Pennsylvania from the Discovery of the Delaware to the Present,* 2d. ed., revised and enlarged, 3 vols. (New York and Chicago, 1905), 2:297–98; "List of Negroes Returned into the Clerk's Office by Peter Grubb, October 21, 1780" [copy], Box 3, GFP, HSP; and Gary B. Nash and Jean R. Soderlund, *Freedom by Degrees: Emancipation in Pennsylvania and its Aftermath* (New York, 1991), 94–95.

9. On gradual emancipation in Pennsylvania and New Jersey, see Nash and Soderlund, *Freedom by Degrees,* 99–113; A. Leon Higginbotham, Jr., *In the Matter of Color: Race and the American Legal Process: The Colonial Period* (New York, 1978), 299–305; Arthur Zilversmit, *The First Emancipation: The Abolition of Slavery in the North* (Chicago, 1967), 124–37, 141–46, 152–53, 159–62, 175–76, 184–89, 192–99; and Graham Russell Hodges, *Slavery and Freedom in the Rural North: African Americans in Monmouth County, New Jersey, 1665–1865* (Madison, Wis., 1997), 113–70.

10. Henry Drinker to Richard Blackledge, Oct. 4, 1786, HDL, 1786–1790, Drinker, HSP, in Thomas M. Doerflinger, "How to Run an Ironworks," *PMHB* 108 (1984): 364; and Drinker to Warner Mifflin, July 15, 1790, HDL, 1790–1793, Drinker, HSP.

11. Drinker to Franklin, Robinson & Co., June 9, 1790; and Drinker to Robert Bowne & Co., Nov. 11, 1790, HDL, 1790–1793, Drinker, HSP. On Drinker's land speculation and his promotion of maple sugar, see Thomas M. Doerflinger, *A Vigorous Spirit of Enterprise: Merchants and Economic Development in Revolutionary Philadelphia* (Chapel Hill, 1986), 321–23; and Alan Taylor, *William Cooper's Town: Power and Persuasion on the Frontier of the Early American Republic* (New York, 1995), 119–26, 130–34, 137–38. On antislavery, capitalism, and humanitarian thought, see the forum with David Brion Davis, John Ashworth, and Thomas L. Haskell in *AHR* 92 (1987): 797–878.

12. Drinker to William Cooper, Jan. 15, 1791, Drinker to Benjamin Walker, Sept. 17, 1790, and Drinker to Robert Bowne, Aug. 14, 1792, HDL, 1790–1793, Drinker, HSP. For more on abolitionist ironmasters, see Joseph M. Paul Daybook, 1800–1821; Paul to Lundy & Garrison, Sept. 21, 1829, and Joseph M. Paul Letterbook, Joseph M. Paul Papers, Acc. #192, HSP.

13. "Return of Negro Servants in Pursuance of the Act Passed 29 March 1788 of the Negro Children of the Minors," Sept. 22, 1788, Box 2, GFP, HSP; Lancaster County Clerk of Courts, "Returns of Negro and Mulatto Children Born After the Year 1780, June 7, 1788–November 13, 1793," Reel #6251, PSA; and Centre County Prothonotary, Birth Returns Negroes and Mulattoes, 1803–1820, Reel #6480, PSA.

14. John Anderson to William Orbison, Mar. 22, 1813, Box 5, Orbison Family Papers, MG–98, PSA.

15. Slaves Folder, Folder 2, Box 1, The Coleman Papers Collection, MG–275, LCHS; George Brooke Papers, PHMC Special Collections, Reel #955, PSA; and Birdsboro Forge Timebook, 1816–1823, Box 229, AISIC, HML.

16. Hopewell Forge Journal J, 1780–1783, Hopewell Forge Journal, 1783–1786, FFR, HSP; and Jasper Yeates to John Hubley, Nov. 28, 1788, Box 3, and "Agreement between Guardians of Estate of Peter Grubb and Cyrus Jacobs," Nov. 6, 1789, Box 2, GFP, HSP. On the Pennsylvania Abolition Society, see Nash and Soderlund, *Freedom by Degrees,* 113–36; and Dee E. Andrews, "Reconsidering the First Emancipation: Evidence from the Pennsylvania Abolition Society Correspondence, 1785–1810," *PH* 64 (1997: Special Supplemental Issue): 230–49.

17. Hopewell Forge Journal J, 1780–1783, FFR, HSP; "List of Negroes Returned into the Clerk's Office by Peter Grubb," October 21, 1780 [copy], Box 3, GFP, HSP; and Hopewell Forge Timebook, 1776–1785, FFR, HSP.

18. "Agreement between the Guardians of the Estate of Burd Grubb and Henry Bates Grubb and John Hare," Mar. 16, 1787; "Agreement between the Guardians of the Estate of Burd Grubb and Henry Bates Grubb and Samuel Jones," Aug. 11, 1788, Box 2; "Indenture of Negro Jasper . . . to John Jones from 4 Feb[y] 1795"; "Bill of Sale between Henry Bates Grubb and George Ege," June 2, 1801, all in Box 4, GFP, HSP.

19. Union Forge Account Book, 1795, Box 1, and Cornwall Furnace Collection, MG–203, PSA; New Market Forge Daybook B, 1795–1799, V–60, Bethlehem Steel Corporation—Archives—Manuscript Collections, Acc. 1699, HML; Charming Forge Journal, 1780–1785, Charming Forge Daybook, 1788–1791, FFR, HSP; and Charming Forge Day Book, 1790–1794, 1821–1822, Charming Forge Day Book, 1794–1800, Library, Historical Society of Berks County, Reading, Pennsylvania. George Ege, Charming Forge's owner, re-

ported nine slaves to census takers in 1790, and seven, plus "4 other free persons except Indians—not taxed" in 1800. See "Population Schedules of the First Census of the United States, 1790," and "Population Schedules of the Second Census of the United States, 1800."

20. Lancaster County Slave Register [transcript], Slave Records of Lancaster County Collection, MG–240, LCHS; and Cornwall Furnace Timebooks, 1776–1785, 1786–1794, FFR, HSP.

21. Cornwall Furnace Timebook, 1776–1785, FFR, HSP.

22. Johann David Schoepf, *Travels in the Confederation [1783–1784]*, trans. and ed. Alfred J. Morrison, 2 vols. (Philadelphia, 1911; reprint, New York, 1968), 1:208.

23. Cornwall Furnace Coal and Cordwood Book, 1776–1790, FFR, HSP. On growing identification of indentured servitude with "slavery" and "blackness," see Robert J. Steinfeld, *The Invention of Free Labor: The Employment Relation in English and American Law and Culture, 1350–1870* (Chapel Hill, 1991), 122–46; and David R. Roediger, *The Wages of Whiteness: Race and the Making of the American Working Class* (New York, 1991), 27–36, 43–60.

24. On white servitude and apprenticeships in the early nineteenth-century Pennsylvania iron industry, see Codorus Forge Daybook, 1818–1819, GFC, HML; and "Agreement between Richard Snowden and Elizabeth Dorsey," July 19, 1819, Box 1, Greensberry Dorsey Family Papers, Acc. # 1627, HSP. Joanne Pope Melish argues that gradual abolition defined concepts of race in *Disowning Slavery.* On immigrant indentured servitude's demise, see Farley Grubb, "The Disappearance of Organized Markets for European Immigrant Servants in the United States: Five Popular Explanations Reexamined," *Social Science History* 18 (1994): 1–30.

25. Cornwall Furnace Timebooks, 1786–1794, 1805–1825, Cornwall Furnace Journal, 1794–1802, Cornwall Furnace Journal, 1798–1802, Cornwall Furnace Journal, 1804–1822, Cornwall Furnace Journal, 1822–1833, FFR, HSP.

26. Cornwall Furnace Journal, 1794–1802, Cornwall Furnace Daybook, 1798–1804, Cornwall Furnace Journal, 1798–1802, Cornwall Furnace Daybook, 1804–1822, Cornwall Furnace Journal, 1822–1833, FFR, HSP; and Workmen's Account Book, 1810–1816, Box 1, Cornwall Furnace Collection, MG–203, PSA.

27. Birdsboro Forge Timebook, 1816–1823, AISIC, HML.

28. Coventry Forge Journal, 1792–1796, FFR, HSP; and Birdsboro Forge Ledger B, 1800–1803, Birdsboro Forge Ledger C, 1804–1808, Birdsboro Forge Collection, MG–258, PSA.

29. Birdsboro Forge Ledger C, 1804–1808 and Birdsboro Forge Ledger B, 1800–1803, Birdsboro Forge Collection, MG–258, PSA.

30. Sergeant and Rawles 351 (1817), 352–55, quoted in Christopher Tomlins, *Law, Labor, and Ideology in the Early American Republic* (New York, 1993), 265. On mobility and free black men, see W. Jeffrey Bolster, "'To Feel Like A Man': Black Seamen in the Northern States, 1800–1860," *JAH* 76 (1990): 1173–99.

31. Birdsboro Forge Timebook, 1816–1823, AISIC, HML.

32. Derick Barnard to Samuel G. Wright, Jan. 9, 1825, Box 1, WFP, HML; and Coventry Forge Journal, 1792–1796, FFR, HSP. On the association of black women with domestic service, see Jacqueline Jones, *American Work: Four Centuries of Black and White Labor* (New York, 1998), 284–85.

33. Cumberland Forge Provision Book, 1828–1835, Box 1, HP, HML. See also Carl D. Oblinger, "Alms for Oblivion: The Making of a Black Underclass in Southeastern Pennsylvania, 1780–1860," in *The Ethnic Experience in Pennsylvania* ed. John E. Bodnar (Lewisburg, Pa., 1973), 94–119; and Oblinger, "New Freedoms, Old Miseries: The Emergence and Disruption of Black Communities in Southeastern Pennsylvania, 1780–1860" (Ph.D. diss., Lehigh University, 1988), 78–81, 160–61; and Joseph E. Walker, "A Comparison of Negro and White Labor in a Charcoal Iron Community," *LH* 10 (1969): 487–97. On exclusion of African Americans from mill and factory jobs in the North, see Jones, *American Work,* 258–64. On maritime labor, see W. Jeffrey Bolster, *Black Jacks: African American Seamen in the Age of Sail* (Cambridge, Mass., 1997), 68–130, 158–232.

34. Gardiner H. Wright to William Potter, Dec. 11, 1831, Box 5, WFP, HML. On worsening racism in the early national North, see James Brewer Stewart, "The Emergence of Racial Modernity and the Rise of the White North, 1790–1840," *JER* 18 (1998): 181–217; Roediger, *Wages of Whiteness*, 43–163; Gary B. Nash, *Forging Freedom: The Formation of Philadelphia's Black Community* (Cambridge, Mass., 1988), 172–279; Jones, *American Work*, 246–97; David Waldstreicher, *In the Midst of Perpetual Fetes: The Making of American Nationalism, 1776–1820* (Chapel Hill, 1997), 294–352; and Melish, *Disowning Slavery*, 119–237.

35. *Martha*, May 18, 1808; and Martha Furnace Diary and Timebook, 1808–1815, Acc. 339, HML.

36. *Martha*, July 10, 1808; June 6, Oct. 1, Oct. 8, 1809; and Jan. 9, 11, 1810.

37. Roediger, *Wages of Whiteness*, 133–63; Noel Ignatiev, *How the Irish Became White* (New York, 1995); and Matthew Frye Jacobson, *Whiteness of a Different Color: European Immigrants and the Alchemy of Race* (Cambridge, Mass., 1998), 15–52. For a skeptical view of "whiteness" as an explanatory device, see Peter Kolchin, "Whiteness Studies: The New History of Race in America," *JAH* 89 (2002): 154–73.

38. Henry Drinker to Richard Blackledge, Oct. 4, 1786, HDL, 1786–1790, Drinker, HSP, in Doerflinger, "How to Run an Ironworks," 364.

39. Drinker to Thomas Stevenson & Co., Apr. 18, 1793, HDL, 1790–1793, Drinker, HSP.

40. Drinker to Embree & Shotwell, Feb. 20, 1787, and Drinker to Pearsall & Pell, Mar. 23, 1787, HDL, 1786–1790; Drinker to Isaac Stoutenburgh & Son, July 15, 1790, and Drinker to Robert Bowne, Aug. 14, 1792, HDL, 1790–1793, Drinker, HSP.

41. Drinker to Isaac Stoutenburgh & Son, Aug. 13, 1789, HDL, 1786–1790, Drinker, HSP.

42. Birdsboro Forge Timebook, 1809–1810, Box 1, Birdsboro Forge Collection, MG–258, PSA; and Birdsboro Forge Timebook, 1816–1823, AISIC, HML. Herbert Gutman casts such behavior as the legacy of a premodern work culture in "Work, Culture, and Society in Industrializing America, 1815–1919," *AHR* 78 (1973): 343–55.

43. Cornwall Furnace Timebooks, 1805–1825, 1821–1832, FFR, HSP; and Pine Grove Furnace Timebooks, 1812–1821, 1822–1837, Pine Grove Furnace Collection, MG–175, PSA. To calculate persistence rates, I compared timebook entries from July for each year and determined how many employees remained one, two, three, four, and five years later. I then averaged persistence rates within five-year spans. Turnover rates at Cornwall and Pine Grove compare favorably to some of those for the textile industry. Jonathan Prude found aggregate annual turnover rates of 100–250 percent between 1814 and 1859 at Massachusetts mills. Anthony Wallace found turnover rates comparable to those of Cornwall and Pine Grove among workers at Rockdale, Pennsylvania: 50 percent in the first two years of employment, and 4–5 percent per year thereafter. See Prude, *The Coming of Industrial Order: Town and Factory Life in Rural Massachusetts, 1810–1860* (New York, 1983), 227–35, 267–70; and Wallace, *Rockdale: The Growth of an American Village in the Early Industrial Revolution* (New York, 1972; paperback, New York, 1980), 63–65.

44. Linda McCurdy, "The Potts Family Iron Industry in the Schuylkill Valley" (Ph.D. diss., Pennsylvania State University, 1974), 181–98; Mrs. Thomas Potts James, *Memorial of Thomas Potts, Junior* (Cambridge, Mass., 1874), 213–26, 247, 249, 260–61, 280–82, 301; GFC, HML; Frederic K. Miller, "The Rise of an Iron Community: An Economic History of Lebanon County, Pennsylvania from 1740 to 1865" (Ph.D. diss., University of Pennsylvania, 1947), 142–48, 158–77; Frederic S. Klein, "Robert Coleman, Millionaire Ironmaster," *Journal of the Lancaster County Historical Society* 64 (1960): 17–33; Ege, *History and Genealogy of the Ege Family*, 85–94; Calvin W. Hetrick, *The Iron King* (Martinsburg, Pa., 1961); and Arthur D. Pierce, *Family Empire in Jersey Iron: The Richards Enterprises in the Pine Barrens* (New Brunswick, N.J., 1964), 3–32.

45. Drinker to Thomas Stevenson & Co., Apr. 18, 1793, HDL, 1790–1793, Drinker, HSP; and William Waples to Samuel G. Wright, July 7, 1821, Sept. 18, 1821, Box 1, WFP, HML.

46. John G. Smith to Samuel G. Wright, Dec. 15, 1821, Box 1, WFP, HML. On cities and labor markets, see Billy G. Smith, *The "Lower Sort": Philadelphia's Laboring People, 1750–1800*

(Ithaca, 1991); and Seth Rockman, "Working for Wages in Early Republic Baltimore: Unskilled Labor and the Blurring of Slavery and Freedom" (Ph.D. diss., University of California, Davis, 1999).

47. Cornwall Furnace Timebook, 1776–1785, FFR, HSP; and "Memorandum of Agreement Between H. B. Grubb & Peter Bright," Jan. 25, 1798, Box 4, GFP, HSP.

48. "Article of Agreement between H. B. Grubb and George Petz," Sept. 28, 1796, GFP, HSP; and Box 1, Haldeman-Wright Family Papers, MG–64, PSA.

49. Workmen's Account Books, 1806–1809, 1810–1816, Box 1, Cornwall Furnace Collection, MG–203, PSA. Even when one excludes individuals with balances over one hundred dollars, Cornwall still owed the average creditor nearly twenty-five dollars. To compile these data, I took the latest balance for each person. For duplicate names, I only used balances when the store clerk distinguished between fathers and sons or listed people of the same name with different occupations.

50. Samuel Laverly to Peter Grubb, June 6, 1785, Box 2, GFP, HSP; Drinker to Daniel Clymer, Aug. 29, 1792, HDL, 1790–1793, Drinker, HSP; and Birdsboro Forge Timebook, 1809–1810, Birdsboro Forge Collection, MG–258, PSA.

51. Derick Barnard to Samuel G. Wright, July 19, 1824, Box 1, WFP, HML.

52. Mount Hope Furnace Journal, 1786–1788, AISIC, HML; Hopewell Forge Daybook, 1797–1805, FFR, HSP; and Codorus Forge Daybook, 1818–1819, GFC, HML.

53. Mount Hope Furnace Journal, 1786–1788, AISIC, HML; Berkshire Furnace Journal, 1787–1789, Box 1, Business Records Collection, MG–2, PSA; Colebrook Furnace Daybook, 1816–1823, Lebanon County Historical Society Collection, MG–182, PSA; Hopewell Forge Daybook, 1797–1805, FFR, HSP; Codorus Forge Daybook, 1818–1819, GFC, HML; and Martha Furnace Diary and Timebook, 1808–1815, Acc. 339, HML. See also Thomas M. Doerflinger, "Rural Capitalism in Iron Country: Staffing a Forest Factory, 1808–1815" *WMQ*, 59 (2002): 20–23.

54. Mount Hope Furnace Journal, 1786–1788, AISIC, HML; Hopewell Forge Journal, 1794–1796, FFR, HSP; and Codorus Forge Journal, 1820–1823, and Codorus Forge Daybook, 1818–1819, GFC, HML.

55. *Martha*, Nov. 7, 1808, June 22, 1811, Aug. 20, 1811, and Jan. 19–20, 1814. According to timebook entries, there were twenty-nine incidents of drunkenness in 1811 before Evans's declaration, ten for the rest of the year, seventeen each in 1812 and 1813, and ten in 1814.

56. *Martha*, Jan. 7, 1809, Jan. 16, 1811; John Campbell to Peter Grubb, Dec. 23, 1783, Box 2, GFP, HSP; New District Forge Daybook, 1799–1801, and Greenwood Iron Works Daybook, 1807–1808, John Pott Employee Records, Acc. 1120, HML; Codorus Forge Journal, 1820–1823, GFC, HML; and Joseph E. Walker, *Hopewell Village: The Dynamics of a Nineteenth Century Iron-Making Community* (Philadelphia, 1966), 412.

57. Barnard to Wright, Nov. 24, 1822, Oct. 29, 1822, Box 1, WFP, HML.

58. Barnard to Samuel Wright, June 21, 1824, Apr. 18, 1825, Nov. 19, 1827, Box 1; Gardiner H. Wright to William Potter, Apr. 17, 1833, Box 5, WFP, HML; and W. J. Rohrbaugh, *The Alcoholic Republic: An American Tradition* (New York, 1979).

59. Barnard to Wright, June 9, 1822, Sept. 2, 1822, Box 1, WFP, HML. On labor conflict at Delaware Furnace, see Stuart Paul Dixon, "Organizational Structure and Marketing at Delaware Furnace 1821–1836" (MA thesis, University of Delaware, 1990), 11–41.

60. Barnard to Wright, July 29, 1822, Box 1, WFP, HML.

61. This paragraph and the next draw from Barnard to Wright, Dec. 23, 1822, Dec. 31, 1822, Box 1, WFP, HML.

62. Barnard to Wright, June 9, 1822, Nov. 24, 1822, and Oct. 29, 1822, Box 1, WFP, HML.

63. *Martha*, Oct. 29, 1811; and David Montgomery, "Wage Labor, Bondage, and Citizenship in Nineteenth-Century America," *International Labor and Working-Class History* 48 (1995): 6–27.

64. *Martha*, June 5–6, 1813. On ethnic tensions, see *Martha*, Dec. 25, 1808. On canal workers, see Peter Way, *Common Labour: Workers and the Digging of North American Canals*

1780–1860 (New York, 1993), 163–99, and "Evil Humours and Ardent Spirits: The Rough Culture of Canal Construction Laborers," *JAH* 79 (1993): 1397–1428.

65. William A. Sullivan, *The Industrial Worker in Pennsylvania, 1800–1840* (Harrisburg, 1955), 59–71. The only four instances at Martha Furnace in which employees united against proprietors were when they complained about provisions. See *Martha*, Aug. 5, 1809, Dec. 6, 1810, July 6, 1811, and Aug. 31, 1811.

66. Barnard to Samuel Wright, July 8, 1823, June 14, 1824, July 12, 1824, Jan. 9, 1825, May 10, 1825, and Sept. 18, 1825, Box 1, WFP, HML.

67. Barnard to Wright, Oct. 25, 1824, Mar. 24, 1827, and Apr. 15, 1827, Box 1; Barnard to Wright, Dec. 26, 1830; William Waples to Wright, Nov. 9, 1830, Box 2, WFP, HML.

68. Samuel McAnnine to Wright, June 26, 1821; Waples to Wright, Dec. 25, 1821; Barnard to Wright, Nov. 15, 1824, all in Box 1, WFP, HML.

69. "Articles of Agreement between Henry B. Grubb and John King and Joseph Klutz," Jan. 26, 1822, Box 4, GFP, HSP.

70. Delaware Furnace Receipt Book, 1823–1827, and Delaware Furnace Receipt Book, 1825–1832, Box 7, WFP, HML.

71. Bela Kingman to Samuel Wright, Aug. 26, 1822, and Waples to Wright, Sept. 8, 1822, Box 1, WFP, HML.

72. Kingman to Wright, Sept. 11, 1822, Oct. 2, 1822, and Oct. 28, 1822, Box 1, WFP, HML.

73. Barnard to Wright, Nov. 19, 1823, and Kingman to Wright, Dec. 8, 1823, Box 1, WFP, HML.

74. Barnard to Wright, Mar. 24, Mar. 29, 1824, Box 1, WFP, HML.

75. Barnard to Samuel Wright, Feb. 18, 1827, Box 1, and Gardiner H. Wright to William Potter, May 17, 1832, Box 5, WFP, HML.

76. James McFadden to Samuel Wright, Nov. 27, 1833, Box 2, WFP, HML.

77. John Lesher to Thomas Wharton, President, Executive Council of Pennsylvania, Jan. 9, 1778, *PA, First Series*, 6:170; and Bining, *Pennsylvania Iron Manufacture*, 122, 125.

78. Lesher to Wharton, Jan. 9, 1778, *PA, First Series*, 6:170.

79. Pierce, *Family Empire in Jersey Iron*, 20–21. Collection of government debts also benefited Pennsylvania ironmasters by creating a larger pool of labor as farmers faced foreclosure on their lands. See Terry Bouton, "A Road Closed: Rural Insurgency in Post-Independence Pennsylvania," *JAH* 87 (2000): 855–87.

80. Bining, *Pennsylvania Iron Manufacture*, 122; Robert Coleman to John Adams [copy], Dec. 4, 1800, Robert Coleman #2, Box 2, Folder 6, The Coleman Papers Collection, MG–275, LCHS; Klein, "Robert Coleman," 25–28; and *Biographical Dictionary of the United States Congress, 1774–1989, Bicentennial Edition: The Continental Congress September 5, 1774 to October 21, 1788 and the Congress of the United States from the First through the One Hundredth Congresses, March 4, 1789 to January 3, 1989, Inclusive* (Washington, D.C., 1989), 955, 1259, 1966. On the Whiskey Rebellion, see Thomas P. Slaughter, *The Whiskey Rebellion: Frontier Epilogue to the American Revolution* (New York, 1986).

81. On ironmasters and the tariff, see Malcolm Rogers Eiselen, *The Rise of Pennsylvania Protectionism* (Philadelphia, 1932), 20–133; F. W. Taussig, *The Tariff History of the United States*, 8th ed., revised (New York, 1931), 46–59; and Bining, *Pennsylvania Iron Manufacture*, 144–46, 148–50. On tariffs and antebellum politics, see Jonathan J. Pincus, *Pressure Groups and Politics in Antebellum Tariffs* (New York, 1977). For adventurers' petitions to Congress regarding tariffs, see *Annals of the Congress*, 42 vols. (Washington, 1834–1856) 4:523–24 (1794); 23:569 (1811); 31:446 (1817). For more on ironmasters and elections, see Walker, *Hopewell Village*, 403–8; and Pierce, *Family Empire in Jersey Iron*, 64.

82. Ronald Schultz, *The Republic of Labor: Philadelphia Artisans and the Politics of Class, 1720–1830* (New York, 1993), 81–84; Waldstreicher, *In the Midst of Perpetual Fetes*, 90–107. On British traditions which viewed the iron trade as a corporate entity, see Chris Evans, *"The Labyrinth of Flames": Work and Social Conflict in Merthyr Tydfil* (Cardiff, Wales, 1993), 83–85.

83. Pennsylvania had one of the most expansive electorates in the new nation, though it denied the vote to free black men beginning in 1838. For more on the general expansion and targeted contraction of the right to vote during the early national and antebellum eras, see Alexander Keyssar, *The Right to Vote: The Contested History of Democracy in the United States* (New York, 2000), 8–67, 340–67.

84. Eiselen, *Pennsylvania Protectionism*, 42; and John Weidman to Jacob M. Haldeman, Oct. 25, 1817, Box 13, HP, HML

85. Taussig, *Tariff History*, 51–52.

86. Bining, *Pennsylvania Iron Manufacture*, 27; and Benjamin Abbott, *The Experience and Gospel Labours of the Rev. Benjamin Abbott: To Which is Annexed, A Narrative of His Life and Death*, 4th ed. (New York, 1813), 130–31. On Rebecca Grace's patronage of the Methodists of Coventry Township, see Potts James, *Memorial*, 390.

87. Elmer T. Clark, J. Manning Potts, and Jacob S. Payton, eds., *The Journal and Letters of Francis Asbury*, 3 vols. (London and Nashville, 1958), 2:697, 729. On the role of women within early Methodism, see Dee Andrews, *The Methodists and Revolutionary America, 1760–1800: The Shaping of an Evangelical Culture* (Princeton, 2000), 99–122. On "republican motherhood," see Linda K. Kerber, *Women of the Republic: Intellect and Ideology in Revolutionary America* (Chapel Hill, 1980); and Rosemarie Zagarri, "Morals, Manners, and the Republican Mother," *American Quarterly* 44 (1992): 192–215.

88. Sermon of Rev. John B. Laman at the Funeral of Catherine Carmichael Jenkins, September 1856, quoted in [Committee on Historical Research, Pennsylvania Society of the Colonial Dames of America], *Forges and Furnaces in the Province of Pennsylvania* (Philadelphia, 1914), 105–6. See also Lori D. Ginzburg, *Women and the Work of Benevolence: Morality, Class, and Politics in the Nineteenth-Century United States* (New Haven, 1990).

89. *Journal and Letters of Francis Asbury*, 1:694; Arthur D. Pierce, *Iron in the Pines: The Story of New Jersey's Ghost Towns and Bog Iron* (New Brunswick, N.J., 1957), 194–95; *Martha*, Jan. 17, 23, and Feb. 8, 10, 1810; Weymouth Works Daybook, 1808–1811, Batsto Manuscript Collection, Batsto Village, New Jersey State Park Service, Hammonton; and Colebrook Furnace Timebook, 1799–1809, FFR, HSP.

90. E. P. Thompson, *The Making of the English Working Class* (London, 1963; paperback ed., New York, 1966), esp. 346–400; Paul E. Johnson, *A Shopkeeper's Millennium: Society and Revivals in Rochester, New York, 1815–1837* (New York, 1978); and Barbara M. Tucker, *Samuel Slater and the Origins of the American Textile System* (Ithaca, 1984), 163–81.

91. John E. Pomfret, *Colonial New Jersey: A History* (New York, 1973), 201; and David Potts to Sarah Potts, Aug. 27, 1826, Box 3, Potts Family Papers, Acc. #520, HSP.

92. Roland Curtin to Judge James Potter, Mar. 7, 1803, quoted in John Blair Linn, *History of Centre and Clinton Counties, Pennsylvania* (Philadelphia, 1883), 261, n.1; and William Henry Williams, *The Garden of American Methodism* (Wilmington, Del., 1984), 159.

93. On evangelical Christianity and metaphors of family, see Christine Leigh Heyrman, *Southern Cross: The Beginnings of the Bible Belt* (New York, 1997; paperback ed., Chapel Hill, 1997), 117–60. On evangelical Christianity and national identity, see Liam O'Boyle Riordan, "Identities in the New Nation: The Creation of an American Mainstream in the Delaware Valley, 1770–1830" (Ph.D. diss., University of Pennsylvania, 1996), 313–92.

94. William Sisson argues that industrial employers and their workers shared values in "From Farm to Factory: Work Values and Discipline in Two Early Delaware and Maryland Textile Mills," *Delaware History* 21 (1984): 31–52, as does Anthony F. C. Wallace in *Rockdale*, 241–397, who ties consensus to evangelization and to mill owners' ability to maintain relatively harmonious relations with operatives.

95. On evangelical Christianity's attractions, especially Methodism's, for working people, see Nathan O. Hatch, *The Democratization of American Christianity* (New Haven, 1989); Andrews, *Methodists and Revolutionary America*, 155–83; Tucker, *Samuel Slater*, 181–84; and Jama Lazerow, *Religion and the Working Class in Antebellum America* (Washington, 1995).

96. For examples, see *Martha,* Jan. 9, 1810, Feb. 18, 1812, and July 5–6, 1813.

97. Weymouth Furnace Timebook, 1820–1821, Weymouth Furnace Timebook, 1825–1826, Weymouth Furnace Timebook, 1826–1827, Weymouth Furnace Timebook, 1827–1828 [photocopy], Weymouth Furnace Timebook, 1830–1831 [photocopy], Atlantic County Historical Society, Somer's Point, New Jersey. See also Anne M. Boylan, *Sunday School: The Formation of an American Institution, 1790–1880* (New Haven, 1988).

98. *Martha,* June 2, 1813; Washington Furnace Timebook, 1814–1818, Batsto Manuscript Collection, Batsto Village, New Jersey State Park Service, Hammonton. On employee liability for accidents at work, see Tomlins, *Law, Labor, and Ideology,* 331–84.

99. Weymouth Works Daybook, 1808–1811, Batsto Manuscript Collection, Batsto Village, New Jersey State Park Service, Hammonton; *Martha,* Feb. 28, 1810, and Dec. 19, 1809.

100. *Martha,* Mar. 20, 1811, Oct. 21, 1811, July 13, 1813, and Aug. 1, 1813.

101. John Weidman to Jacob M. Haldeman, Oct. 25, 1817, HP, HML.

102. Gordon S. Wood argues that voluntary associations, especially those of evangelical Christianity, made possible the rise of a market society and democratic capitalism in "The Enemy Is Us: Democratic Capitalism in the Early Republic," *JER* 16 (1996): 307–8.

103. *Address of the Friends of Domestic Industry, Assembled in Convention, at New-York, October 26, 1831, to the People of the United States* (Baltimore, 1831), 20, 39.

104. Here I borrow from studies of British industrialization, especially Patrick Joyce, *Work, Society, and Politics: The culture of the factory in later Victorian England* (Brighton, UK, 1980); and Evans, *"Labyrinth of Flames,"* 137–44.

Conclusion

1. Julian Ursyn Niemcewicz, *Under Their Vine and Fig Tree: Travels through America in 1797–1799, 1805 with some further account of life in New Jersey,* trans. and ed. Metchie J. E. Budka (Elizabeth, N.J., 1965), 228–29.

2. *Ibid.,* 227.

3. See, for example, Anthony F. C. Wallace, *Rockdale: The Growth of an American Village in the Early Industrial Revolution* (New York, 1972; paperback ed., 1978); Teresa Anne Murphy, *Ten Hours' Labor: Religion, Reform, and Gender in Early New England* (Ithaca, 1992).

4. On the comparison of "wage slavery" to "white slavery," see David R. Roediger, *The Wages of Whiteness: Race and the Making of the American Working Class* (New York, 1991), 65–87. On southern critiques of free labor, see Jacqueline Jones, *American Work: Four Centuries of White and Black Labor* (New York, 1998), 219–32.

5. Walter Johnson, *Soul by Soul: Life Inside the Antebellum Slave Market* (Cambridge, Mass., 1999). See also Jeffrey Robert Young, *Domesticating Slavery: The Master Class in Georgia and South Carolina, 1670–1837* (Chapel Hill, 1999), 91–234.

6. Charles B. Dew, *Ironmaker to the Confederacy: Joseph R. Anderson and the Tredegar Iron Works* (New Haven, 1965); Ann Kelly Knowles, "Labor, Race, and Technology in the Confederate Iron Industry," *Technology and Culture* 42 (2001): 1–26; Jones, *American Work,* 232–45; Frederick K. Miller, "The Rise of an Iron Community: An Economic History of Lebanon County, Pennsylvania, from 1740 to 1865" (Ph.D. diss., University of Pennsylvania, 1948), 202–12; Mrs. Thomas Potts James, *Memorial of Thomas Potts, Junior* (Cambridge, Mass., 1874), 288; and Wallace, *Rockdale,* esp. 459–71.

7. See, for example, Iver Bernstein, *The New York City Draft Riots: Their Significance for American Society and Politics in the Age of the Civil War* (New York, 1990); Eric Foner, *Reconstruction: America's Unfinished Revolution, 1863–1877* (New York, 1988); David Montgomery, *The Fall of the House of Labor: The Workplace, The State, and American Labor Activism, 1865–1925* (New York, 1987); and Gary Gerstle, *American Crucible: Race and Nation in the Twentieth Century* (Princeton, 2001), 3–127. On how immigrants invoked and tried to redefine "free labor," see Gunther Peck, *Reinventing Free Labor: Padrones and Immigrant Workers in the American West* (New York, 2000).

8. On white working-class racism and exclusion of African Americans from industrial employment and from unions, see Bruce Nelson, *Divided We Stand: American Workers and the Struggle for Black Equality* (Princeton, 2001); and Jones, *American Work*, 258–355. On freedmen, see Foner, *Reconstruction;* and Julie Saville, *The Work of Reconstruction: From Slave to Wage Laborer in South Carolina, 1860–1870* (New York, 1994).

9. Foner, *Reconstruction;* Jacquelyn Dowd Hall, James Leloudis, Robert Korstad, Mary Murphy, Lu Ann Jones, and Christopher B. Daly, *Like a Family: The Making of a Southern Cotton Mill World* (Chapel Hill, 1987); Allen Tullos, *Habits of Industry: White Culture and the Transformation of the Carolina Piedmont* (Chapel Hill, 1989); John W. Cell, *The Highest Stage of White Supremacy: The Origins of Segregation in South Africa and the American South* (New York, 1982); Michael K. Honey, *Southern Labor and Black Civil Rights: Organizing Memphis Workers* (Urbana, Ill., 1993); and Jones, *American Work*, 301–39.

10. Montgomery, *Fall of the House of Labor;* and Lizabeth Cohen, *Making a New Deal: Industrial Workers in Chicago, 1919–1939* (New York, 1990). See also Herbert G. Gutman, "Work, Culture, and Society in Industrializing America, 1815–1919," *AHR* 78 (1973): 531–88.

11. See, for example, Randall Robinson, *The Debt: What America Owes to Blacks* (New York, 2000); Dalton Conley, "The Cost of Slavery," *New York Times*, Feb. 15, 2003, p. A25; Tatsha Robinson, "Reparations for Slavery: Old Idea Goes Mainstream," *Boston Globe*, Apr. 4, 2002, p. A1.

12. For a critique of the dominant model for global economic development, see Joseph E. Stiglitz, *Globalization and Its Discontents* (New York, 2002).

INDEX